Springer Aerospace Technology

AF772877

Jens Eickhoff

Simulating Spacecraft Systems

With 234 Figures and 9 Tables

Springer

Dr.-Ing. Jens Eickhoff
Säntisweg 28
88090 Immenstaad
Germany

ISBN 978-3-642-01275-4 e-ISBN 978-3-642-01276-1
DOI 10.1007/978-3-642-01276-1
Springer Heidelberg Dordrecht London New York

Springer Series in Aerospace Technology ISSN 1869-1730 e-ISSN 1869-1749

Library of Congress Control Number: 2009932687

© Springer-Verlag Berlin Heidelberg 2009

This work is subject to copyright. All rights are reserved, whether the whole or part of the material is concerned, specifically the rights of translation, reprinting, reuse of illustrations, recitation, broadcasting, reproduction on microfilm or in any other way, and storage in data banks. Duplication of this publication or parts thereof is permitted only under the provisions of the German Copyright Law of September 9, 1965, in its current version, and permission for use must always be obtained from Springer. Violations are liable to prosecution under the German Copyright Law.
The use of general descriptive names, registered names, trademarks, etc. in this publication does not imply, even in the absence of a specific statement, that such names are exempt from the relevant protective laws and regulations and therefore free for general use.

Cover picture modified from original in ISSN 1866-7376, Issue 2.0.
Original, by Sabine Leib EADS DS and Jens Eickhoff.

Printed on acid-free paper

Springer is part of Springer Science+Business Media (www.springer.com)

Foreword

Satellite development worldwide has significantly changed within the last decade and has been accelerated and optimized by modern simulation tools. The classic method of developing and testing several models of a satellite and its subsystems with the aim to build a pre-flight and finally a flight model is being replaced more and more by a considerably faster and more inexpensive method. The new approach no longer includes functional test models on entire spacecraft level but a system simulation. Thus overall project runtimes can be shortened. But also significantly more complex systems can be managed and success oriented tests on integration and software level can be realized before the launch.

Applying modern simulation infrastructures already during spacecraft development phase, enables the consistent functionality checking of all systems both in detail and concerning their interaction. Furthermore, they enable checks of the system's proper functionality, their reliability and safety / redundancy. But also analysis regarding aging and lifetime issues can be performed by simulation. Project-related simulations of operational scenarios, for example with remote sensing satellites, and the checking of different operational modes are of similar importance. On the whole, risk is reduced significantly and the satellite can be produced in a considerably more cost efficient way, with higher quality and in shorter periods of time.

Therefore "Simulating Spacecraft Systems" - the title of the present book - is an important domain of modern system engineering, which meanwhile has successfully established a position in many other sectors of industry and research, too.

For this reason it was a matter of particular concern for the Universität Stuttgart, Faculty of Aerospace Engineering, to offer this subject as a lecture held in two semesters for prospective engineers. The goal was to achieve a close relation to industrial satellite development and include demonstrations in the MDVE, (Model based development and verification environment), laboratory of the Institute. It is a big asset for the faculty that Dr. Jens Eickhoff from EADS Astrium GmbH - Satellites is engaged on this topic. He has combined theory, industrial experience and research in this very modern sector into his lectureship for many years now.

The present book results from several years of lectures, has been consistently complemented and practical examples have been added. This work is the first book of its kind, guiding the reader from simulator application overview, adding detail sequentially as it teaches the student simulator development, numerics, software technology etc. The book is equally applicable for students as well as experts of many engineering disciplines. It is suitable for introduction and reference in modern system engineering.

Stuttgart, spring 2009

Prof. Dr. Hans-Peter Röser
Institute of Space Systems
Universität Stuttgart

Preface

This book results from the author's lectures at the Universität Stuttgart. The idea comes from a visit by the Institute of Space Systems management team to EADS Astrium GmbH - Satellites in Friedrichshafen, a key European satellite developer.

Astrium has been developing a complex system simulation infrastructure for several years. However, it is difficult for industry to find graduates which are not only well trained in space engineering but also have adequate knowledge in simulator software development. The idea was to address this shortfall by giving lectures and workshops by the author at the Institute of Space Systems, (IRS). These lectures meanwhile have evolved comprising industry infrastructure visits, tutorials etc. From University side they are based on according teaching assignments whilst Astrium authorized the project as an agreed sideline task of the Author.

The decision to write a book on system simulation from the lecture notes originates because there is indeed lots of technical literature on simulation, but all with deficiencies for the target audience. There are many books on simulation in control engineering, however tackling almost exclusively special development tools. The literature on process engineering simulation again mostly concentrates on specific tools like flowsheet applications. All books known to the author put only little emphasis on how the simulator development is interwoven with engineering pathway of the to be simulated target system. Therefore application examples in this book address this deficit and always explain simulation in the context of the engineering process towards satellites, space probes and rocket stages.

Another important deficit is, very few books considering, that most interested students are beginners in the simulation domain. Such students need to be guided to receive a proper introduction. This results in a requirement on the author to guide the reader on their way from spacecraft system engineering topics to the system simulation case, and beyond to the modeling of the system inside the simulator. Finally arriving at the deeper topics of simulator coding, whilst addressing all the caveats along the journey.

Students' responses to the lectures, and the demand for study, diploma and doctoral theses topics since the beginning of this IRS / Astrium cooperation, clearly show great interest in this fascinating subject. I hope this book contributes to imparting background knowledge to the student, enabling them to begin professionally in the simulation domain.

Immenstaad, May 2009 Jens Eickhoff

Acknowledgements

This book covers a broad spectrum of simulation technology aspects and would not have become so educative without availability of significant material from industry. Therefore first of all I am greatly indebted to my superior, Eckard Settelmeyer, "Director Earth Observation & Science", EADS Astrium - Satellites. He granted me approval to use the presented figures and photographs from Astrium's spacecraft and the development infrastructure of EADS Astrium GmbH - Satellites.

Further photographs and figures from industrial companies and institutes were provided to me by courtesy of:

- EADS Airbus S.A.S., Toulouse, France
- Institute of Space Systems, Universität Stuttgart, Germany
- RUAG Aerospace Sweden AB, Göteborg, Sweden
- Oerlikon Space AG, Zürich, Switzerland
- Satellite Services B.V., Katwijk, Netherlands
- ScopeSET GmbH, Fischbachau, Germany
- Northrop Grumman LITEF GmbH, Freiburg Germany
- Thyssen-Krupp Stahl AG, Duisburg, Germany

All figures used from these industrial providers are cited with the according source and copyright information. Figures from ESA and NASA Internet pages are used according to the copyright and usage conditions cited there, e.g. multimedia@esa.int, and also are cited with according copyright owner information.

This book is derived from a lecture at Universität Stuttgart and I want to express my gratitude to Prof. Dr. Hans-Peter Röser at the Institute of Space Systems for engaging me in 2003 as visiting lecturer and for his support to derive this book from my lecture material accumulated over the years. He also initiated the contact with Springer-Verlag GmbH and permitted his scientific assistants to support me in translating all the material.

This directly leads me over to thank Michael Fritz, Claas Ziemke and Alexander Brandt, for taking over parts of the translation work to keep me in schedule towards the manuscript delivery duedate. And finally I am very much obliged to Dave T. Haslam who accomplished the entire translation proofreading as native English speaker.

At Springer-Verlag GmbH I was very well supported by Mrs. Carmen Wolf and Dr. Christoph Baumann concerning all the questions about layout and other topics typically arising for a newcomer to book publishing.

Finally I want to thank my family and especially my wife for her encouragement and motivation, and for bearing me spending many evenings in front of the computer during the last months before manuscript submission.

Grateful for all the support I received,
Jens Eickhoff

Contents

Part III Advanced Technologies

List of Abbreviations

General Abbreviations

a.m.	above mentioned
cf.	confer
e.g.	example given
i.e.	Latin: id est $\Rightarrow$ that is
w.r.t.	with respect to

Technical Abbreviations

AI	Artificial intelligence
AIT	Assembly, integration and testing
ANSI	American National Standards Institute
AOCS	Attitude and orbit control system
AR	Acceptance review
ASIC	Application specific integrated circuit
ATV	Automated Transfer Vehicle
BIOS	Basic I/O-System
CAD	Computer aided design
CADU	Channel access data unit
CASE	Computer aided software engineering
CCD	Charge-coupled device
CCS	Central Checkout System
CCSDS	Consultative Committee for Space Data Systems
CDF	ESA / ESTEC's Concurrent Design Facility
CDR	Critical design review
CLTI-system	Continuous, linear, time invariant system of differential equations
CLTU	Command link transmission unit
Cmd	Command
CNES	Centre National d'Études Spatiales
CPU	Central processing unit
Ctrl	Control
DA-system	Differential-algebra system of equations
DB	Database
DD	Design document
DDR	Detailed design review
DEQ	Differential equation
DLR	Deutsches Zentrum für Luft- und Raumfahrt e. V.
DORIS	Doppler orbitography and radiopositioning integrated by satellite
DSP	Digital signal processor
dtd	XML document type definition file
ECLSS	Environment control and life support systems
ECSS	European Cooperation for Space Standardization
EEPROM	Electrically erasable PROM
EFM	Electrical functional model

EGSE	Electrical ground support equipment
EMC	Electromagnetic compatibility
EMF	Eclipse Modeling Framework
ESA	European Space Agency
ESOC	ESA Space Operations Center
ESTEC	European Space Research and Technology Center
FAA	Federal Aeronautics Association
FAR	Flight acceptance review
FCL	Foldback current limiter
FDIR	Failure detection, isolation and recovery
FEEP	Field emission electrical propulsion
FEM	Finite element method
FOG	Fiber-optic gyroscope
FPGA	Field programmable gate array
FVB	Functional Verification Bench
GEO	Geostationary Earth orbit
GPS	Global Positioning System
GSOC	German Space Operations Center
GSWS	Galileo Software Standard
HITL	Hardware in the loop
HW	Hardware
I/O	Input / output
IABG	Industrieanlagen-Betriebsgesellschaft mbH
ICD	Interface control document
ICD	Interface control document
ICU	Instrument control unit
IDE	Integrated development environment
IEEE	Institute of Electrical and Electronics Engineers
IF	Interface
IRR	Integration readiness review
IRS	"Institut für Raumfahrtsysteme" or "Institute of Space Systems", Universität Stuttgart, Germany
ISO	International Standardization Organization
ISS	International Space Station
ITT	Invitation to tender
JPL	NASA Jet Propulsion Laboratory
LAN	Local area network
LCL	Latch current limiter
LEO	Low Earth orbit
LEU	Load emulator unit
MDA	Model Driven Architecture
MDVE	Model-based Development & Verification Environment: (A system simulation and verification infrastructure from EADS Astrium GmbH - Satellites)
MEO	Medium Earth orbit
MGM	Magnetometer
MJD	Modified Julian date time format
MMI	Man-machine interface
MOF	Meta Object Facility
MRR	Mission requirements review

NASA	United States National Aeronautics and Space Administration
OAV-triple	Object-attribute-value triple
OBC	On-board computer
OBCP	On-board control procedure
OBSW	On-board software
OBT	On-board time
OCL	Object Constraint Language
ODE	Ordinary differential equation
OMG	Object Management Group
PA	Product assurance
PCDU	Power control and distribution unit
PDC	NASA / JPL's Product Design Center
PDE	Partial differential equation
PDR	Preliminary design review
PIM	Platform Independent Model
PPS	Pulse per second
PROM	Programmable ROM
PRR	Preliminary requirements review
PSM	Platform Specific Model
PSP	Product structure plan
PST	Polling sequence table of an on-board software
PUS	ESA Packet Utilization Standard
Pwr	Power
QR	Qualification review
RAM	Random access memory
RF	Radio frequency
ROM	Read-only memory
RPN	Reverse polish notation
RPT	Report
RTCA	Radio Technical Commission for Aeronautics Inc.
RWL	Reaction wheel
S/C	Spacecraft
SCOE	Special checkout equipment
SDAI	Standard data access interface
SDO	Astrium Satellite Design Office
SEDB	System engineering database
SM	State machine
SMT	Simulated mission time
SOA	Service oriented architecture
SPARC	Scalable processor architecture
SRD	Software / system requirements document
SRR	System Requirements Review
SRT	Simulation run time
STB	Satellite Testbed or System Testbed
STEP	Standards for Exchange of Product Model Data
STR	Start tracker
SVF	Software Verification Facility
SW	Software
SysML	Systems Modeling Language
TAI	Temps Atomique International: Terrestrial time reference from the International Bureau of Weights and Measures (BIPM).

TC	Telecommand
TCL	Tool Command Language (Generic open source Script Language)
TDRS	Tracking Data & Relay Satellite
TINA	Timeline Assistant: Mission Planning Tool of Astrium GmbH - Satellites
TM	Telemetry
TM/TC-FE	Telemetry / Telecommand-Frontend
TMS	Truth maintenance system
TN	Technical note
TRR	Test readiness review
TTR	Telecommand / telemetry responder
TUHH	Technische Universität Hamburg-Harburg, Germany
UMAN	User manual
UML	Unified Modeling Language
URD	User requirements document
UTC	Universal Time Code
VHDL	VHSIC (Very High Speed Integrated Circuits) hardware description language
W3C	World Wide Web Consortium
WGS	World Geodetic System
XML	Extensible Markup Language
xsd	XML schema definition file

Notation of Variables and Symbols

General Notation used in Mathematical Deductions:

u	an input parameter of a component
w	an output parameter of a component
y	a state variable of a component
x	a variable
$\bar{x}$	a vector
$\bar{\bar{x}}$	a matrix
z	Geometric location inside a component
∇	Tensor of geometric derivatives

Specific Notation used in Physical Formulae:

A	Area / surface area
c	Compound chemical concentration
D	Diffusion coefficient
h	Enthalpy
H	Momentum
m	Mass
M	Arbitrary variable in balance equations
N	Torque
P	Pressure
q	Quaternion
Q	Heat
r	Position of a spacecraft or celestial body
R	Specific gas constant
S	Source / sink term
t	Time
T	Temperature
v	Velocity
V	Volume
W_t	Technical work done
$\bar{\bar{\delta}}$	Unit tensor
θ	Matrix of moments of inertia
ρ	Density
τ	Viscosity of a fluid
φ	General rate flow over the system boundary
ω	Rotational rate

Further Notation

Explanations on graphical notations used in

- Unified Modeling Language for software design and in
- System Modeling Language for system modeling

can be found in chapters 8.4 and 10.1 respectively.

Introduction

As mentioned in the foreword, this book was written to serve as a reference material for the attendees of the author's lectures at the Institute of Space Systems, (IRS), Universität Stuttgart, Germany. These lectures cover the topic of "System Simulation in Satellite Development", parts 1 and 2, and the seminars on "Functional analysis and on-board Software Design", treating topics also based on simulation tools.

The lecture "System Simulation in Satellite Development", covers two semesters. This accompanying book contains all information for both sections, parts 1 and 2, with exception of the tutorials. The summer term lectures cover classical engineering process for spacecraft, and describes various characteristics of simulation based design verification, and the applied test tools. The focus is on the functional system simulation, with selected simulation and test tools. These topics are reflected in this book's part 1 with the chapters 1 to 5.

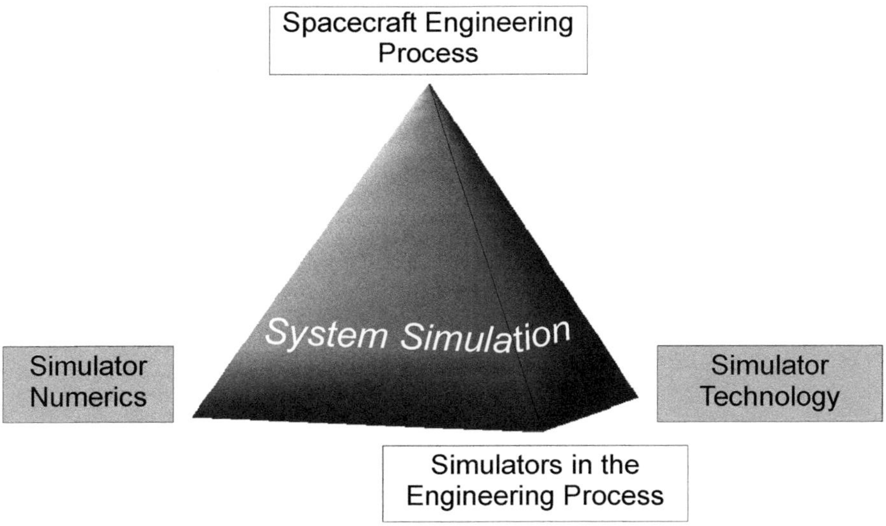

Figure 1: Simulator technology and the system engineering process.

The winter semester specializes in simulator numerics as well as implementation and software technologies for such simulation systems. Furthermore, outlined will be software architecture technologies for the development and verification of complex simulators. These topics are reflected in this book's part 2, which comprises the chapters 6 to 10. The book recapitulates formal functional notation methods like UML and explains the steps from the identified function set via software requirement documents down to the potential, and the limitations, of spacecraft on-board software verification and validation via ground based simulation.

Finally in part 3, including the chapters 11 to 14, some advanced and research topics in the field of spacecraft simulation respectively simulator technology are treated.

Aside from attendees of the lectures themselves, the book addresses lecturers and students looking for a consistent summary on the state of the art in modeling, simulation and spacecraft testing in the context of overall spacecraft system engineering.

The presented infrastructure examples are based either on the "Model-based Development and Verification Environment", (MDVE), from Astrium GmbH, or alternatively on the open source simulation tool *OpenSimKit*. The latter is an open source tool freely available to download from the Internet [23]. The original version was coded by the author and it has been enhanced by the developers community and diverse theses from students of various universities. Code examples from this source are used due to the compactness and simplicity of *OpenSimKit*[1] code compared to professional simulator implementations.

OpenSimKit

http://www.opensimkit.org

[1] *OpenSimKit* is a registered trademark of the author.

Only few know, how much you need to know,
for to know, is to realize how little one knows.

Werner Heisenberg

Part I

Simulation Based System Development

1 Complex Systems in Spaceflight

Ariane V164 © ESA / Arianespace

J. Eickhoff, *Simulating Spacecraft Systems*, Springer Aerospace Technology 1,
DOI 10.1007/978-3-642-01276-1_1, © Springer-Verlag Berlin Heidelberg 2009

Complex systems require detailed system engineering for their design, construction, verification, and finally for testing their completion and final validation. For many years system engineering has been supported by computer based system simulation techniques. In fact, as early as the Apollo program, NASA and its contractors applied such methods. However with today's significantly more powerful computers and sophisticated software tools, one can derive much greater performance from simulation infrastructures.

Such simulation techniques in principle are used in every industry sector from space, through the automotive industry to plant manufacturing. In every domain of use, special requirements concerning the design and verification tools are applicable. This book introduces the techniques for system simulation in the context of "Model-based System Engineering". Provided real world examples mainly originate from the field of satellite development. However, the introduced underlying steps of system design, verification and the provided software methods are of universal use.

The following are some examples of complex systems originating from the field of space applications, which require system simulation for their development. To maintain the analogy from ground into space firstly all launch vehicles will be addressed.

Rocket Launchers

Figure 1.1: Cutaway of an Ariane 5 rocket. © ESA

Figure 1.2: Soyuz launcher. © ESA

In the field of launch vehicles such as the Ariane 5, one must consider more than a significant number of detailed simulations needed for system development. Very complex simulations of the entire system as a whole are required for the verification of overall operational behavior. Effects to be simulated range from rocket engine simulation to the modeling of solid rocket boosters down to the functionality of trajectory control, stage separation, on-board software and orbit propagation.

Launcher Stages and their Subsystems

A subgroup of the launch vehicles is made up by the different launch vehicle stages. In the case of deep space probes, the uppermost stage also can be a part of the probe itself. The aspects, which have to be modeled respectively for simulation, range from very complex calculations of the engine's fluid dynamics to the functionalities of the turbo pumps, the propellant chemistry, thermodynamics, to the trajectory control and hardware / software thus needed.

Launcher stages with cryogenic propellants, used for direct insertion of spacecraft into desired orbits, (e.g. telecommunications satellites), differ from upper stages used for insertion into more complex orbits such as polar. The latter type of stages typically using reignitable hypergolic propellants.

Figure 1.3: Ariane 5 upper stage L10.
© ESA

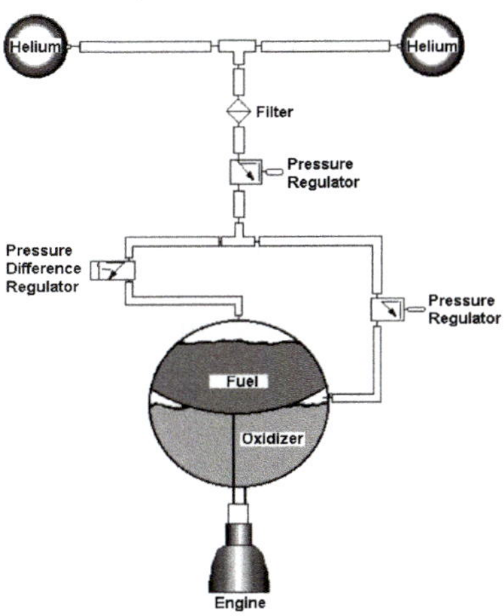

Figure 1.4: Medium energetic propulsion system.

Similar stages also are typically used for deceleration maneuvers of space probes after long coast phases or insertion into required orbit around the target celestial body / planet.

Apart from classical expendable launch vehicles, next reusable launch and space transportation systems shall be addressed.

Space Transportation and Supply Systems

Figure 1.5: Phoenix. © Astrium

Space shuttle systems, as spaceships and carrier capsules, in addition have to comply with requirements concerning safe reentry in to Earth's atmosphere. Manned shuttle systems furthermore have to be equipped with complex life support systems.

Freight container systems like the modern "Automated Transfer Vehicle", (ATV), are part of this supply systems group also. ATV has its own power supply system and is equipped with a fully autonomous docking system in order to couple itself to the "International Space Station", (ISS).

Figure 1.6: Automated Transfer Vehicle. © ESA

Although the current version of the ATV is a pure cargo transportation system, a future manned version capable of transporting astronauts to the International Space Station and back to Earth is envisaged.

Figure 1.7: Columbus launch with Shuttle. © NASA / ESA

This directly leads to the next – extremely demanding – category of spacecraft in a wider sense space stations, respectively the station modules.

Manned Space Laboratories and Auxiliary Systems

Figure 1.8: Columbus module in Space Shuttle bay. © Astrium

Systems like the International Space Station, (ISS), are of such complexity that it is not possible to simulate them as a whole. For preliminary calculations and the analysis of dynamical nominal and failure value ranges in operations, specific subsystem simulations are necessary, which have to be correlated at a system level. This has to be performed in conception and design phases as well as later for monitoring the operational conditions.

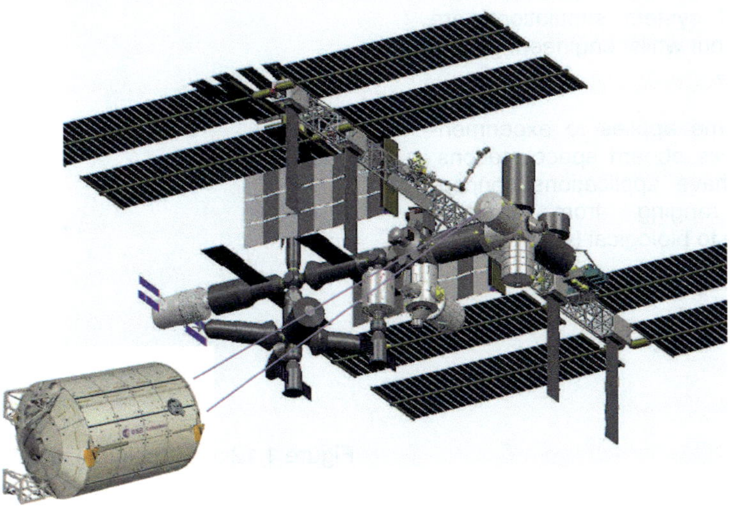

Figure 1.9: Columbus module of the International Space Station. © Astrium

Typical separately modeled and simulated subsystems are laboratories and auxiliary systems, as well as life support, power supply and attitude and orbit control systems.

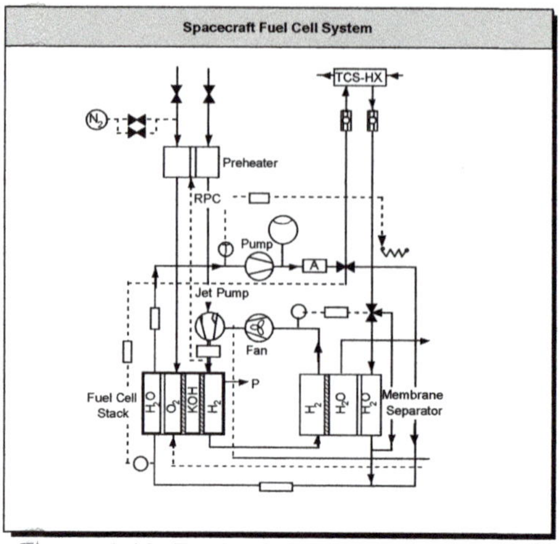

Figure 1.10: Fuel cell power subsystem for spacecraft.

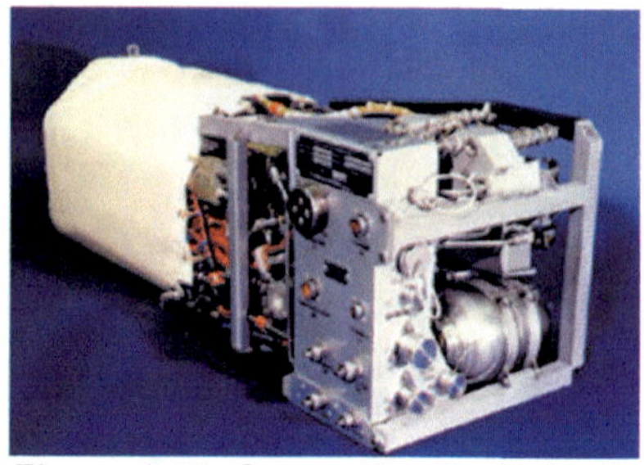

Figure 1.11: Space Shuttle fuel cell. © NASA

The power supply systems of manned spacecraft in most cases are based on fuel cells, while the ones of space stations are supplied by a combination of solar arrays, battery systems and if necessary, fuel cell / Sabatier reactor systems. During the conception of such systems not only complex cybernetic, but also physical / chemical effects have to be modeled. As a result a multitude of system simulations are carried out whilst engineering such systems.

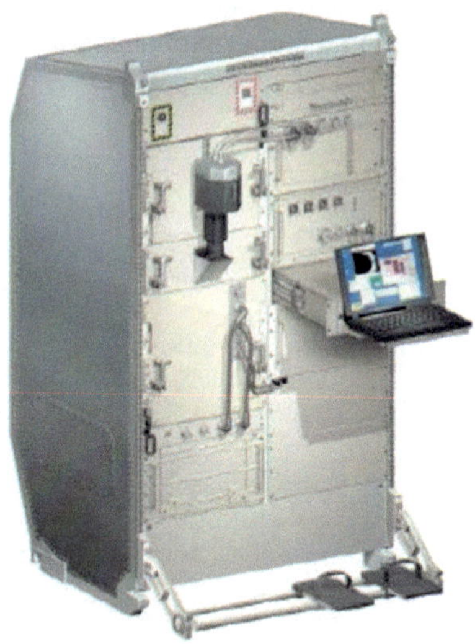

Figure 1.12: Fluid science laboratory. © ESA

The same applies to experimenting racks aboard space stations, which have applications conceptions ranging from material science to biological laboratories.

The European space laboratory, Columbus, is already equipped with a multitude of experiment racks of various types.

Figure 1.13: Columbus inside view. © Astrium

Finally among the technical infrastructure of space stations, besides the basic infrastructure like "Attitude and Orbit Control Systems", (AOCS), power supply systems and "Environmental Control and Life Support Systems", (ECLSS), also infrastructural elements for the external maintenance are found – e.g. robotic arms and other assembly support systems. Today these are also highly complex, programmable and highly automated functional elements which require detailed calculations and system simulation during their development.

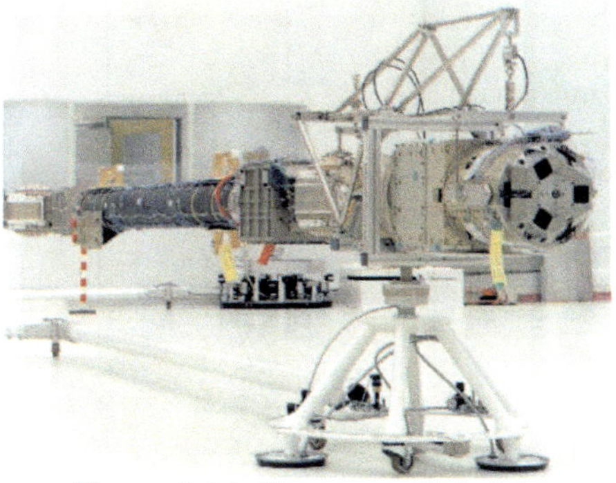
Figure 1.14: European Robot Arm. © Astrium

Finally the essential category of spacecraft, which make up the largest number, should not be overlooked: The research, telecommunications and military satellites, and ultimately space probes which explore foreign planets and the remote parts of the Solar System.

Satellites and Space Probes

Satellites are equipped with complex attitude and orbit control systems which, depending on the mission, may have extreme requirements regarding their accuracy. The payloads of satellites range from radar systems through optical systems to special applications such as

Figure 1.15: MetOp. © Astrium

gradiometers. Space probes, in addition, may have mission specific subsystems such as radio thermal power generators, or landers which are to be placed safely on a remote celestial body.

This book concentrates on the system simulation for spacecraft and illustrates most of the facts and coherences using examples from the domain of satellite development – the author's field of work. Nevertheless systems of comparable complexity can be found in numerous engineering domains, in the field of aviation, plant manufacturing, powerplant construction, automotive, and medical engineering. In all such domains, system simulations are applied for development support and for system testing.

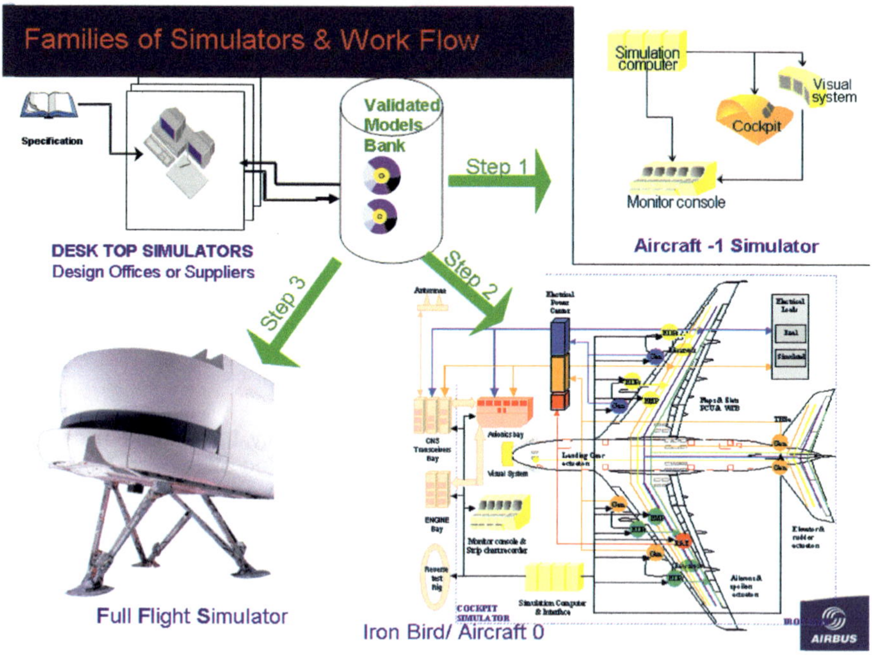

Figure 1.16: System simulation in airplane development. © Airbus S.A.S.

Before describing the simulation technology in more detail, the term "simulation" should be clarified by a definition:

> Simulation is an approach for analyzing a dynamic system for gaining an insight to its dynamic behavior. Simulation implies conducting experiments on a model of the system. In the context of simulation the term "simulated system" refers to the real world system, while the term "simulation model" refers to an abstraction of the real world system.

2 System Simulation in System Engineering

Metop © Astrium

J. Eickhoff, *Simulating Spacecraft Systems*, Springer Aerospace Technology 1,
DOI 10.1007/978-3-642-01276-1_2, © Springer-Verlag Berlin Heidelberg 2009

2.1 Development Process Phases for Spacecraft

The system development of spacecraft is divided into four developmental phases, plus an operational phase and - if necessary - a disposal phase, as depicted in the figure below. The system manufacturer, e.g. of an entire satellite or a subunit, usually participates in the first four phases as well as in the start up at the beginning of phase E. Established during this development process are some important milestone reviews with the customer (which for a spacecraft usually is a space agency or a commercial contractor, for a subsystem it is the spacecraft prime contractor).

The typical milestones, their position within the spacecraft development process together with their abbreviations are outlined in the figure below, too.

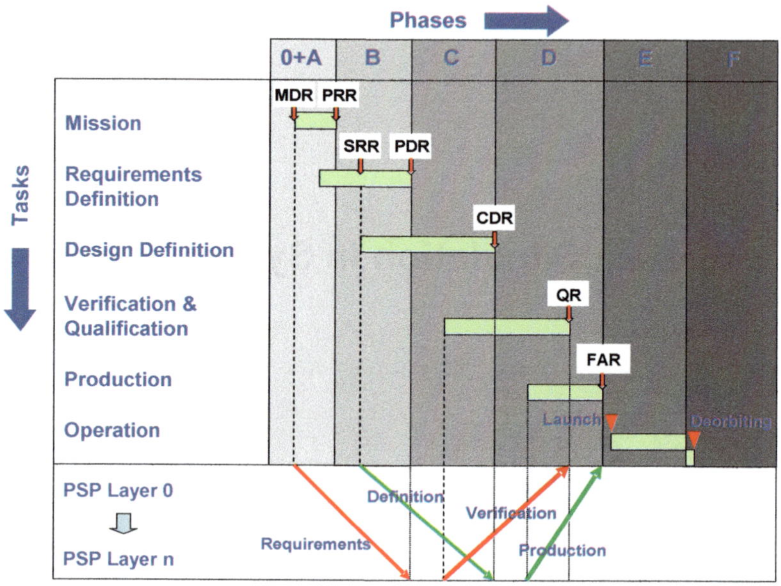

Figure 2.1: Phases and milestones in space projects. Source: ECSS-M30A

Phase A, sometimes including a previous conceptual phase 0, is carried out as a study. During this phase the spacecraft manufacturer analyzes the requirements for a satellite in order to accomplish a specific mission with particular quantitative results. One example is the analysis of requirements for orbit parameters and characteristics in order to achieve the designated resolution and revisit cycles with a certain payload. At this point with the "Mission Requirements Review", (MRR), definition of requirements starts for the the overall system level, which is Level 0 of the "Product Structure Plan", (PSP). This implies design requirements for the satellite itself concerning power supply for the payload, data transfer to ground, attitude and orbit control and requirements for the power and thermal control systems. This analysis is initially limited to pure budget analyses, (e.g. necessary battery capacity on board, memory capacity etc.). The only exceptions are detailed orbit and ground station contact simulations. Phase A is finished by the "Preliminary Requirements Review", (PRR).

The work contracts of phase A are usually assigned to two or more competitors simultaneously, so that the customer, e.g. the space agency, receives at least two different, independent analyses and concepts worked out for the planned mission. The customer chooses the best of the received phase A concepts and submits an "Invitation to Tender", (ITT), for the development phases B, C and D. The phase B/C/D development is awarded in most cases as one contract to the winner of the B/C/D tender. This contract usually includes the support of the satellite operations from the manufacturer at the start of phase E.

During phase B the requirements for the components of a satellite are worked out, for example

- algorithmic requirements for attitude and orbit control,
- qualitative and quantitative requirements on equipment components and their design,
- qualitative and quantitative requirements on the entire system design regarding structure, thermal and power control functionality,
- functional and performance requirements on payload and its control,
- and, last but not least, technical requirements concerning the on-board software.

After the adequate specification of requirements for orbits, system, operational and payload functionalities etc., the "System Requirements Review", (SRR), with the customer takes place. The design definition on system level now begins:

- Initial attitude / orbit control algorithms are developed.
- The exact system topology is specified as product structure.
- First CAD drawings and electrical block diagrams are created.
- Furthermore, thermal and mechanical calculations are performed for the first time.

Phase B is finished with the "Preliminary Design Review", (PDR).
After this review milestone the invitations to tender for equipment subcontractors are submitted subsequently for development and manufacturing of elements on the lower PSP levels.

Phase C is in fact the real definition phase. The design on system level is consolidated once again. Components and subsystems are defined at the level of subcontractors (PSP levels 1 to n). Phase C is completed with the "Critical Design Review", (CDR). For standard components on subsystem level also first equipment verifications take place on hardware breadboard or engineering models.

The subsequent phase D is the production phase, which is finished by the completion of an operational system, e.g. the satellite. The final acceptance milestone is the "Flight Acceptance Review", (FAR). The complete production must be finished by then. For the spacecraft prime contractor however, mostly more critical is the previous milestone, namely the "Qualification Review", (QR). This review marks the successful completion of all equipment verification tests, integration tests and system verification tests. The latter also comprise complex verification in a thermal vacuum chamber and mechanical vibration tests on a shaker. For QR also all parts must be space qualified, (e.g. electronic components such as application specific integrated circuits), as well as all applied manufacturing processes for electronics, soldering, bonding etc.

After phase D, the system is taken into operation (e.g. through launch of a satellite). And within the operational phase E, the system manufacturer still is bound to support the spacecraft operating agency during the commissioning phase and the on-orbit characterization and calibration of payloads etc. For completeness, the disposal phase F shall be mentioned, which takes place after the operational phase E and comprises shutdown and eventual de-orbiting of the spacecraft.

2.2 A System, its Control Functions and their Modeling

A system, except for a few of its passive elements, typically can be abstracted to control functions and controlled physics. This applies for both entire spacecraft as well as subcomponents, for example, a radar payload, a rocket stage etc. Examples for entirely passive elements are, for example, the central structure of satellites - without deployable antennae - or sunshields for optical instruments and so forth. The design analysis for such parts shall not be topic of this book.

Instead the focus of this volume is on system functionalities and control functions which will be formally analyzed and modeled. The design and verification of such functional systems these days is mostly performed through applying system simulation technologies. In this scope both the physics of the system functions, (w.r.t. electrics, mechanics, thermodynamics, fluid dynamics etc.), are to be modeled as well as the specifications of the system controllers. These might range from pure mechanical controls to software based applications. For this sort of integrated engineering approach for system physics, plus control technology, typically system simulations are applied on various levels of detail. The technical criteria for such simulations are focusing on

- analysis and simulation of the interaction of all system components,
- resulting in the simulation of the complete system as a whole, achieved by modeling of:
    - ◇ System components and their functionality,
    - ◇ Component interfaces and interactions through such connections,
    - ◇ The system's external environment throughout operation.

The level of detail and the complexity of system modeling are driven by questions raised from the domain of system engineering. Simulation models only reflect the real equipment functionally, which means, for example, concerning the equipment's communication protocols, its operational modes or power consumption. In functional simulations the goal is not to exactly reflect the internal design of an equipment component. Rather, the level of detail in modeling, the modeled effects and on the other side the resulting simplifications in the simulation are adapted to the requirements and the requested precision. These requirements are driven by the type of system analyses to be performed with the simulator.

System simulation is characterized by application in different development phases of the project. It is an integrated task inside the domain of system engineering. The resulting questions from system engineering - such as system performance verification or system internal failure management verification - impose the boundary conditions onto the applied simulation techniques in a project. Simulation technologies nowadays are very advanced, so that entire complex applications like aircraft, satellites etc. can be developed purely based on simulation techniques. The former development philosophy to implement, for example a separate mechanical prototype for a satellite, (see figure 2.7), and a separate thermal model before assembling the real flight model, is outdated. The old approach also no longer can be financed according to the shrinking mission budgets of private and institutional customers. The reduced number of models however may not endanger mission success. Risk mitigation therefore is achieved via

- a system design based on standardized components as far as possible,
- the simulation of elaborated configurations before start of hardware manufacture,
- and an extensive use of simulation techniques to support all important steps of system design verification, and of "Assembly, Integration and Testing", (AIT).

This technology approach is called, "model-based development and verification". CryoSat 1 was the first European Space Agency satellite project in which the satellite engineering model prototype was replaced entirely by simulation.

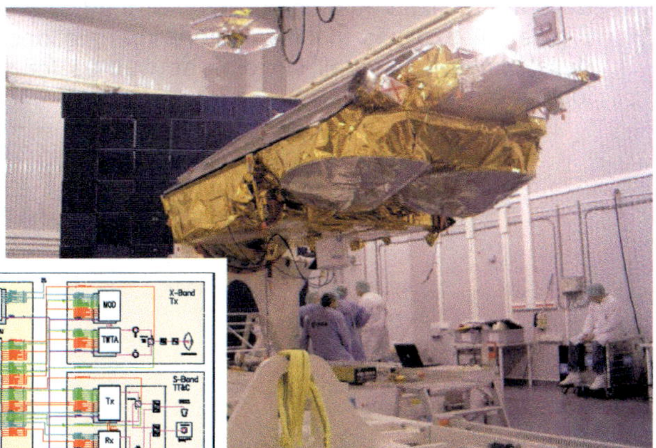

Figure 2.2: CryoSat 1 © ESA

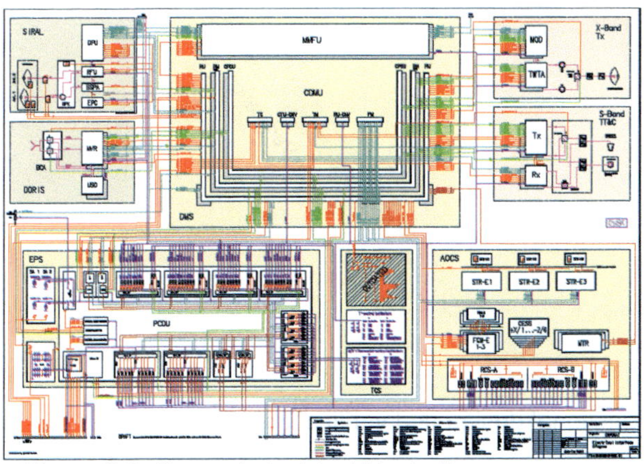

Figure 2.3: Cryosat electrical block diagram. © Astrium

At this point, discussing simulation based system verification, it is relevant to precisely define terms "Verification" and "Validation":

- Verification means checking that all defined system requirements, which are laid down in a formal requirements document, are fulfilled. Proof can be achieved by analysis computation, simulations, tests and inspections - the appropriate method to be chosen according to the requirement type.
- Validation is to check that the system, all in all, performs as originally expected - e.g. for a satellite payload, that it provides the images with desired spacial respectively spectral resolution.

The next paragraphs will start explaining the concepts of system physics simulation and functions simulation. Thereafter the discussion will lead over to verification and validation topics and will explain the verification and validation tasks in overall system engineering that can be based on simulation technology, and tasks which need support by further means.

2.3 Algorithms, Software and Hardware Development and Verification

The application and embedding of system simulation within system engineering is discussed on the basis of the figure below. Modern complex systems, such as spacecraft, automobiles, airplanes, power plants or other machinery installation these days are realized as a combination of hardware equipment and software for control. Figure 2.4 explains the elementary interrelations which are applicable both for the development of overall systems as well as for subsystems, such as satellite payloads.

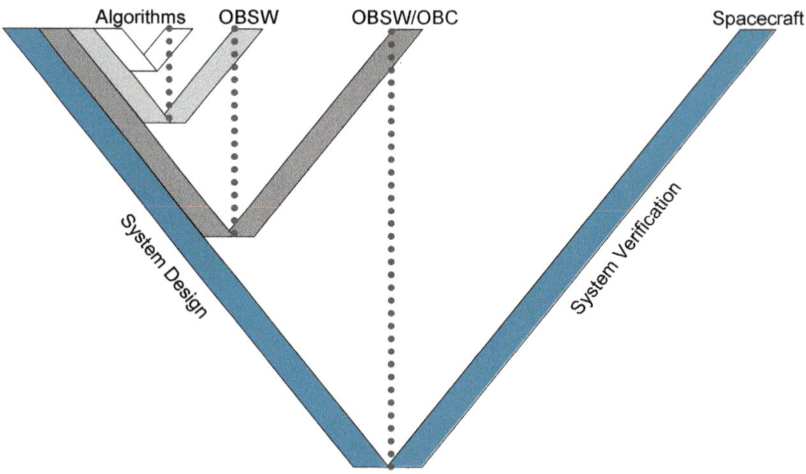

Figure 2.4: Functional design and verification.

The figure depicts the development process as a classic V-model, consisting of the main branches, design and verification. This V-Model however, in the engineering of an entire system, breaks down to a set of interlinked V-steps as shown in figure 2.4. It can be read as follows - from right to left:

- To be able to verify the detailed requirements on the system (here a spacecraft) by system testing, a suitable target computer based control software has to be available.
- This means, hardware (HW) / software (SW) integration has to be completed for this step.
- To verify the proper HW / SW integration, already a pre-verified control software must be available. In spacecraft engineering this is on-board software.
- And finally to be able to pre-verify the on-board software, for the integrated control algorithms - e.g. for attitude control - reference data must be available. The latter come from an algorithm verification campaign, which itself has to be already completed at that point.

The majority of project management literature only tackles, in a simplified way, design and verification of system and subsystem, (PSP levels in figure 2.1), as a semi-two layer problem. The interdependencies from algorithm design down to system verification as shown in figure 2.4 are not addressed. These interrelations however, are essential for understanding which participant in a project at which time is dependent on which input results. Or, formulated vice versa, who in the project will be pushed onto the critical path in development by delayed input.

Figure 2.4 furthermore points to an important fact concerning system modeling and verification infrastructures. For verification of any functionality test, infrastructure is needed, independent from the level being of simplest algorithm verification up to extremely complex top level system tests. So in a system development, a design and test environment must be foreseen

- for the control algorithms,
- for transformation of the algorithms into on-board software code (before being integrated with target hardware),
- for hardware / software integration and finally,
- for the entire system - here spacecraft - including its integrated on-board computer.

In the ideal case, all these infrastructures should be based on a common toolkit suite and the results should be portable from one development step to the next. The requirements, functionalities and implementation examples for such an overall infrastructure are presented in chapter 3. The verification concept for a spacecraft including its on-board software, as shown in figure 2.4, already depicts a stepwise development principle, which also is applied in many other industries:

- In an initial step, the physics of the system is modeled - in software or as a test stand - and the developed control algorithms are integrated to control the system. The algorithms mostly are not yet implemented in the target

programming language, neither on targeted hardware. This type of test is called "Algorithm in the Loop".
- Secondly, the algorithms are coded in software in the target language. The now available control software is loaded onto the - eventually modified - test stand, again, to control the system. This type of tests is called "Software in the Loop".
- The third step is to load the control software onto a representative target computer, which now controls the hybrid test stand. The final software on the target computer now has simulated system physics. This principle is called "Controller in the Loop". The first step of tests is the software integration on the target computer.
- The fourth and final step of system testing now aims to make the control software on the target hardware now control the real system, and no longer the test stand's system simulation. This deployment phase is called "Hardware in the Loop", (HITL). In spacecraft development simulators here are required for computations of stimuli parameters to reflect gravity-free space conditions.

For all these steps, each time the requirements documentation for the "Item under Test" - the algorithm, the on-board software, etc. - must be written. The principal verification approaches are to be documented, the design of the item under test is to be documented and finally test plans for the verification are to be generated and results are to be collected and analyzed in test reports.

This kind of simulation based system development approach requires fundamentally new workflow processes to be applied, both concerning applied technology as well as with respect to project organization and distribution of responsibilities. In brief, what is to be managed can be summed up as

- the integration of engineering disciplines such as mechanics / kinematics, electrics, thermal and system operations / data handling,
- the allocation of a simulator infrastructure responsible to the project. This role also is called "simulator architect", and the task is to manage the in-time and requirement compliant development and qualification and installation of the simulation infrastructure.
- The tasks to be managed furthermore comprise the consistent application of simulators, system models, configuration databases and test procedures over all project phases, (B, C / D, E).
- The system design, simulation and verification environment has to be standardized as much as possible to reuse qualified elements in the next space project.
- And finally a consolidated work process is needed for the integrated development and test of
  - ◇ satellite hardware and software,
  - ◇ system simulator,
  - ◇ check-out software and equipment.

It has to be kept in mind that the functionality of the simulation based test stands has to be defined, implemented and verified following an analogous process as explained for the spacecraft itself above.

2.4 Functional System Validation

At end of a system development, the goal is to have a validated system design. This means - see definition of the term, "validation", beforehand - that at system delivery, launch, series production or plant commissioning it is assured, that the system functions as expected. And this is regarding functionality, reliability and especially performance.

For this validation, either multiple test stands are required - not to be mixed up with the verification infrastructures discussed in the previous section, since for verification of system requirements a test stand eventually only needs to comprise parts of the overall system. Alternatively, also a test operation of the real system can be performed. In the automobile industry, an example for performance validation would be testing a new car prototype on a roller rig following a specified load cycle to validate the fuel consumption. A real system test could, for example, be a road test on a test track, a drive under polar or desert climate conditions.

However, here exactly are where the specific problems arise for spacecraft. An automobile prototype which on the test track does not accelerate, brake as desired, shows insufficient steering control response or otherwise performs inadequately can be reoptimized before series manufacture starts. E.g. drive train geometry can be perfected.

Spacecraft - and especially Earth observation satellites and deep space probes - themselves are prototypes. A validation of single major components, e.g. of a radar payload in an EMC chamber, usually is achievable. Tests comprising the entire system under real space conditions impose a huge effort. System tests which typically are still carried out before launch, based on "test stands" are thermal tests in a thermal vacuum chamber,

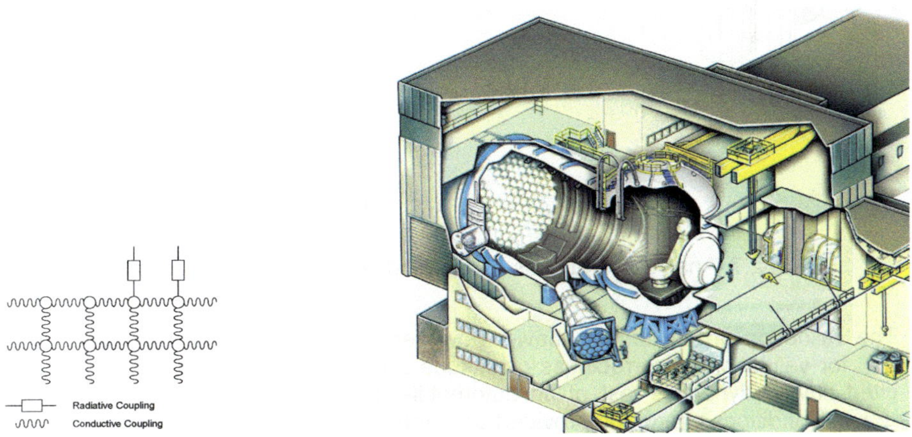

Figure 2.5: Analyses and tests of thermal design. Cutaway © ESA

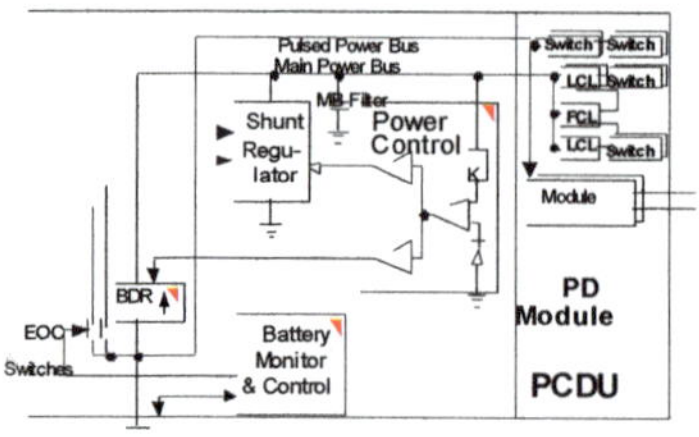

Figure 2.6: Analyses and tests concerning electromagnetic compatibility. © Astrium

tests concerning electromagnetic compatibility, (EMC), of the overall system in an EMC chamber as well as tests concerning structural mechanics compatibility to the load spectrum of the foreseen launcher. These tests are performed with the spacecraft on a shaker.

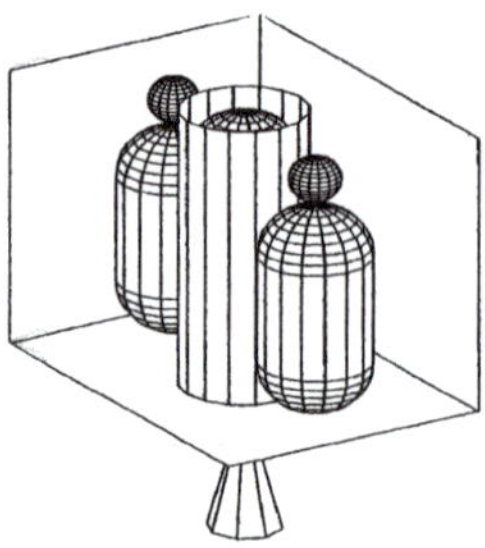

Figure 2.7: Analyses and tests concerning structure mechanics. Photo © ESA

The validation of functional behavior however, turns out to be a very difficult topic, e.g. for validation of correctness of attitude and orbit control - as an analog to the cited driving dynamics test of an automobile. For a spacecraft, the essential zero gravity conditions cannot be provided on Earth. Therefore it theoretically is necessary to work with test stands where attitude and position of the spacecraft can be modeled geometrically. On 3-dimensional turn table test stands with mounted spacecraft sensors, e.g. Sun and Earth visibilities for each sensor formerly were simulated by optical and infrared lamps. Similar setups existed for optical injection of star positions into star trackers. On these turn tables, at the same time, the angular rate sensors of the spacecraft could be stimulated.

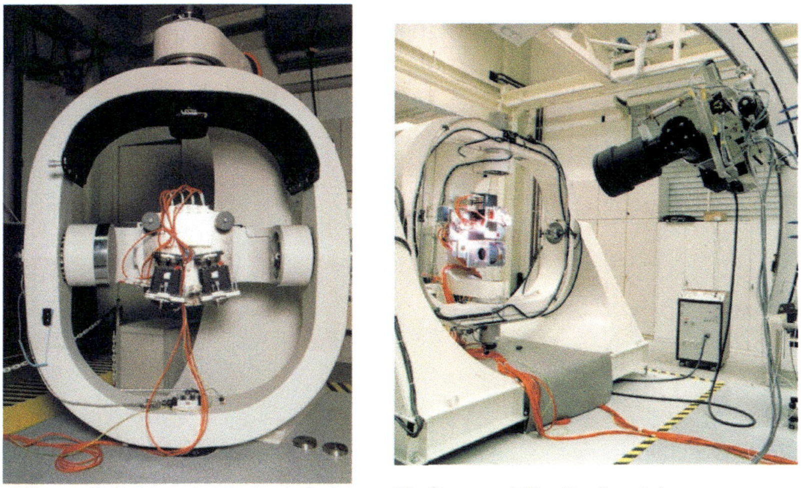

Figure 2.8: Turn table installation with Sun and Earth simulator. © Astrium

From the figure of the turn table installation shown above - which is limited to stimulation of Sun and Earth sensors of a satellite attitude control system - the test bench complexity already can be estimated. The complexity even increases if a comparable approach is to be followed for additional attitude sensors, rotational rate sensors and finally also the for satellite's actuators - eventually even the pyrotechnic ones. Including all this hardware equipment in a closed-loop verification test stand is far out of financial mission budgets today.

For this reason, at least for unmanned spacecraft, a more pragmatic approach is followed based on simulation technologies as they are treated in this book. The approach comprises

- the limitation of validation tests on component tests,
- to verify the entire system against its specifications as soon as possible,
- the verification of the on-board software, (OBSW), in a consistent approach from "Algorithm in the Loop", down to runs on target hardware against real system components and,
- to limit the system validation on the test stand types for thermal, mechanical and EMC as depicted in figures 2.5 to 2.7.

This mitigates the risk of an entirely non functional system, to an acceptable level and the final performance validation is carried out during the on orbit commissioning part of the operational phase E. In theory, not before these final tests, the system design validation is closed and the same applies for the design of all applied test infrastructures which were used for intermediate verification steps.

3 Simulation Tools for System Analysis and Verification

TanDEM-X © Astrium

J. Eickhoff, *Simulating Spacecraft Systems*, Springer Aerospace Technology 1,
DOI 10.1007/978-3-642-01276-1_3, © Springer-Verlag Berlin Heidelberg 2009

The main development tasks within the different development phases of spacecraft are grouped together by phase in the following table. From these tasks the requirements for the spacecraft development and verification infrastructures can be derived directly.

Table 3.1: Main tasks in spacecraft development phases.

Phase 0/A	Phase B	Phase C	Phase D	Phase E
• Mission analysis • Development of system concept and configuration alternatives • Analysis of these concepts and configurations, "system-trade-offs" • Development of standardized documentation for the selected variant	• System design refinement and design verification • Development and verification of system and equipment specifications • Functional algorithm design and performance verification • Design support regarding inter-faces and budgets	• Subcontracting of component manufacturing • Detailed design of components and system layout • EGSE develop-ment and test • On-board software development and verification • Development and validation of test procedures • Unit and subsystem tests	• Software verification • System integration and tests • Validation regarding operational and functional performance • Development and verification of flight procedures	• Ground segment validation • Operator training • Launch • In-orbit commissioning • Payload calibration • performance evaluation • Prime contractor provides trouble shooting support for spacecraft

Phases 0/A aim toward the development of a system concept. Basic system characteristics are elaborated and defined here. For a satellite, these comprise the orbit definition and decisions on system configurations such as body mounted or deployable solar panels, etc. These tasks basically are design tasks. Design refinements are the main task of phase B, which is completed by finishing the design at system level and the requirements completion up to the level of detailed components. During these phases for all physical design elements - such as spacecraft structure - as well as for the functions to be implemented technical requirement specifications are created.

With current technology one collects these requirements as text modules in databases. Later in the project test, documents - such as test plans, procedures and reports - can be assigned to these requirement text modules. The requirements specify which level they need to be verified on component level - by unit tests, on integration level - via integration tests, or on system level - via system tests. Through these test documents it can be demonstrated how the requirements are to be verified.

The subsequent development phases C and D focus on construction and verification of components and on their integration right up to the complete system. Finally phase E tasks focus on in-orbit verification of the spacecraft with respect to functionality and performance.

Here it can be identified that in phases 0/A/B design tools and design simulators are required. Phases C/D in contrast require tools, which support verification tasks. As depicted in the table on the next page, seven system design and verification infrastructure types can be determined - where in the ideal case one should evolve from the previous.

Table 3.2: Steps of system development and according infrastructure setups.

1) Conceptualization infrastructure: Design Offices.

Mission concept analyses, budget analyses etc. based on spreadsheets. Orbit analyses and simulations based on commercial tools.

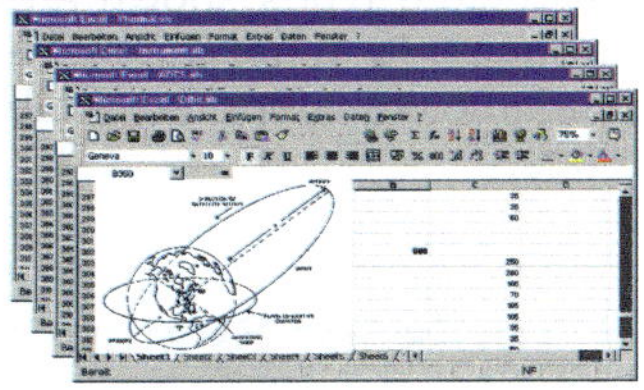

2) Design infrastructure: Discipline specific tools.

For design of AOCS algorithms, for control engineering, thermal design, electrical design, structure analysis.

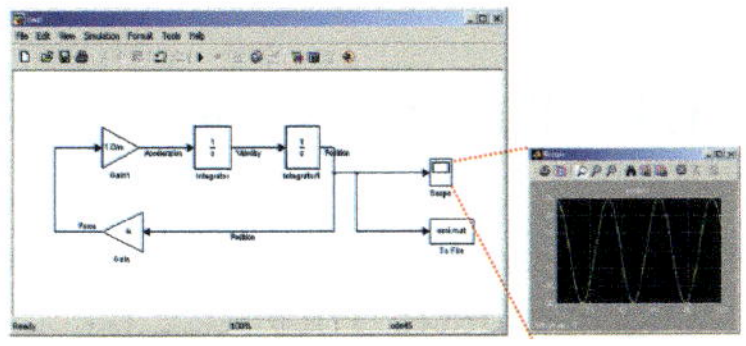

3) Algorithm verification infrastructure: "Algorithm in the Loop", (FVB).

First complete functional modeling of satellite and space environment. Test of control algorithms from 2) within the simulation reference.

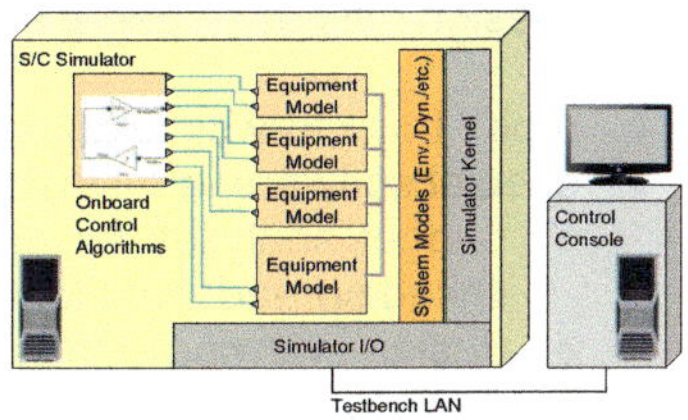

4) Software verification infrastructure: "Software in the Loop", (SVF).

Detailed on-board computer, (OBC), simulation, allowing on-board software, (OBSW). Binary execution in an OBC model within a simulated satellite.

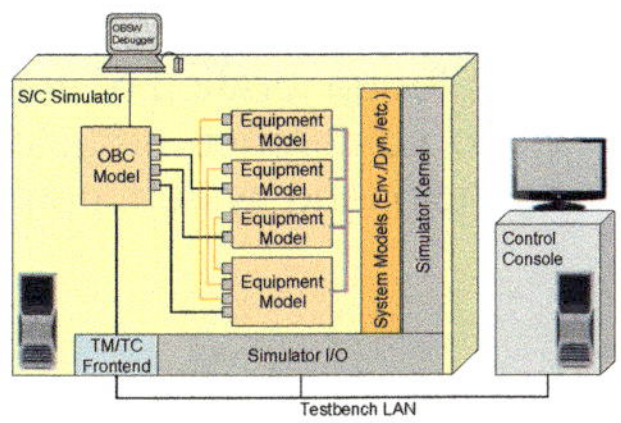

5) Hybrid verification infrastructure: "Controller in the Loop", (STB).

OBC available in hardware. Test of compatibility between OBC software with hardware and test of OBSW with simulated satellite.

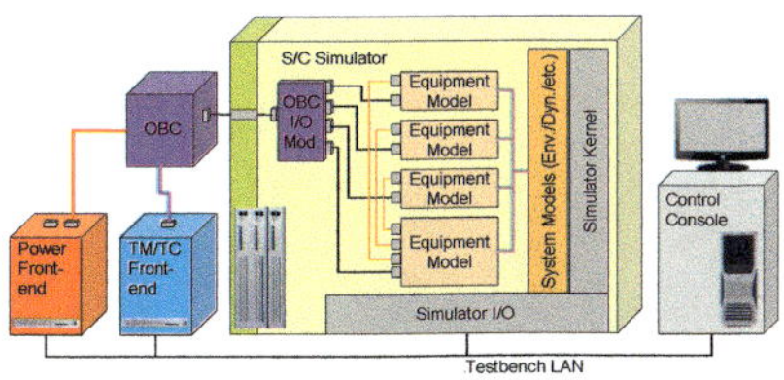

6) Hybrid verification infrastructure: "Hardware in the Loop", (EFM).

"Hardware in the Loop" extension up to the complete HW/SW compatibility with all equipment.

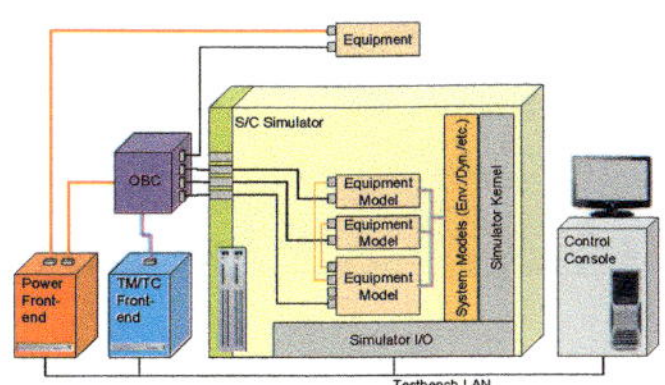

7) Simulator for operations support:

Detailed simulation of the satellite. Connected to the ground station for operations support.

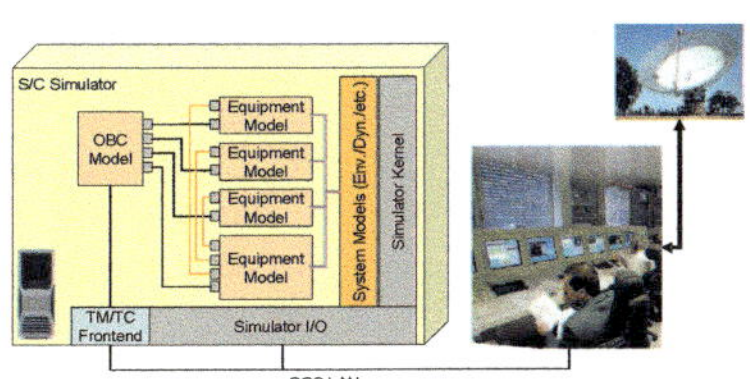

Simulation technology in each setup plays a significant but changing role. The first two infrastructure types shown are conception and design infrastructures, whereas those belonging to steps 3-7 are based on a common verification infrastructure. All these infrastructures depicted in table 3.2 are explained in more detail in the following subsections.

3.1 Tools for System Design and Dimensioning

3.1.1 Tools for System Predesign and Conception

Phase A analyses nowadays are performed based on an optimised, semi-concurrent, working approach. They take place partly in classic distributed work - every domain specialist on his own, in his own office - and partly in integrated team sessions in so-called "Design Offices", open-plan offices with

- orbit / trajectory simulation tools,
- special software infrastructure for
    - ◇ budget analyses for all spacecraft system engineering domains, applying linked spreadsheet tools and
    - ◇ design viewing infrastructure like beamers or even 3D visualization devices,
- databases for analysis results management,
- tools for generation of analysis reports,
- video conference infrastructure for meetings between customer, (e.g. space agency), and system prime contractor.

Figure 3.1: Satellite Design Office (Astrium Friedrichshafen). © Astrium

Such infrastructures are available both at prime contractor sites as well as in space agencies. Their naming varies accordingly. Some examples are listed here:

- Astrium "Satellite Design Office".
- ESA "Concurrent Design Facility".
- NASA / JPL "Project Design Center".

Budget analyses and orbit analyses respectively, orbit simulations are performed in such Design Office meetings for conceptual mission design. However, there are no functional spacecraft simulations performed in this phase of spacecraft development, since neither the system detailed topology nor any system control functionalities are yet quantitatively defined.

Figure 3.2: ESA Concurrent Design Facility. © ESA / ESTEC

The following figure outlines the the logic of a total process, which belongs to a phase A system study. The chosen example is taken from the Astrium Satellite Design Office.

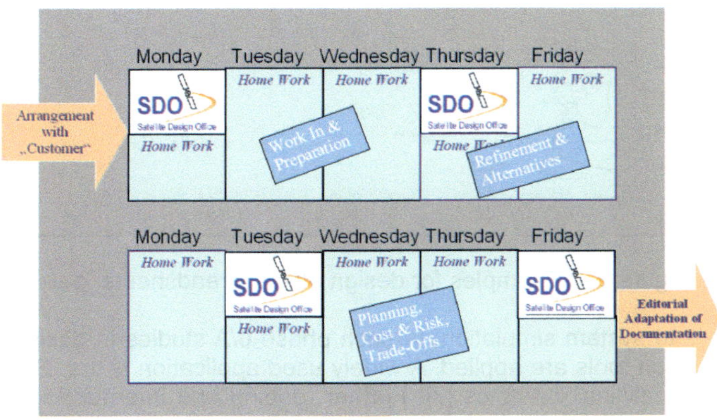

Figure 3.3: Study example for the Satellite Design Office. © Astrium

In this example, the system concept design takes two weeks including several Design Office meetings. Non-technical tasks like risk analyses and report writing are also proceeded in the frame of these time slots. The participating engineering disciplines typically are:

System	RF communication
Payload	Power supply
Operations and on-board software	Mechanical configuration / structure
Orbit analysis	Thermal design
Electrical system and on-board data handling	Scheduling and ground operations
AOCS	Cost and risk analysis
Propulsion subsystem	+ SDO sessions moderator

Essential for this approach is the coordinated work process concerning development logic and sequence and including systematic documentation of all evaluated system design alternatives and mission concepts which came up throughout such a process. This documentation is essential for archiving the design variant selection decisions, selected concepts and discarded alternatives to build up a knowledge-base for similar successor projects. Furthermore, the concerted sessions improve process efficiency for these, mostly scarcely, financed phase 0/A studies.

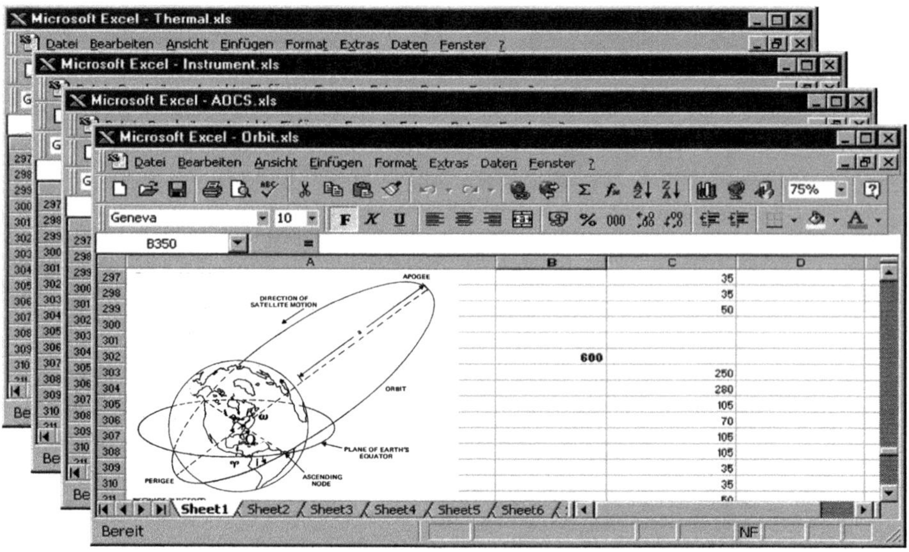

Figure 3.4: Examples for design office spreadsheets © Astrium

With respect to system simulation, in such phase 0/A studies in design offices, only orbit propagation tools are applied. A widely used application is the "Satellite Toolkit", (STK), from Analytical Graphics [7]. Further reading and Internet pages concerning Design Offices are listed in the according subsection of this book's references annex.

3.1.2 Functional System Analysis Tools for Phase B

In phase B simulation tools are applied for the first time in the system engineering process for overall analysis and subsystem design analysis. Besides tools for functional simulation, further analysis tools are applied in the disciplines of structure mechanics, thermal engineering and electric design. These include finite element method tools, (FEM), thermal network solvers etc. The functional tools applied in phase B typically are commercial toolboxes which comprise special libraries for control engineering and system dynamics engineering. The most widespread infrastructure of this kind is Matlab[2] together with the add-on toolboxes Simulink and Stateflow. An open source competitor is SciLab and a semi open tool is Modelica, which also is largely used in automotive industry.

Basic tool technology and an overview on the functionality of such tools is shown here through the example of Matlab / Simulink / Stateflow. Further reading and Internet pages on these type of simulation toolkits for system design analysis can be found in the according subsection of the references annex.

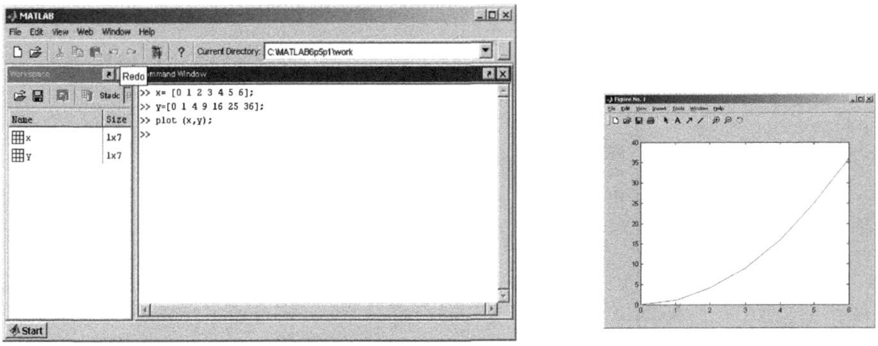

Figure 3.5: Matlab user interface.

Matlab is a mathematical interpreter for matrix and vector algebra, ("Matrix-Laboratory"). Based on this tool diverse types of mathematical computations can be carried out. Matlab provides

- global variable definitions,
- extensive functions for standard mathematical operators, especially for vector and matrix algebra,
- substantial graphical visualization features,
- the interactive computation considering user input and commands,
- the possibility of script-based computations using so-called *.m-Scripts,
- the possible extension of its function scope by shared libraries (Mex), coded in C/C++,
- and finally, it can be complemented by the simulation and automata toolboxes Simulink and Stateflow.

[2] Matlab, Simulink and Stateflow are registered trademarks of The MathWorks Inc.

Simulink is a simulation kit allowing graphical modeling of dynamic systems. The functional elements of a signal flow chain can be assembled from element blocks via a mouseclick. As well as mathematical operator blocks, also digital and analog visualization blocks can be added to the overall system graph.

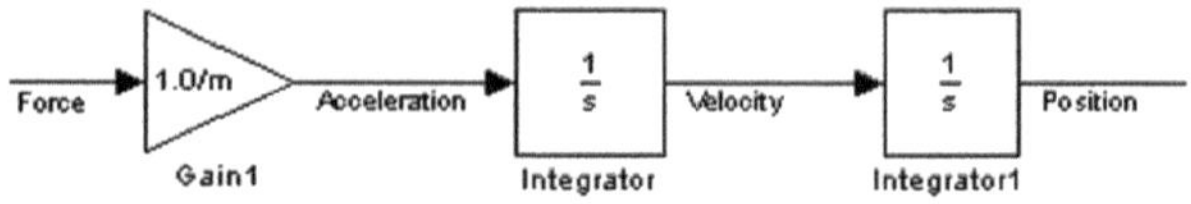

Figure 3.6: Assembly of Simulink blocks.

The vast commercially available libraries with standard elements for system modeling, control theory and engineering are extendible by self programmed function blocks, so-called "S-Functions". These can be implemented in FORTRAN or C. S-Functions can be also grouped and stored in shared libraries. Simulink allows for the time domain state space system modeling as well as for modeling in the domain of Laplace transformed signals, the frequency or S-domain.

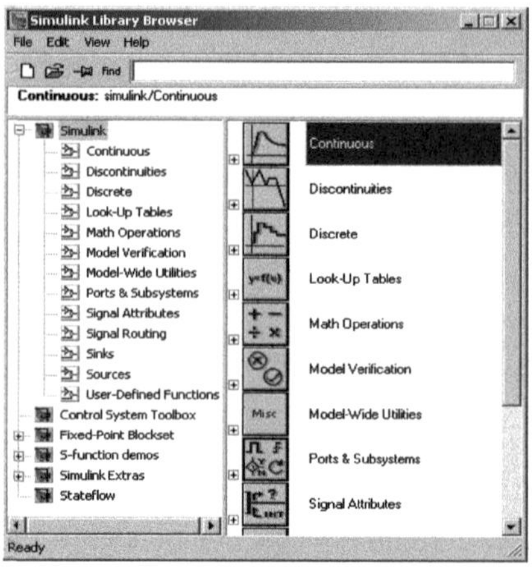

Figure 3.7: Simulink module libraries.

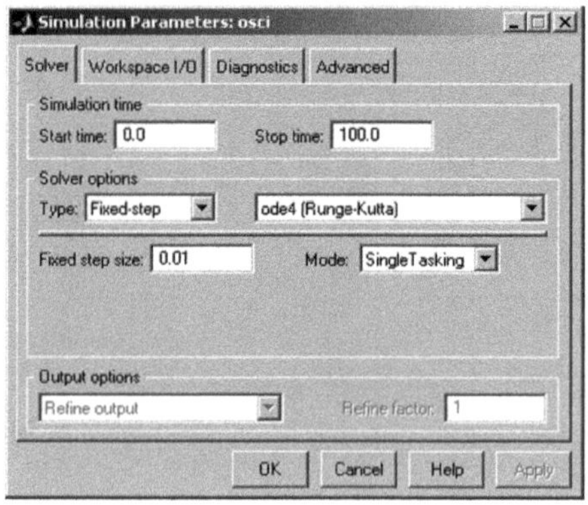

Figure 3.8: Selection of numerical solver.

The state integration of the overall equation system is carried out by integrated solvers. Simulink provides a large variety of them. Some of these solvers even are suited for systems with algebraic closed loops.

A further feature of Simulink is the possibility of partitioning systems into subsystems and within hierarchical nesting. A block diagram for a very simplified satellite model is depicted in figure 3.9.

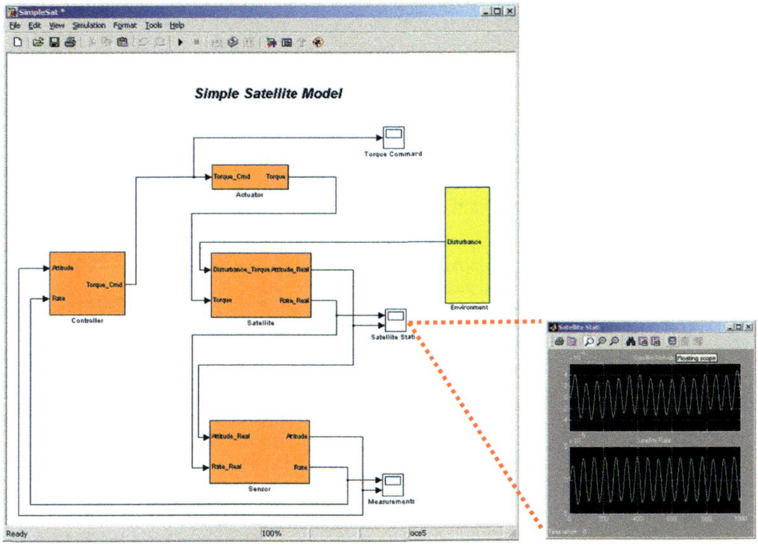

Figure 3.9: Aggregation of functions in higher level building blocks.

Last but not least Stateflow is a functional extension to Matlab and Simulink, which supports the modeling of system components as finite state machines. These state machines also can be hierarchically nested like Simulink models. Stateflow represents the modeled equipment as state machines of Harel type which covers also transition times, nesting and allow to model multiple cross-functional state diagrams within one statechart.

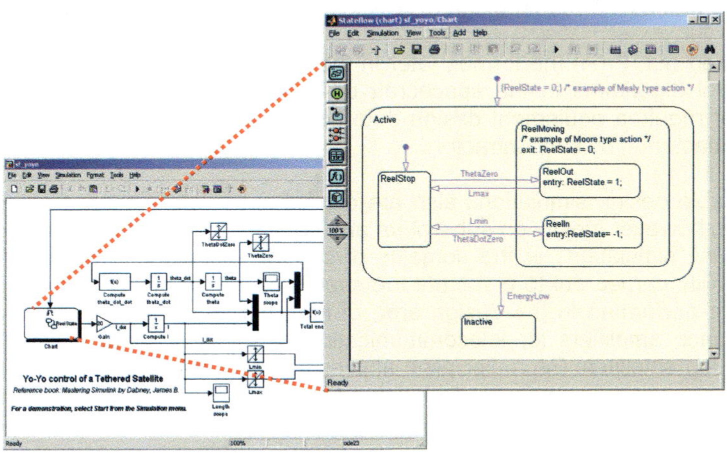

Figure 3.10: Typical modeling granularity for phase B models. © The MathWorks Inc.

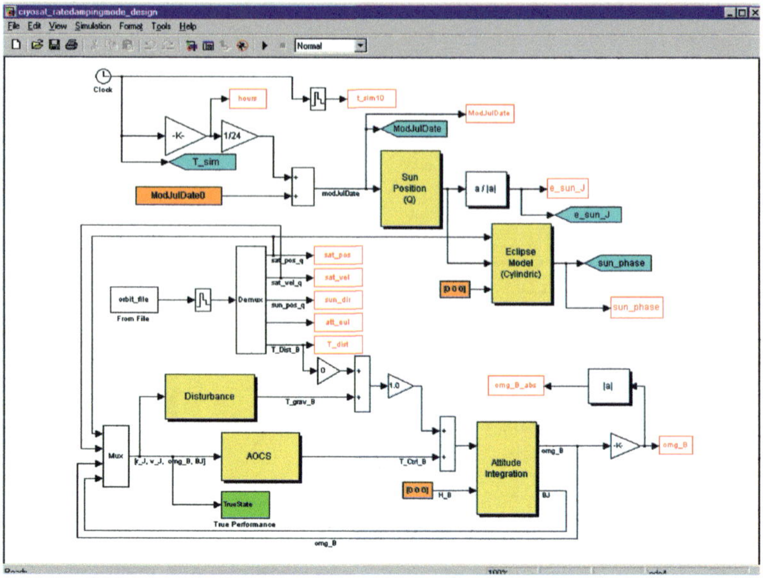

Figure 3.11: AOCS Simulink functional model - toplevel layer. © Astrium

The figure above shows the top layer of a satellite attitude and orbit control system for one operational mode. It is assembled from Simulink and Stateflow blocks. The explained system analysis tasks performed with tools of this type are classic design tasks within system engineering. In the above example

- AOCS controller algorithms and their quantitative characterization
- as well as the required characteristics for the AOCS sensors and actuators

are worked out. From the former, later in the engineering process the specifications for control algorithms in the spacecraft on-board software are derived. The latter serve for required equipment design data flowing into the component specifications for manufacturing by subcontractors.

Comparable design simulations also are carried out in other spacecraft engineering domains, however mostly on subsystem and equipment level. For example for power control unit simulations in the domains of control engineering and electromagnetic compatibility carried out to elaborate the specifications for power bus control. Similar tasks are performed in the design area of high frequency equipment, e.g. for high performance amplifiers for telecommunication satellites or radar payloads of Earth observation satellites. Each discipline applies its specific design tools such as AOCS engineering using the presented Matlab / Simulink / Stateflow Suite [8], electronics design often uses PSpice[3] [11], the designers of fluid systems apply Sinda / Fluint [12], the designers of ECLS systems use EcosimPro[4] etc.

[3] PSpice is a trademark of Cadence Design Systems.
[4] EcosimPro is a trademark of Empresarios Agrupados A.I.E.

Besides these example tools for the functional system simulation and design, quite a significant number of further analyses and dimensioning computations are performed. However not all of them are focused on the functional system behavior, but rather for best-case / worst-case scenarios. These comprise for example, thermal analysis, (e.g. with the tools ESARAD / ESATAN / FHTS), structural mechanics, (e.g. applying ANSYS[5], NASTRAN[6]), and other fields. Also such best-case / worst-case analyses can cover dynamic load cases such as the finite element analysis of behavior of a satellite structure under shock and vibration loads from the launcher.

3.2 System Verification Tools

After these systems, controllers and equipment have been designed, the first steps of verification in the system development process are taken. These range from lowest equipment and algorithm verifications up to the top level of system verification, which might concern a complete spacecraft. An AOCS algorithm and a satellite OBSW are chosen again as an example: The first step would actually be the test of the AOCS algorithms, which have been designed in the previous chapter.

As the algorithms are part of the OBSW and are designed to control the spacecraft, they should be tested with the real spacecraft. This would result in a typical "Algorithm in the loop" configuration. As there still is no real spacecraft available at this point in time, the algorithm can only be pre-verified against a spacecraft simulation. The required spacecraft simulator is the first one of a whole chain of verification simulators. The next one is related to verification of the OBSW after coding. The OBSW then contains the AOCS algorithms from the previous step. The OBSW has also to be verified with a real satellite, which usually is not yet available at this point. Thus, again a simulation is used to verify the OBSW with a "Software in the Loop" concept. The next step is the OBSW running on an on-board computer core ("Controller in the loop") with an appropriate satellite model as periphery. The OBC core is finally replaced by a full functional on-board computer with all interfaces ("Hardware in the loop"), if applicable with real connected satellite components.

The system simulators applied in all these four scenarios should ideally have a common technical basis to reduce the delta developments from one simulation scenario to the next to a technical minimum. An example for such a modular simulator infrastructure covering all verification phase setups red framed in table 3.2 is depicted in the following figure.

It is the so-called "Model-based Development and Verification Environment", (MDVE), of Astrium GmbH - Satellites. The successor system following the same principle, being applied internationally over all european Astrium - Satellites sites and in addition featuring a better performing simulator core with additional configuration databases, is called "Functional Verification Infrastructure", (FVI).

[5] ANSYS is a trademark of ANSYS Inc.

[6] NASTRAN is a trademark of The MacNeal-Schwendler Corporation (MSC).

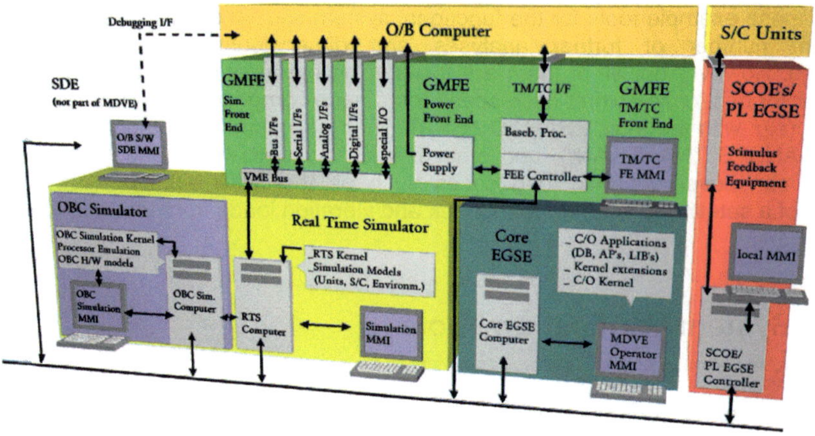

Figure 3.12: Model-based Development & Verification Environment. © Astrium

The analogies to the components of steps 3 to 7 in table 3.2 are obvious. The core of the MDVE infrastructure is the system simulator, which is called a "Real-Time Simulator" here. A subassembly of the system simulator is the on-board computer simulator, "OBC simulator", which models the on-board computer of a satellite[7]. The complete system is controlled by a control console, the so-called "Core EGSE"[8]. There is an interface between Real-Time Simulator and integrated satellite hardware, (e.g. real on-board computer instead of simulation), for hybrid bench configurations. This interface is called "Generic Modular Frontend Equipment" at Astrium. It performs the data transfer between integrated hardware and simulation, is responsible for the power supply of integrated hardware and for routing of telecommands and telemetry between integrated hardware in the loop, namely the on-board computer, and the control console.

Such a simulator infrastructure is similar to a "LEGO" construction kit and can have different peculiarities in its configuration for algorithm, software, controller and "Hardware in the Loop" tests. These characteristics are explained in the following subsections. Therefore first it is necessary to explain, that all these simulator configurations typically are based on a standardized kernel - which for real-time capable applications in most cases will be implemented in C or C++ programming language. The models which represent the system equipment are coupled to the kernel. Using the satellite as example again, the models represent on-board computers, sensors, actuators, payload, solar arrays etc. Furthermore, the equipment is modeled as occurrences in an object oriented manner. If a satellite has three star trackers, three star tracker model occurrences will be available in the simulation and will be coupled to the simulator kernel. By applying this object oriented design concept, each of the three star trackers models can have individual values for the same design parameter, e.g. for the mounting position. The same concept applies for modeling the functional interfaces, such as data interconnections,

[7] In the MDVE infrastructure the on-board computer model is a separate simulation application and not an equipment model like all others. In the more modern FVI all equipment models are treated equally.

[8] EGSE stands for "Electrical Ground Support Equipment".

power connections etc. The real interconnections in the spacecraft also are represented as interconnection occurrences of the according type in the simulator. Besides this the simulator kernel comprises simulation control functions, contains mathematical integrators and is equipped with functions to load in configuration files at start-up, provides interconnection to the control console and functions result logging.

3.2.1 Functional Verification Bench (FVB)

A first simulator of such a kind is depicted in the following figure, showing the configuration used for algorithm verification. This first configuration level with the embedded algorithms to be tested usually is called the "Functional Verification Bench", (FVB). For most spacecraft projects, this bench is limited to algorithm verification of the attitude and orbit control system, (AOCS), of the satellite.

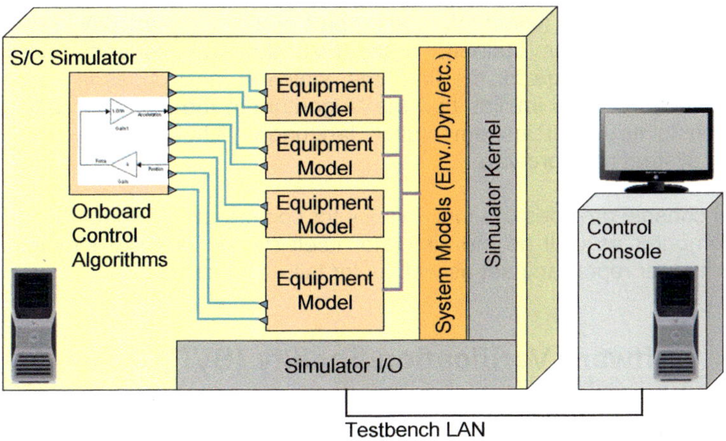

Figure 3.13: Functional Verification Bench – FVB.

Once more using the example of a satellite, this infrastructure consists of

- a simple functional on board computer model into which the controller functions - originally developed on Simulink or similar tools - are embedded as C code,
- functional models of satellite equipment,
- a numerical model of space environment for the calculation of external influences onto the simulated spacecraft, such as thermal loads, radiative pressure, magnetic fields, gravitational fields etc., and a dynamics propagator in order to integrate attitude and position of the satellite over time.

Data interconnections between simulated OBC, (control algorithms) and equipment models are established by the transmission of parameters - expressed in engineering units - via the component interfaces in the FVB. These interfaces do not yet match with the later system cabling lines, as there are still no specifications for equipment hardware interfaces at this point. For example, two analog signals can be sent from an equipment to the OBC in one design alternative via two separate lines later when spacecraft design is frozen. They could also be transmitted as calibrated packets via a serial interface or a MIL-STD-1553B bus in another design. At time of algorithm verification on FVB, the spacecraft electrical design is not yet frozen to a level where data interface types are selected.

In the FVB, mathematical models of system equipment for the first time run on the simulator reference environment. Therefore, they have to be converted from tools used in the functional design phase like Simulink to the corresponding object oriented simulator code where necessary - mostly into C++. In the case of Simulink this can partly be achieved by using the so-called "Real-Time Workshop", (RTW), developed by the Simulink producer The MathWorks Inc. The RTW allows transformation of Simulink models to C code which then can be embedded into the overall FVB Simulator C++ architecture.

Furthermore, controller algorithms, (being part of the on-board SW later), do not run in Simulink any longer. Instead, in the FVB they run converted to the OBSW implementation language for the first time, which in most cases is either Ada or, in newer projects, C. This implies that the "Algorithm in the Loop" already is compiled and tested in its target implementation language, although not yet on the target operating system nor hardware.

A control console provides the functionality for commanding of both the simulated on-board computer as well as the simulator itself (for actions like simulator start and stop, parameter report and set as well as for failure injection etc.).

3.2.2 Software Verification Facility (SVF)

In the development process of a satellite, the FVB is usually followed by implementation of a so-called "Software Verification Facility", (SVF). The programmed on-board software of the satellite is pretested on this testbench for the first time (cf. figure 3.14). Considering an on-board computer with a classical computer architecture, the on-board software consists of

- the operating system of the on-board computer and,
- the spacecraft system control code, including all control algorithms,
- all interface drivers for input/output interfaces between on-board computer and satellite equipment,
- and functions necessary to receive telecommands from ground and to generate satellite telemetry for the ground station.

These software elements have to be tested extensively. In contrast to the FVB, such an SVF is equipped with a detailed model of the on-board computer, (OBC). In the SVF's OBC simulation model, all components, including the microprocessor or "central processing unit", (CPU), are exactly reflected w.r.t. their functionality. The on-board software, (OBSW), compiled for target CPU and periphery hardware architecture can thus be loaded directly into the SVF OBC model and can control the simulated satellite. This is also applicable for the OBC's basic I/O-system, (BIOS), boot software and the operating system of the on-board computer. The SVF infrastructure thus corresponds to the "Software in the Loop" implementation step in the development process.

Furthermore, the SVF is not only used for on-board software tests in the scope of attitude / orbit control as it is often the case for the FVB in satellite development. The SVF should enable control and monitoring of all functions of the simulated spacecraft. For satellites this includes AOCS, platform and payloads - all controlled by OBSW in the loop. For this reason, numerous additional models have to be added and functional model upgrades are to be implemented compared to the previous FVB. This also concerns modeling of equipment interconnections. In FVB for example, a control algorithm and a sensor model are exchanging data in engineering units. In the SVF, the OBSW in the OBC communicates with the enhanced model of the same sensor via the real protocol which later in spacecraft hardware will be transmitted over a real wired connections.

Since the loaded OBSW in the SVF thus controls sensors, actuators, platform and payload components of the spacecraft via binary data protocols, the "functional interfaces" reflecting interconnections of real hardware in the spacecraft in the simulator, have to be functionally enhanced compared to their representation in the FVB. This implies the data protocols on simulated lines have to be modeled, including aggregation of parameters exchanged via equipment to protocol packets and the modeling of calibration of raw engineering value data to protocol binary formats - e.g. conversion of a simulated temperature from Kelvin to 16 bit binary.

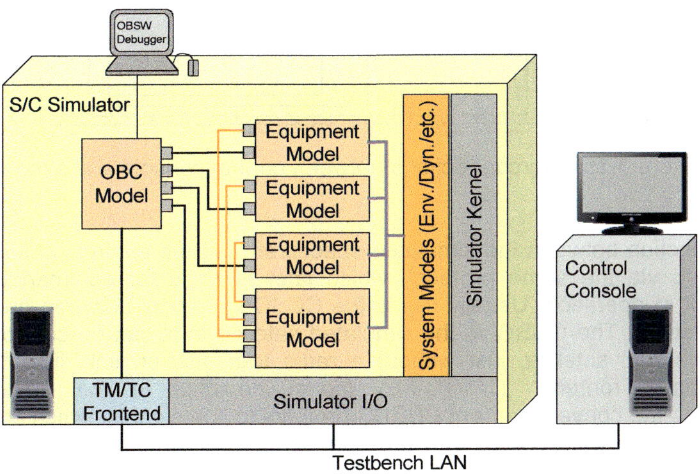

Figure 3.14: Software Verification Facility - SVF.

The connections between non-OBC equipment models also have to be reflected. Their functions and their lines not connecting to the OBC, e.g. power lines, have to be enhanced to the full system functional representation - compare figure 3.13 and 3.14 for the cross-linking between all types of equipment models.

Today, such "Software in the loop" simulations enable the running of a complete functional satellite simulation including on-board computer model and spacecraft dynamics propagator on a single powerful PC or laptop. However, for this system simulation the SVF has to be connected to an adequate control console. This results from the on-board software being commanded in the simulated system like in the real spacecraft, for example the real satellite from the ground. Thus if the focus of simulation scenarios is on the OBSW overall system tests, the control console replaces the ground station. However, if the focus is on unit or integration tests, especially the software developers' control console has to provide further features like debuggers.

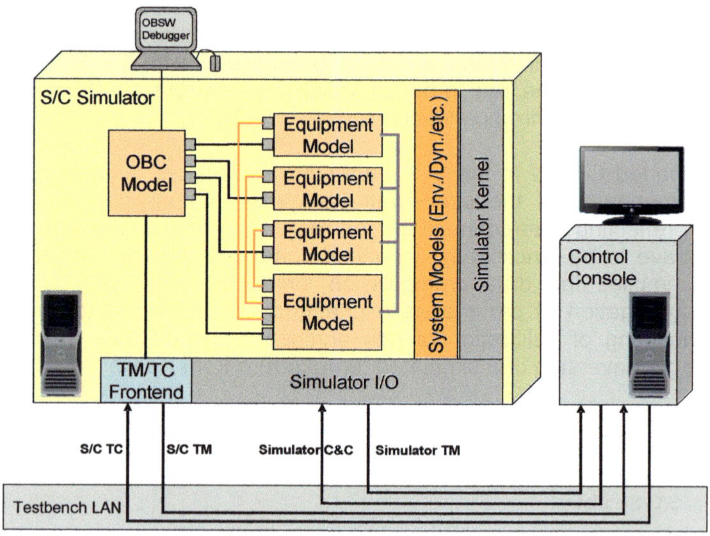

Figure 3.15: Interconnections between control console and simulator.

The connection between the simulated spacecraft and the control console is usually established via the same protocol, which later is used for commanding the real spacecraft in operation. Usually this is the CCSDS protocol between ground station and spacecraft. The OBSW in the simulated satellite thus can be controlled like the one in the real satellite, just without a radio transmission link. The "Telemetry / Telecommand-Frontend" (TM/TC-FE) model depicted in the following figure implements the conversion from CCSDS protocol to line signals, which correspond to those received by the OBC from the spacecraft transponder. The control console sends CCSDS protocol to the TM/TC-FE model which replaces the real spacecraft transponder and sends the converted data via simulated lines to the OBC model's transponder interface. Telemetry to ground passes vice versa from OBC model via TM/TC-FE model to control console. The protocol definition used for spacecraft and simulator command specifies the

- binary packet structure,
- packet header / trailer structure,
- addresses of receiver and sender, (e.g. OBC model or simulator kernel),
- binary coding of raw data in the packet, (with checksums etc.).

In analogy to commanding simulator and simulated spacecraft it is useful to mention that not only the on-board software of the simulated spacecraft can send telemetry data "to ground" respectively to the control console. Also simulator telemetry packets normally can be defined via configuration files. These packets are automatically evaluated with the control console processing tools. For each packet type of such simulator telemetry packets, it can individually be defined which simulator parameters are included in a packet. The transmission frequency in most cases is individually selectable for the different packet types, too. For example, a simulator telemetry packet can be sent from the "Power Control and Distribution Unit", (PCDU), with its simulated relay positions to the control console every 10 seconds by definition. A packet with thermal data might be sent only every minute. The packet length can differ too - limits are specified by the CCSDS standard.

The control consoles usually applied for detailed OBSW tests by the developers, mostly are based on scripting languages which are easy to handle, like Java. Definitions for satellite telecommands and telemetry and the included parameters as well as their binary calibrations can be imported into the control consoles mostly via XML files[9] from the project's telecommand / telemetry reference database. Simple test procedures for the OBSW can be implemented based on these satellite TCs and Java as interpreted scripting language.

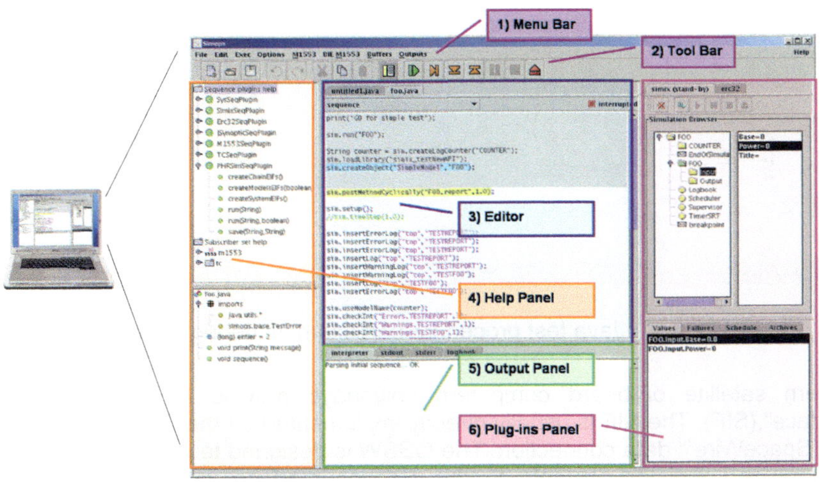

Figure 3.16: SVF with Java-based control console running on a laptop. © Astrium

The following screenshot depicts a section of an OBSW test procedure. Without discussing the syntax of the commands and constructs here the reader who is

[9] XML is a markup language for files. Further details are explained in chapter 8.5.

familiar with Java will identify the programming language immediately. This system setup also enables OBSW debug output if the OBSW was compiled with a debug option for the test. If so, in advanced simulation infrastructures a debugger can be connected to the OBC simulator model inside the simulator and thus the output can be displayed and logged into files and OBSW internal variables are accessible for tracking and manual override.

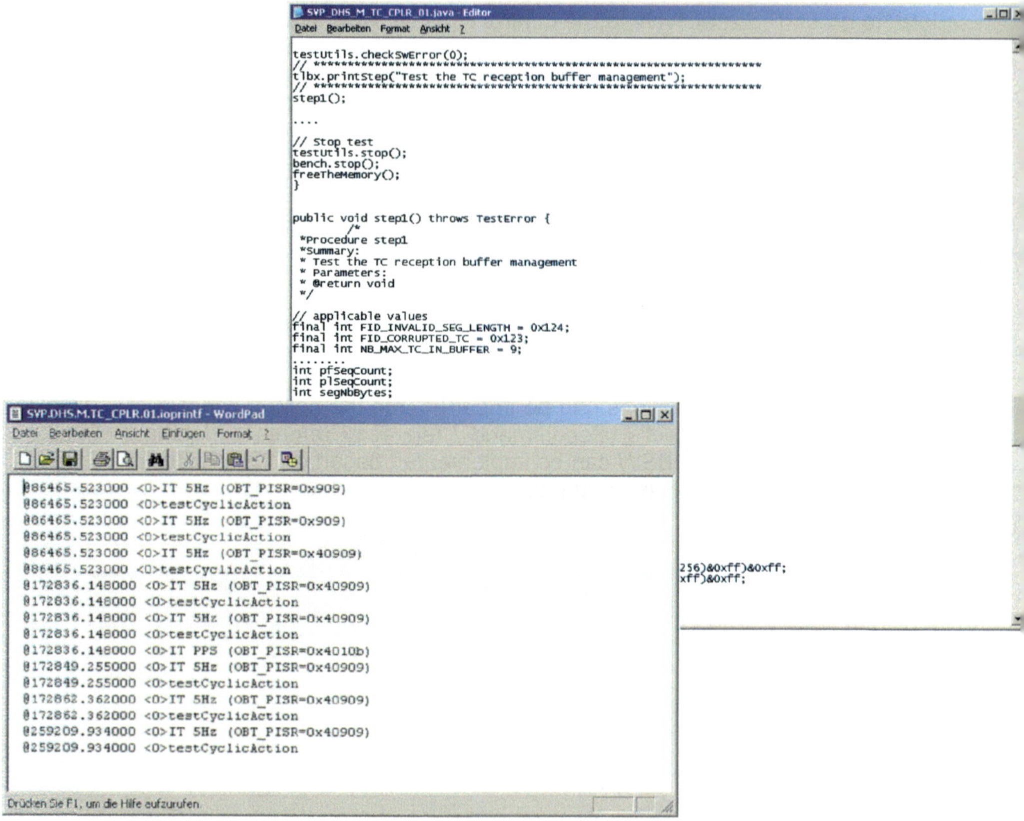

Figure 3.17: Java test procedure and example debug output. © Astrium

Modern satellite on-board computers nowadays provide a so-called "Service-Interface",(SIF). The SIF is usually directly implemented on the CPU board facilitated by a SpaceWire[10] data connection. The OBSW is designed to cyclically send specific memory and register contents to this Service-Interface independent of whether anybody is tracking the output or not. This SIF output is also generated in flight conditions of the spacecraft later. Under ground conditions, this data can be received, displayed and logged as hex dumps. Experts can get some central information on the

[10] SpaceWire is a standard for high-speed links and networks for use onboard spacecraft. It is defined in the European Cooperation for Space Standardization ECSS-E-ST-50-12A standard.

OBSW operating status from these dumps. This Service-Interface access is even available on the launch pad for last minute tests before launch. The electrical connectors are disconnected just before the start. The code instrumentation for data output to the Service-Interface remains on board. A modern SVF can display Service-Interface outputs in the control console, too. As these outputs directly address dumps of specific on-board computer memory and register sections, the direct output however is very cryptic. The following screen dump shows such a SIF output.

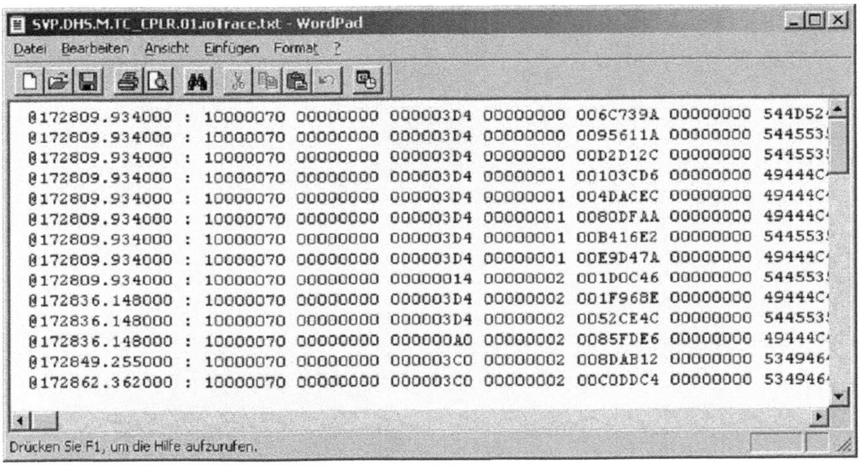

Figure 3.18: Service Interface hex dump resulting from an OBC test run. © Astrium

A close relative to the SVF for OBSW developers as described above is the SVF for tests of entire system operation scenarios - see figure 3.19. The central difference here to the OBSW development SVF is to preferably apply the same control console as used in real spacecraft operations and control later on. Using the example of satellites, this is the control software which is also applied in the ground station including a complete definition set of all satellite telecommands and telemetry packets, the definition of contained parameters and their calibrations. Such control consoles additionally enable the definition of synoptic displays, (e.g. one in order to display all thermal parameters of a satellite, one for required power supply data etc.).

In the frame of hybrid system testbeds described in the following sections, such a control console, besides system simulator and the OBSW, has to control further "Electrical Ground Support Equipment", (EGSE). This central or core console of EGSE therefore also is commonly called Core-EGSE. The Core EGSE enables the definition of control scripts by use of scripting languages which are ideally compatible to the language used in the control station of the spacecraft later during the mission.

This SVF configuration level with a Core-EGSE as control console is used for the verification of complex operation scenarios of the entire satellite such as

- closed-loop attitude control scenarios,
- closed-loop orbit control tests,

- power control tests modeling supply, charging and discharging cycles etc. over several orbits,
- thermal control verification tests,
- payload control verification tests.

The following figure shows such an infrastructure in operation:

Figure 3.19: SVF with Core EGSE. © Astrium

3.2.3 Hybrid System Testbed (STB)

The hybrid testbench configuration is used for "Controller in the Loop" tests. For this purpose it often is extended stepwise. The following examples show such a configuration, again using the example of a satellite simulation and a conventional on-board computer. This type of testbench is often called "System Testbench", (STB).The first stage of expansion is used for hardware / software compatibility tests. This means that in such a test setup, for the first time in the development process, the on-board software is loaded onto real OBC hardware. Elementary functions like booting the on-board software on the real hardware, switching between redundant hardware instances and telecommand and telemetry handling are tested on this type of bench. A very important aspect here is that the previously applied SVF incorporates a functional model of the on-board computer, which was built purely based on OBC documentation from the supplier. The SVF's OBC model characteristics such as timings of ASICS are only theoretical data since at SVF build time, only approximations of the later supplied real hardware characteristics were available or could be taken from OBC prototypes. During the hardware / software testing on an STB, it is evaluated for the first time whether the OBSW is interacting properly with the ASICS and controller chips in the real hardware OBC.

The following two figures show the principle testbench setup for such HW / SW compatibility tests as well as an example of a testbench from the Galileo-IOV project. The photo shows the OBC breadboard box in the middle, the controller monitors and the rack for the Power-Frontend and TM/TC-Frontend on the very right side and a PC to upload the OBSW to the OBC on the left side. The laptop serves as a MIL-STD-1553B bus tracer / responder to test the outputs and inputs of the OBSW to and from the main connection of equipment implemented in a later stage of the development. The control console is located on the right side out of camera sight.

Figure 3.20: STB for HW / SW compatibility tests.

Figure 3.21: Galileo-IOV STB, first stage of extension. © Astrium

After the compatibility of the OBSW and the hardware was successfully tested the loop is closed, leading to the "Controller in the loop" configuration. The OBC-Core is then connected to the simulator which is simulating the satellite's equipment, just like in the SVF. Additionally the still missing analog, digital, serial, pulse and bus interfaces of the OBC are simulated. Now complex controller functions of the OBSW can be tested while running partially on the real hardware and connected with the simulated equipment. The following figure shows such a setup.

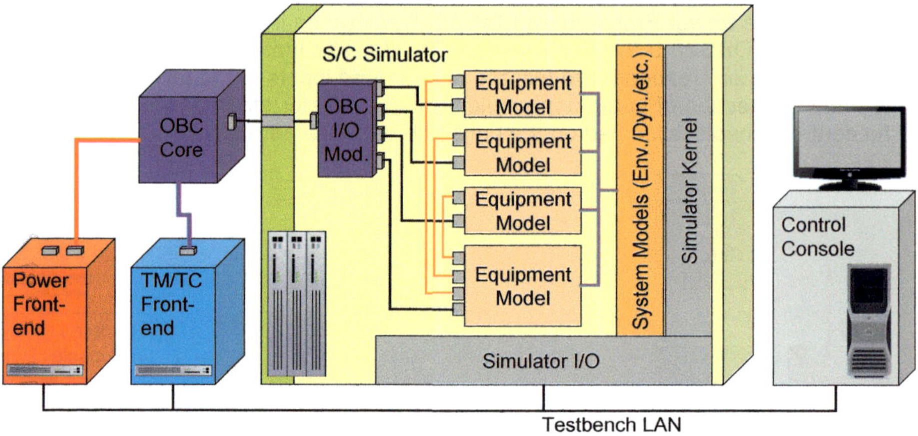

Figure 3.22: First "Controller in the Loop" setup.

If at this stage in the project a fully equipped OBC with an I/O-module already is available, this setup can be skipped. A setup consisting of the real OBC and simulated equipment can be implemented directly as shown in figure 3.23.

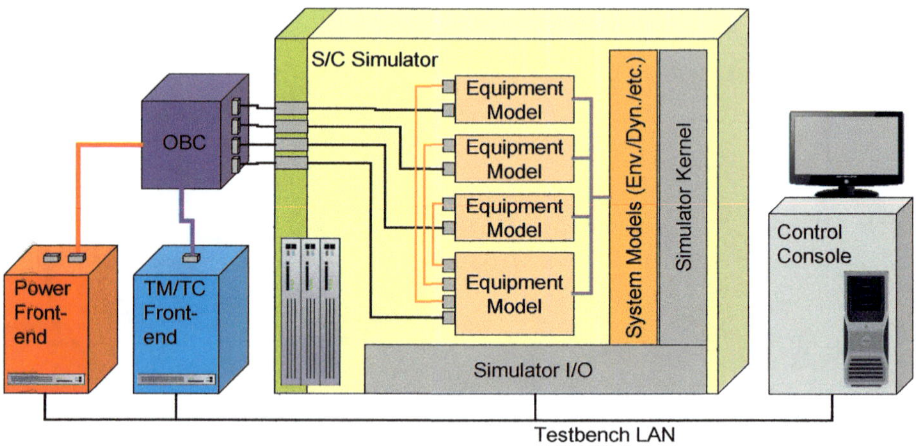

Figure 3.23: STB with complete OBC in the loop.

This configuration has the advantage that all the I/O-interfaces of the OBC including their handling by the OBSW can be tested directly, in contrast to the setup of figure 3.22 where the interfaces are only simulated. On the other hand significant effort has be made to provide all wirings from the on-board computer to the simulator, (test harness), and to provide, configure and test all the simulator interface cards. The interface equipment marked in green in figure 3.23 above, only depicts symbolically four connections. In a real OBC / simulator interconnection as with a satellite, the amount of cabling rapidly increases as shown in the figure below. The photos in figure 3.24 show the OBC of CryoSat 1, (black box), and the wiring to the simulator rack which here still is a relatively simple spacecraft configuration.

Figure 3.24: CryoSat 1 STB ©Astrium

In the case of more complex spacecraft systems with lots of interfaces, the test harness and the interface hardware – the Simulator-Frontend – also can be much more complex, as demonstrated in figure 3.25 below. Visible, from left to right are:

- The TC/TM-Frontend next to the window, (partially hidden)
- The rack with the two slots of the breadboard OBC-Core and the OBC I/O-module
- The interface rack with the cables from the OBC. This Simulator-Frontend rack also incorporates the load emulator for the current driven interfaces and the processor boards of the system simulator
- To the very right the Power-Frontend, the electrical power supply of the whole installation can be identified

Besides the effort to verify the test harness and interface card drivers for both setups figure 3.22 and 3.23 , it has to be mentioned that the system simulator's numerics now has to serve the interfaces in real time. This means that all data protocols of all interfaces have to be processed in parallel which requires sufficient processing power, a corresponding real time operating system, (normally VxWorks or a real time capable Linux distribution), and a real time capable data bus system, (in most cases VMEbus).

Figure 3.25: System testbed (Aeolus project). © Astrium

Because in the STB configuration all satellite components except for the OBC are still simulated, this "Controller in the Loop" setup is fully closed-loop capable. This means that any closed-loop operations scenario of the satellite can still be simulated with this setup. Many system tests which have already been verified on the fully developed SVF typically are replicated on such setups during the “Assembly, Integration and Testing“, (AIT) program of the satellite.

3.2.4 Electrical Functional Model (EFM)

The "Controller in the Loop" setup described in the preceding chapter now can be extended step by step with further "Hardware in the Loop" components leading to a fully deployed testbed, which in most cases is called the "Electrical Functional Model", (EFM).

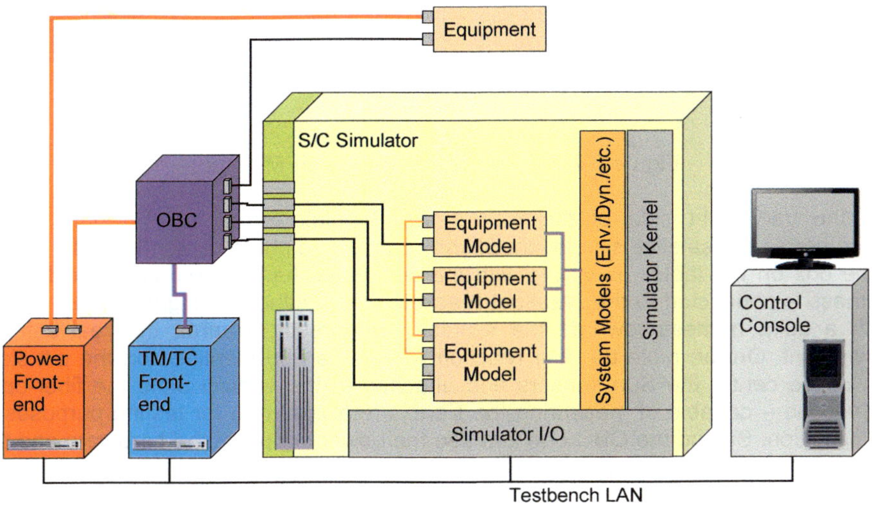

Figure 3.26: Infrastructure configuration EFM.

The EFM evolves from the hybrid testbed STB by integrating more and more satellite hardware. Corresponding to the integration of the real hardware, the simulated equipment model occurrences are removed from the simulator. To command this additional spacecraft hardware or to externally stimulate it, eventually additional equipment may be needed. For example stimulation equipment may be required for Sun or Earth sensors or star trackers. This additional infrastructure as a whole is called "Special Checkout Equipment", (SCOE). It is intended to command the SCOEs from the same control console so that the satellite's hardware, the simulated system components and the stimulation of the hardware in the loop via the SCOE can be commanded by means of test scripts from a single source.

Such configurations of OBC, additional hardware in the loop, cabling and stimuli equipment often are integrated and tested as an arrangement on a table, thus in satellite engineering they are named "FlatSat". The following panorama image arrangement shows such a setup for the avionics subsystem of the Galileo-IOV satellites. The system components of the other subsystems, (Power, Propulsion, Payload), are still simulated in this configuration.

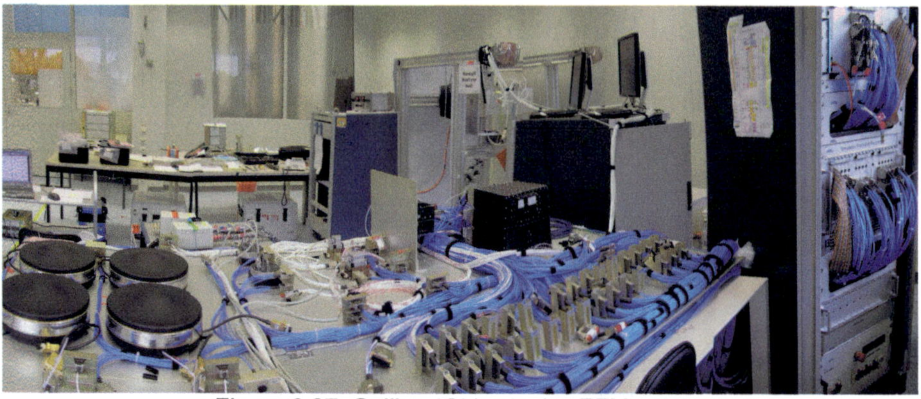

Figure 3.27: Galileo-IOV avionics EFM setup.

On the very right side of the picture the Simulator-Frontend is visible, which incorporates the satellite simulation. Slightly right of the photo center the OBC - black cube box on the table - is visible with the connected harness on its rear side. The harness is connected to the breakout brackets where either the signal can be routed with a delta harness towards the simulator or directly towards the real satellite equipment. On the table the four reaction wheels can be identified on the left side, left of the center the Sun sensors and attached to the vertical plate the fiber-optic gyroscope – capable of measuring the Earth's rotational rate for testing purposes in this position. Behind the OBC one can see the power supply of the hardware and the Test-SCOE for the Earth sensor stimulation.

Figure 3.28: CryoSat 1 FlatSat assembly. © Astrium

An example of an EFM setup for a whole satellite including payload electronics and so on is given in the figure 3.28 originating from the CryoSat 1 project above. After successfully passing all tests in the EFM configuration, the satellite components are integrated into the prepared satellite structure.

Figure 3.29: Mounting functional equipment from FlatSat into satellite structure.
© Astrium

Finalizing integration work is done afterwards, integration tests are conducted at entire system level and the thermal isolation is completed. The functional system verification according to figure 2.4 is completed with these steps.

Figure 3.30: CryoSat 2
Final assembly steps.
© Astrium

The physical design validation follows. For satellites, this normally comprises the thermal vacuum tests, (cf. figure 2.5), the electromagnetic compatibility tests, (cf. figure 2.6), and the structure and mechanics tests, (cf. figure 2.7). The infrastructure needed for such tests normally is not available at the spacecraft manufacturing premises but is located in special facilities owned by the space agencies or specialized technical service providers, (for example ESA / ESTEC, IABG, CNES). Therefore the satellite has to be shipped to the corresponding facilities in order to conduct the final design validation tests.

Figure 3.31: CryoSat 2 preparations for transport.
© Astrium

3.2.5 Spacecraft Simulator for Operations Support

Detailed complete system simulators resulting from the development chain described above can finally be applied to support spacecraft system operations. The SVF configuration is the most appropriate setup. It can be modified such that the control console is replaced by the flight operations system installed in the spacecraft ground station. The SVF simulator's interfaces and the data protocols between the simulator and the control console already are implemented to be compatible with the control system of the ground station as was described in 3.2.2. The resulting simulator setup in the ground station can be used for

- training of the spacecraft operations staff, and for,
- tests of OBSW patches and bug fixes on the simulator before they are uplinked to the real spacecraft.

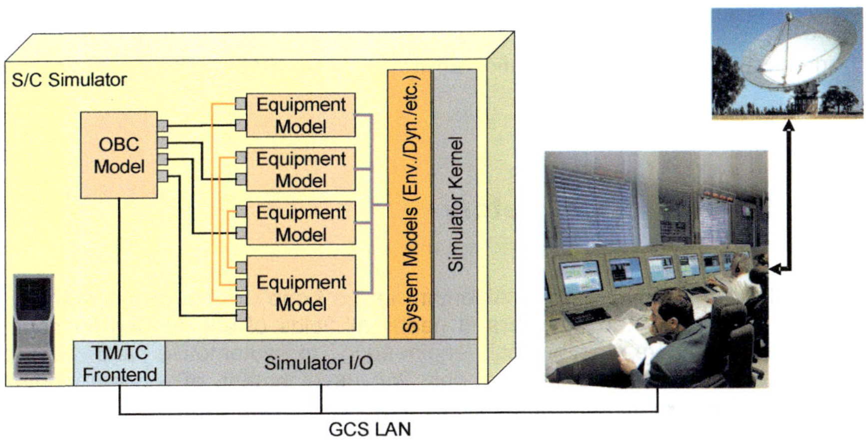

Figure 3.32: System simulator for spacecraft operations support.

The acceptance of such simulators originating from spacecraft system development largely varies from space agency to space agency. Some are using the system simulators, (like in case of the TerraSAR-X MDVE applied by DLR / GSOC for satellite operation), with the argument that such a simulator has already passed a comprehensive verification process and has a very good validation quality.

The "ESA Space Operations Center", (ESOC), typically does not accept system simulators from the development cycle because of their philosophy to use only tools for operations support which are independently developed. This approach minimizes the risk of potential inherent development process errors and such errors can be spotted during operation. For this reason the ESOC has developed its own system simulation infrastructure called SIMSAT - see e.g. [20].

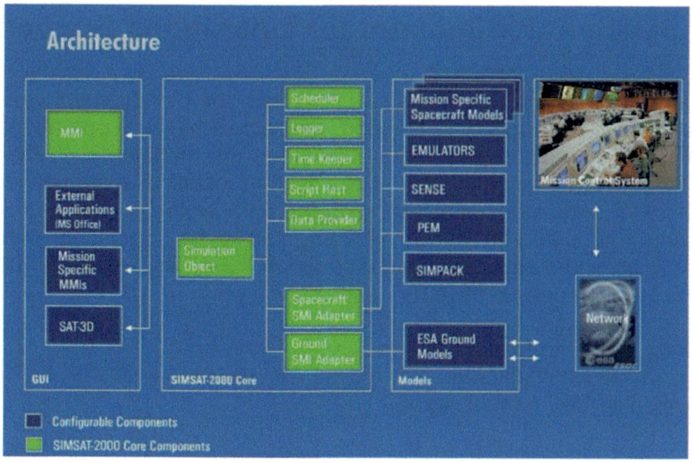

Figure 3.33: The system simulation environment SIMSAT by ESOC. © ESA / ESOC

3.3 Infrastructure History

The development principle from "Algorithm in the Loop" to "Hardware in the Loop" has been state of the art for years in nearly all fields of engineering for controller development – from mechanical engineering to automotive and aerospace engineering. However this applies only for development of system controllers themselves. The functional design and verification of entire systems as a whole is a fairly new application for simulation infrastructures, especially in the field of spacecraft development. This is due to the fact that suitable techniques to model entire systems within a simulation software had to be developed first and the numerics and performance of the computers had to be available for simulating complex systems. In the European space industry, Astrium has established itself in this area with technically up-to-date solutions over the last 10 years. The first satellite simulator was developed from 1998 to 1999 in the ESA case study "System Simulation & Verification Facility", (SSVF), in a cooperation between VEGA Space Systems Engineering, Darmstadt and Daimler-Benz Aerospace, Friedrichshafen (today Astrium GmbH). The simulator was based on a commercial FORTRAN tool and was used in STB configuration for the NASA "Grace" project from 1999 to 2001. This was the first time AOCS closed loop test where conducted as real-time tests with hardware in the loop and simulated spacecraft. Later in 2002, this tool was developed further and was used for "Grace" operations support at NASA/JPL and GSOC. Based on the lessons learned, confirming the technical approach but pinpointing the insufficiency of FORTRAN77 in respect to simulator design, based on the findings of a dissertation at the TUHH[11] and the "Small Satellite Simulator" case

[11] *ObjectSim 2.0* today is Open Source → *http:/www.OpenSimKit.de*, see [120] and [23]

study of the DLR, the MDVE simulator kernel was implemented by Astrium. The software design was performed based on the Unified Modeling Language, (UML), and the code platform since then is C++. These software technologies will be described in later chapters in more detail. The infrastructure was first deployed in the ESA project CryoSat-1 from 2001 to 2003. For CryoSat-1 testbench installation types covered SVF, STB and EFM FlatSat. This project was the first ESA satellite project where no spacecraft engineering model on system level was implemented. Based on this milestone the model-based system development technology found its way into more and more follow-up projects, a fact which is illustrated in the following figure. The figure also illustrates the development of the testbench design from project to project. Up to now 11 satellite projects, (including SSVF / GRACE), of Astrium GmbH were based on the model based simulation and verification technology. The Astrium in-house developed toolkit was called "Model-based Development and Verification Environment", (MDVE). In the meantime this has been transnationally harmonized with a complementary technology “SimWare” [17] from Astrium S.A.S. applied in the CNES Pleijades project. The resulting next generation tool suite called “SimTG” [19] is for the first time applied in the ESA Mercury Science Mission "Bepi Colombo". In the meantime this technology also is used in other areas such as in space transportation and similar infrastructures like "Eurosim" [18] and have been developed by other companies.

	FVB	SVF	STB	EFM	SC-Sim	Struct./ Thermal
ESA SSVF / GRACE			X			
First Simulation and Verification Infrastructure (FORTRAN Tool)						
ESA CryoSat		X	X	X		
“Model-based Development and Verification“ Infrastructure (UML/C++)						
EU Galileo IOV	X	X	X	X	X	
Additional OBSW Verification Setup						
ESA Virtual S/C Study	X	X	X	X	X	X
Optimized digital Spacecraft Modelling Infrastructure						

Figure 3.34: Gradual extension of the simulation technology.

For a first simplified overview it can be summarized that in satellite engineering the model and simulator based development technology has successfully replaced the historic approach of building an engineering model on spacecraft level. This technology is characterized by:

- Dynamical modeling of spacecraft for operations simulation on a detailed level

- Simulation based development forcing the development team, (including subcontractors), to freeze system component interfaces and the operational behavior in an early stage and in a consistent way

- An appropriate concept of the simulation infrastructure which enables:
  - ◇ The early functional system simulations – “Algorithm in the Loop” – before spacecraft HW and SW are available.
  - ◇ To provide a comfortable platform for development and testing of the OBSW (Control software + Operating System + BIOS).
  - ◇ Early tests of the OBSW, (prior to the availability of OBC hardware)
  - ◇ To a large extent the verification of the OBSW on an SVF, (without need for access to the flight hardware while testing).
  - ◇ Tests of the hardware and software interaction can be conducted on the HW as soon as it is available.
  - ◇ Reuse of the simulator in the ground station for operator training and as a software maintenance facility is possible if requested by the customer.

Further reading describing verification testbeds in more detail is provided in the according subsection of this book's references annex. The same applies for literature on use of simulation infrastructure in ground stations.

4 Testbench Components in Detail

On-board Computer Model in a Testbed © Astrium

J. Eickhoff, *Simulating Spacecraft Systems*, Springer Aerospace Technology 1,
DOI 10.1007/978-3-642-01276-1_4, © Springer-Verlag Berlin Heidelberg 2009

The detailed description of test and simulation facility components is the subject of the following section. It begins with the more non-central elements to subsequently lead over to the central simulator kernel, the models and to numeric topics. In this flow the control consoles of are first treated.

4.1 Control Consoles

Already in chapter 3.2.2 it was explained that simulator infrastructures are applied for various tasks, namely on-board software unit and integration tests as well as for complete system tests. As was further explained, the control console in such setups has to operate both the spacecraft on-board computer and the simulation, which means it must provide interfaces to both of them, (cf. figure 3.15). The required OBC service interface and the necessity for visualization of OBSW log data output synchronous to general OBSW telemetry respectively with simulator telemetry also have already been mentioned. In "Controller in the Loop" and respectively "Hardware in the Loop" scenarios the console moreover has not only to manage the OBSW on the OBC and to control the simulator. It must in addition be able to command other special checkout equipment, (SCOE), like Power-Frontend, TC/TM-Frontend, sensor stimulation equipment, eventually measurement data recording systems for actuators or the like. The communication between control console and connected spacecraft, simulator or SCOE equipment typically is established all via the same communication protocol - the one which later also applies for communication between ground and spacecraft. This is the international standardized packet oriented CCSDS[12] protocol. ESA / DLR projects apply the "Packet Utilization Standard", (PUS), which is a service based system control technique based on top of the CCSDS packets. In most cases either special Core EGSE tools are used as the testbench control console, or the testbench is commanded by the same tool later applied for spacecraft flight control, i.e. a mission control system which is suitable also for application as Core EGSE.

A Core EGSE system has to provide all required functions for sequential control of tests, (execution of test procedures), for result and data visualization and for test documentation / logging, (data base). The following figure 4.1 shows the conceptual internal structure of such a machine. At the top of the schematic, the user interface or "Man Machine Interface", (MMI), is depicted with its different kinds of display types (cf. figures 4.2, 4.3 and 4.4). Referring to figure 4.1 commands - including parameter values - can be sent from the user interface passing several processing steps down the right yellow path and proceeding via the LAN towards the addressee, (spacecraft, simulator, SCOE, etc). The telemetry of these units is also received via LAN and processed up through the left yellow path. Incoming packets are sorted by type. Then their parameter values are read out and forwarded to the MMI displays. Besides pure visualization TM is checked against specified limits and the command procedure executor in the control console also can react to telemetry parameter values. All known telecommand and telemetry packets, their structure and parameters are contained in the database, (bottom middle of figure 4.1). Based on the TC and TM packets defined in this database, the user can write test scripts in a script language

[12]CCSDS is the Consultative Committee for Space Data Systems (CCSDS).

which enable one to send commands and analyze incoming telemetry. Script processing is performed by the test executor marked in orange in figure 4.1.

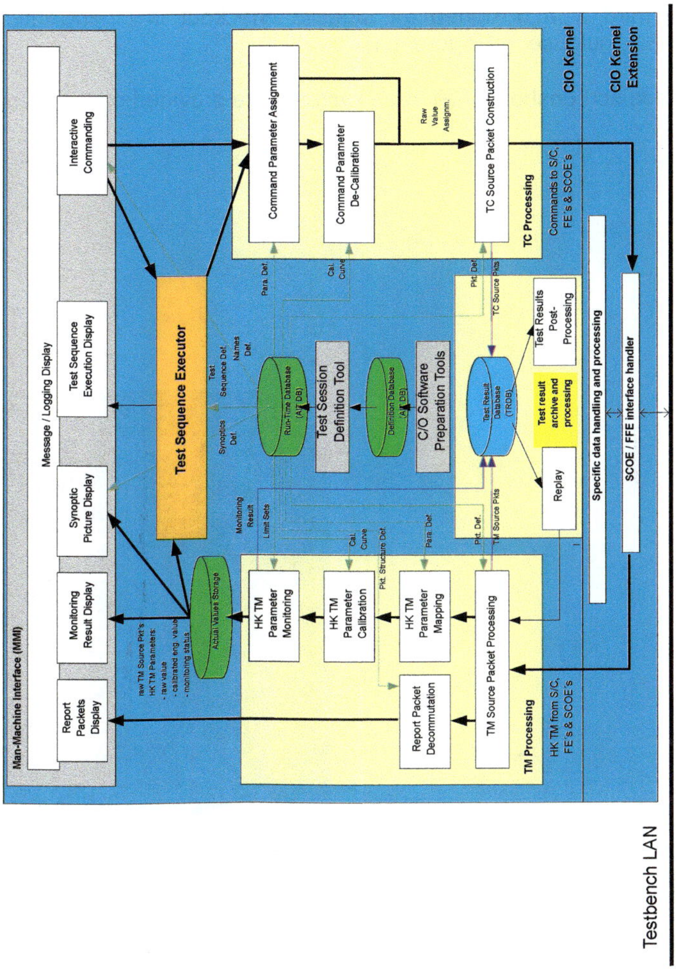

Figure 4.1: Internal architecture of a Core EGSE. © Astrium

Three major systems have established as Core EGSEs in Europe. These are:

- Diverse variants of the Spacecraft Control and Operation System SCOS 2000, ESA / ESOC, Darmstadt, (cf. figure 4.2).
  The basic version, as "Spacecraft Control & Operation System", is free for use in projects in ESA member states, however for full functionality in spacecraft checkout and testbenches it requires commercial upgrades such as those in the projects TerraSAR-X and Galileo-IOV.

- The second one to be discussed is the Central Checkout System, (CCS), formerly called Columbus Ground System, (CGS) of Astrium GmbH, Bremen, (cf. Figure 4.3).
- The third type designed for satellite checkout is OpenCenter, by Astrium S.A.S., Toulouse, (cf. figure 4.4).

User interface screenshots of all three systems are provided on the following pages.

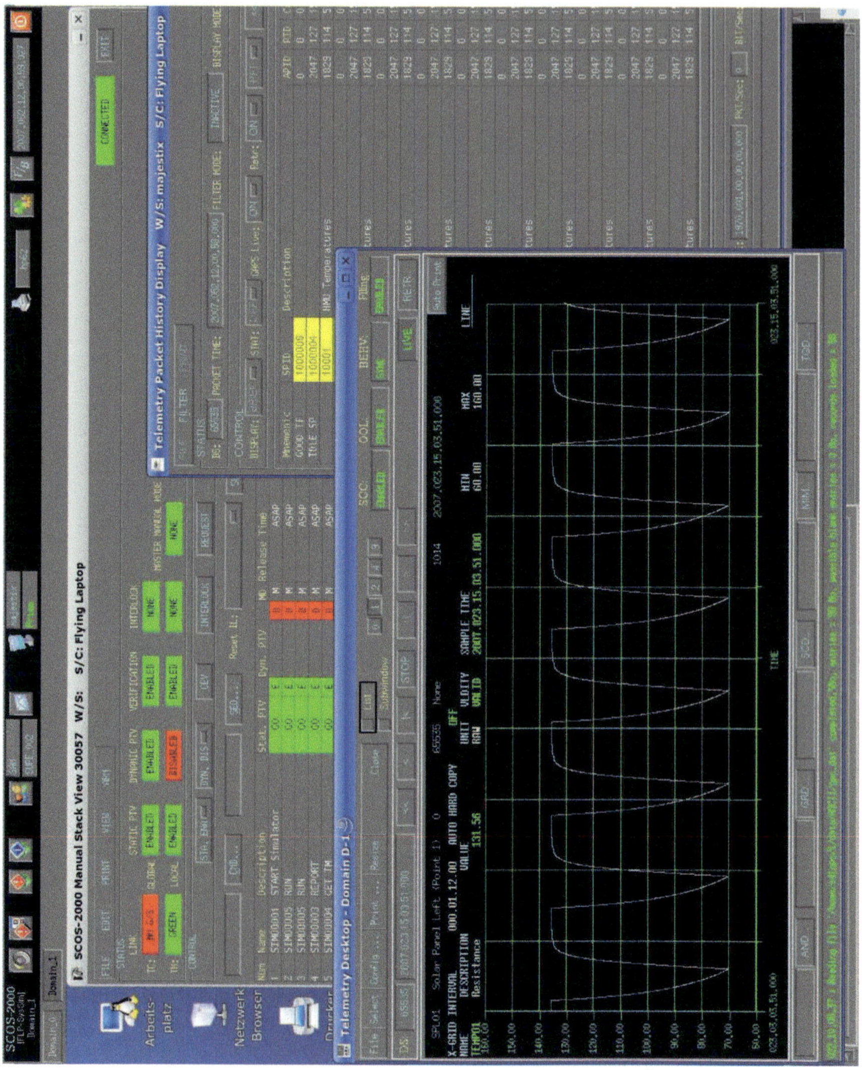

Figure 4.2: Control Console MMI: SCOS 2000.
Application example: IRS, Universität Stuttgart

Figure 4.3:
Control Console MMI:
Example CCS.
© Astrium

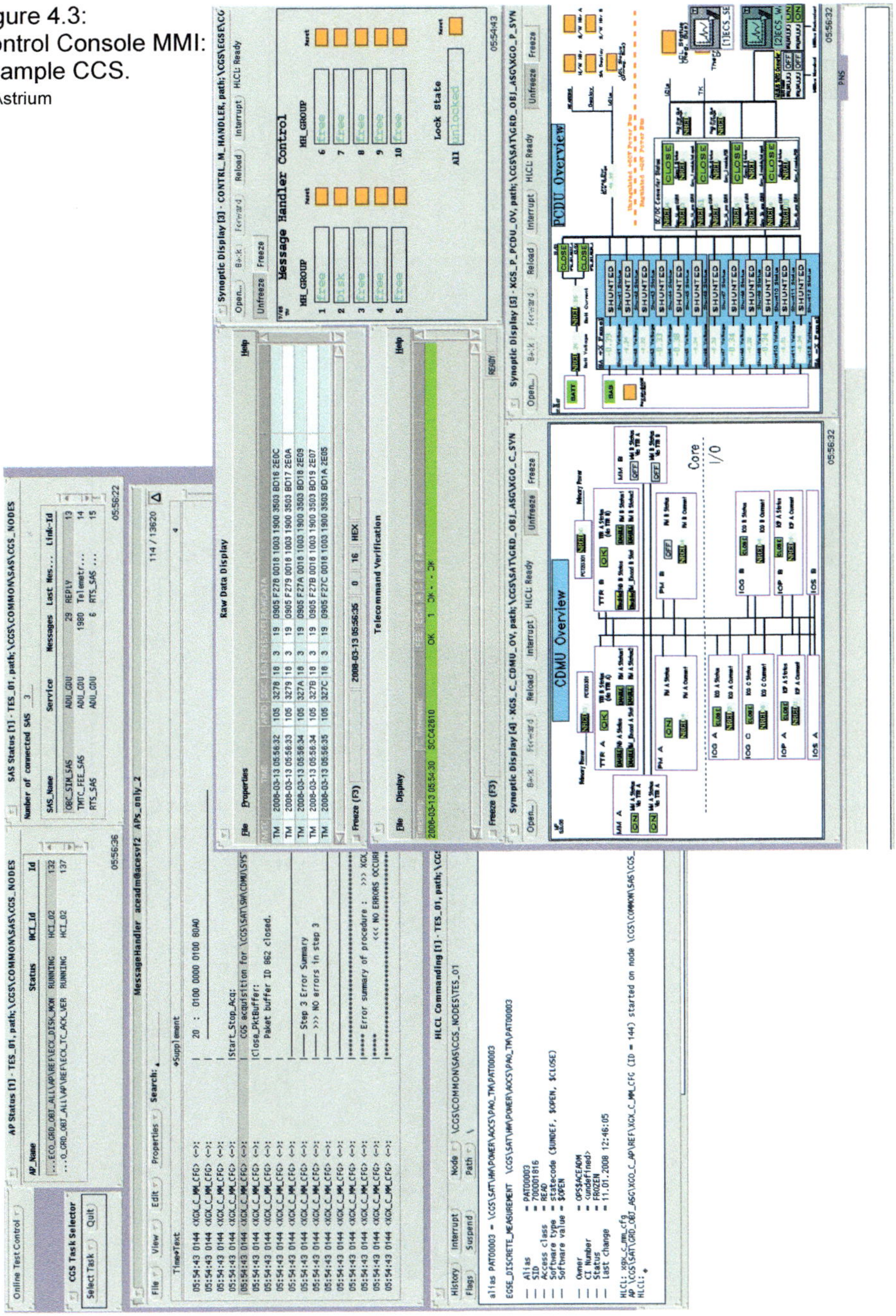

Figure 4.4:
Control Console MMI:
Example Open Center.
© Astrium

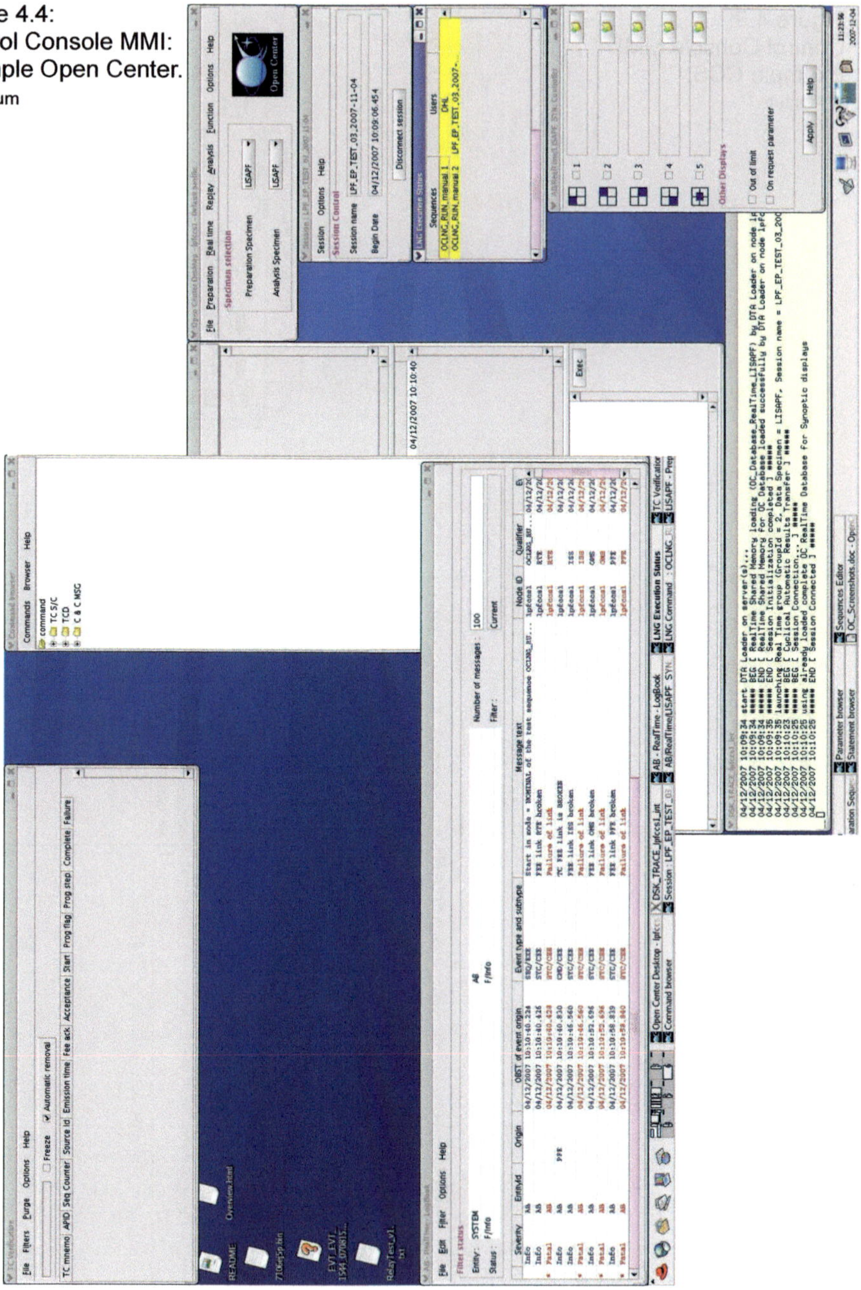

Almost all Core EGSE systems support the user by graphical editors to create displays for visualization of requested parameter values. The display types also are selectable, e.g.

- curve plot displays,
- alphanumeric displays,
- bar charts.

Such visualization windows are called "synoptic displays" or "synoptics". One of these displays is usually created per spacecraft subassembly. For examples please refer to the CCS screenshot provided in figure 4.3.

4.2 Test Procedure Editors and Interpreters

Directly related to these Core EGSE systems are test procedure editors and test procedure interpreters which are needed for procedure definition and execution. The most important test procedure languages in Europe are:

- UCL - command language of CCS,
- ELIZA - command language of OpenCenter,
- TCL[13] - command language of SCOS-2000 / with the TOPE procedure interpreter, and
- Pluto - command language of SCOS-2000 / with the Apex procedure interpreter.

The required functional scope leads to languages which fulfill all necessities for control of tests, reaction to incoming telemetry, reaction to user interactions etc., but on the other hand it makes most languages similarly challenging to handle as a conventional programming language. Therefore, graphical tools, e.g. the "Manufacturing and Operations Information System", (MOIS) [132]), have appeared on the market which

- allow to edit test procedures graphically via flowcharts,
- to provide for each procedure step input masks for the user to add diverse detail information using a text-based view,
- allow direct generation of the test procedure code in the desired test language, (UCL, TCL, ELIZA, Pluto), and
- provide an integrated procedure execution engine which can be coupled to the Core EGSEs.

The TM and TC packet definitions have to be defined in the control console's database and the test procedure engine then allows to subsequently execute the steps of a test procedure inside the control console via an appropriate interface. In the following figures some screenshots can be found demonstrating editing and execution of a test procedure with the MOIS tool from Rhea Group.

[13] The "Tool Command Language", (TCL), used by one SCOS-2000 version is widespread and also is used by many other applications [131].

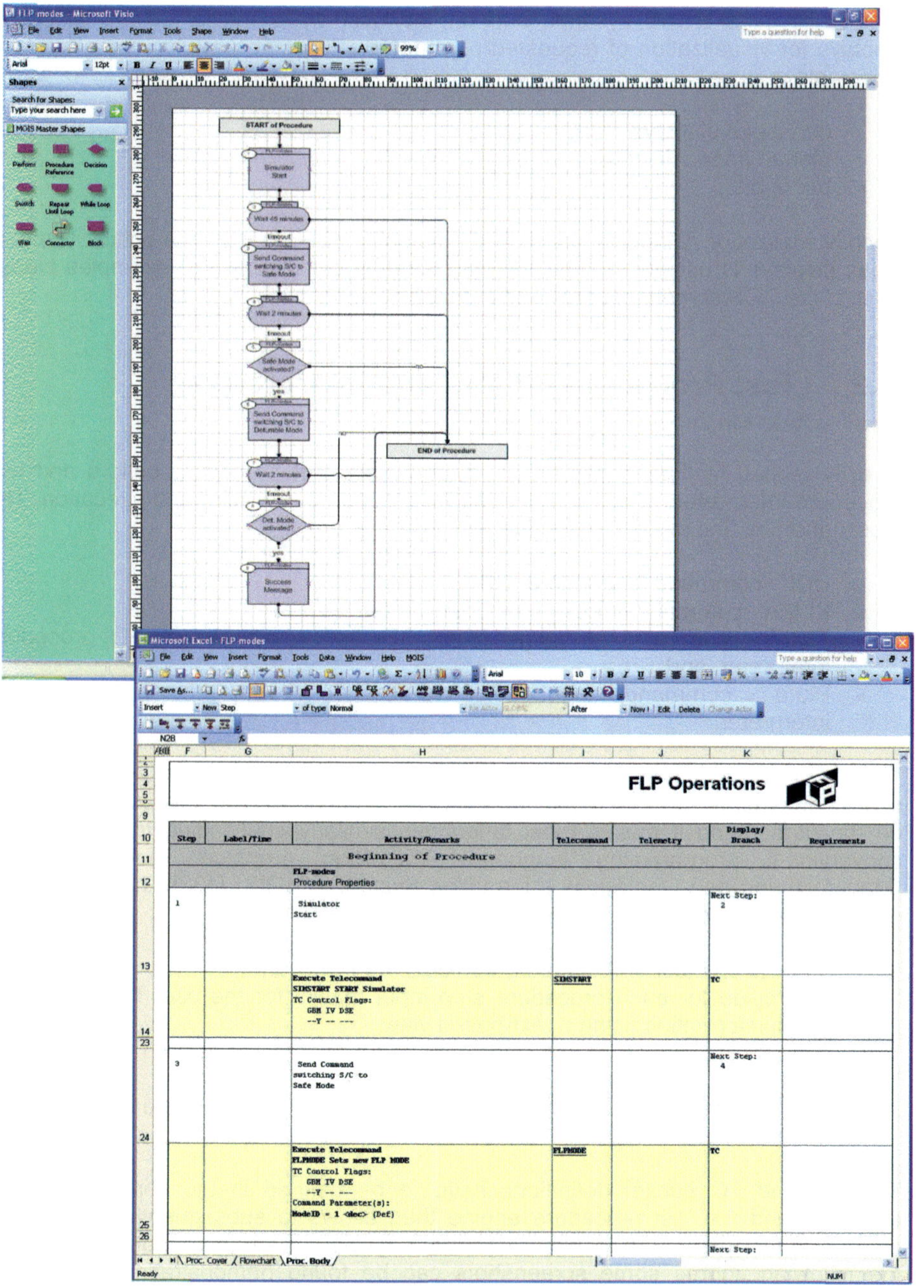

Figure 4.5-A: Definition of procedure logic and details with a flowcharter.
Application example: IRS, Universität Stuttgart.

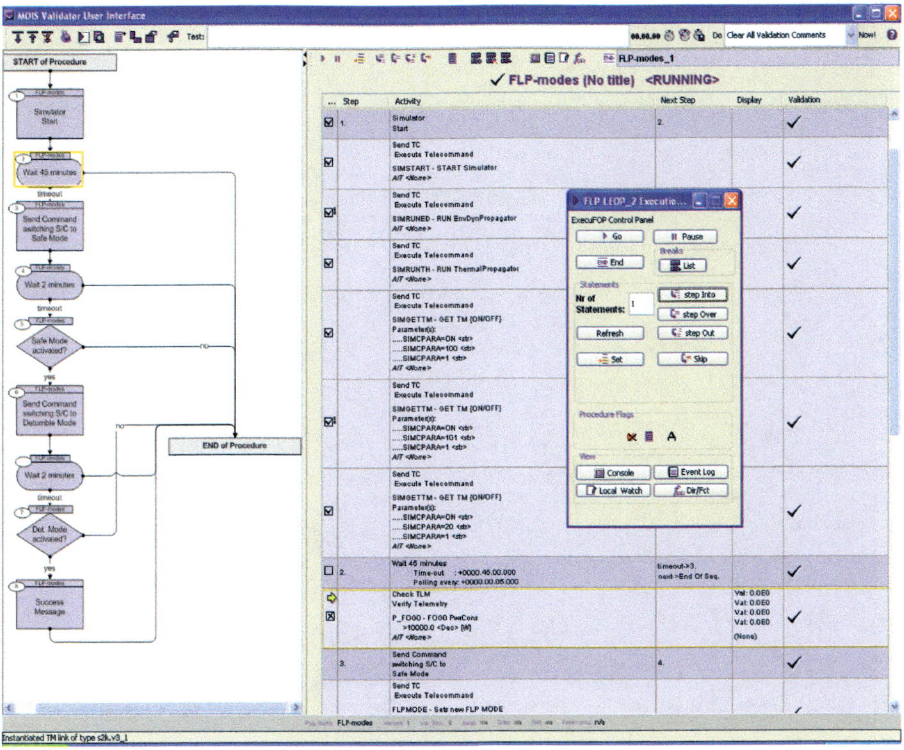

Figure 4.5-B: Tracing of procedure execution with a "validator".
Application example: IRS, Universität Stuttgart.

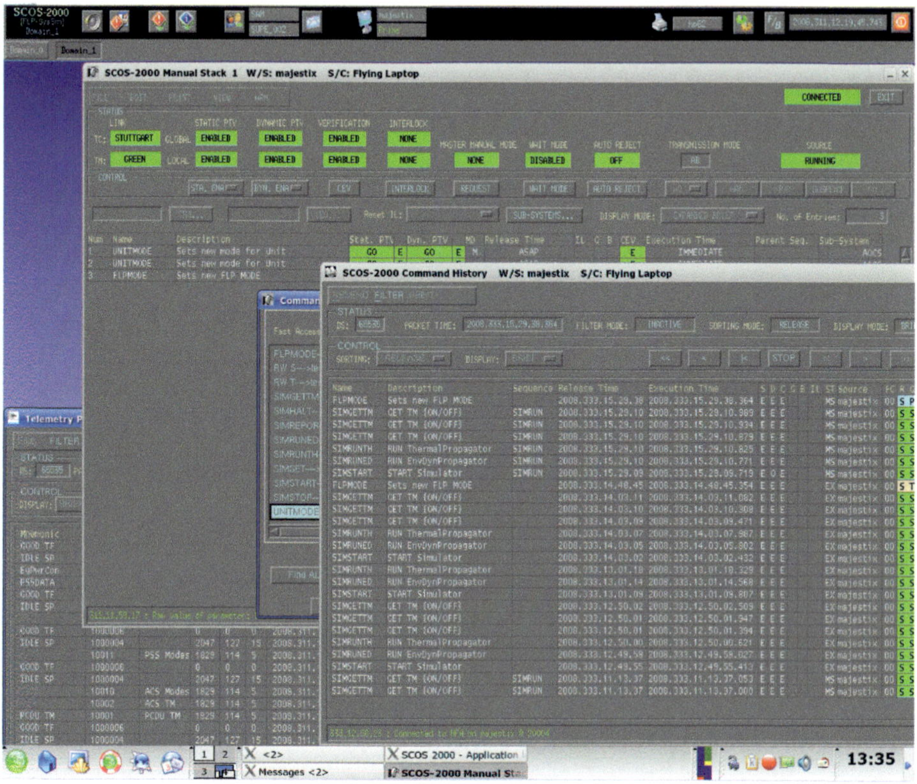

Figure 4.6-A: Command log of simulator and simulated spacecraft.
Application example: IRS, Universität Stuttgart.

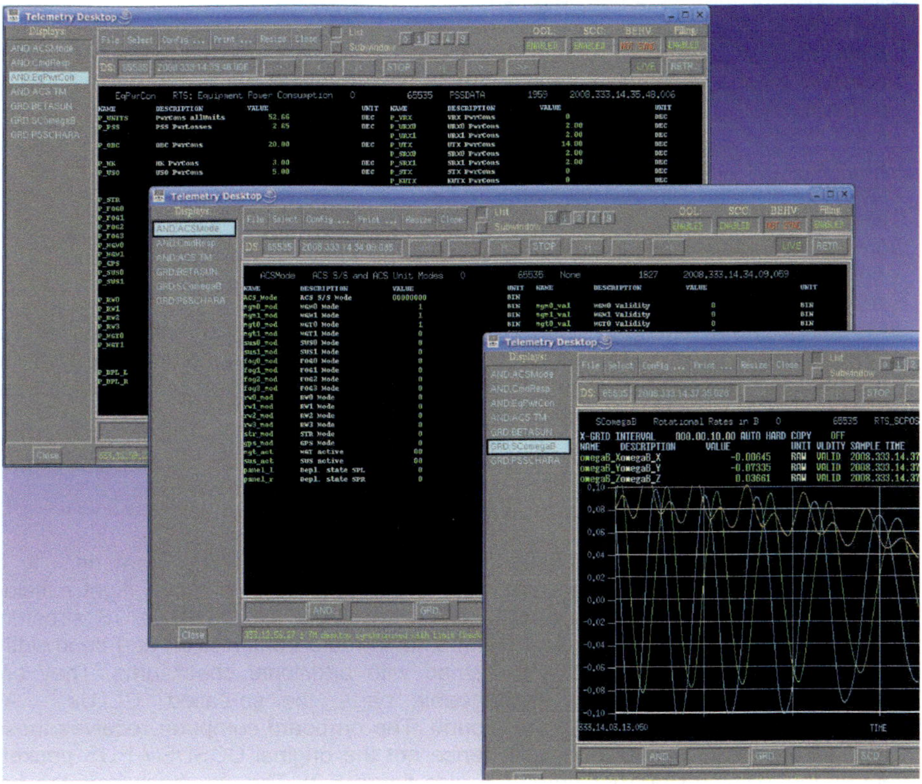

Figure 4.6-B: Parameter values of simulated spacecraft.
Application example: IRS, Universität Stuttgart.

4.3 Special Checkout Equipment

SCOE Example: Power-Frontend

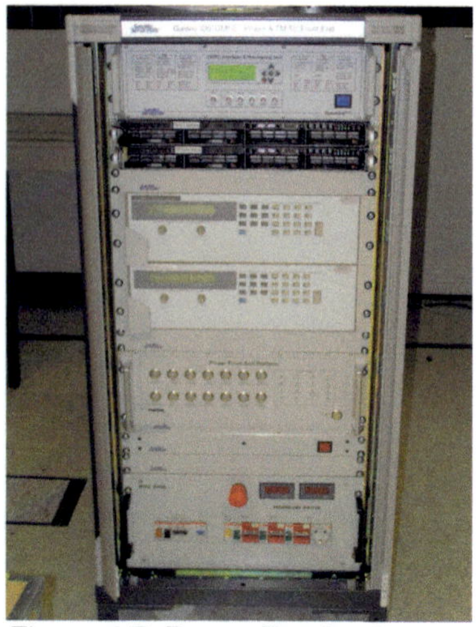

Figure 4.7: Power Frontend.
© Satellite Services B.V.

A typical "Special Checkout Equipment", (SCOE), of a hybrid testbench is a power supply system. To differentiate from standard laboratory power supplies, the testbench supply usually is called the "Power-Frontend". It supplies all “Hardware in the Loop” units of the testbench, e.g. an OBC with power. Depending on the project complexity such a Power-Frontend is either controlled manually or can be commanded by Core EGSE, which is the usual case.

The example shown on the right represents a complex Power-Frontend as used in the Galileo-IOV project. It is commanded via Core EGSE and has 14 direct current regulated outputs with 50V respectively 28V. Each circuit is individually monitored concerning overvoltage and undervoltage and provides overcurrent protection via latch-up current limiters. The settings are also controllable by commands from the Core EGSE. The output currents are precisely filtered and optimally decoupled from main power network influences by complex circuitries.

SCOE Example: TM/TC-Frontend

Another standard SCOE equipment is the so-called "TM/TC-Frontend". It is necessary since telecommands are sent to the spacecraft in a real flight mission from the Mission Control Center as packets following the CCSDS/PUS structure mentioned previously. For technical constraints in radio transmission and bandwidth, these packets are segmented and framed with additional checksums. They are transmitted to the satellite in manageable parts, the so-called "CLTUs"[14]. An acknowledgment is sent back to the ground. The on-board computer receives these CLTU data from its transponder and hence not the original CCSDS / PUS packets sent by the Mission Control Center, (cf. also figure 5.2). The way back from the on-board computer is similar. Frames are transferred to the transponder, which sends so-called "CADUs"[15] to the ground. The CADUs are reassembled to segments and packets there and handed to the Mission Control Center.

In a testbench, the Mission Control Center is replaced by the Core EGSE. Both are compatible to the same protocol, (CCSDS / PUS). Both cases involve a real on-board

[14] CLTU stands for Command Link Transmission Unit.
[15] CADU stands for Channel Access Data Unit.

computer. However, what is missing between the Core EGSE and the OBC is the RF-link and the transponder equipment. The TM/TC-Frontend here replaces these two missing links in the TC and TM chain and performs the required telecommand and telemetry reformatting between OBC and Core EGSE. Furthermore, it enables logging and debugging of data exchange between on-board computer and Core EGSE via various editors and debugging functions.

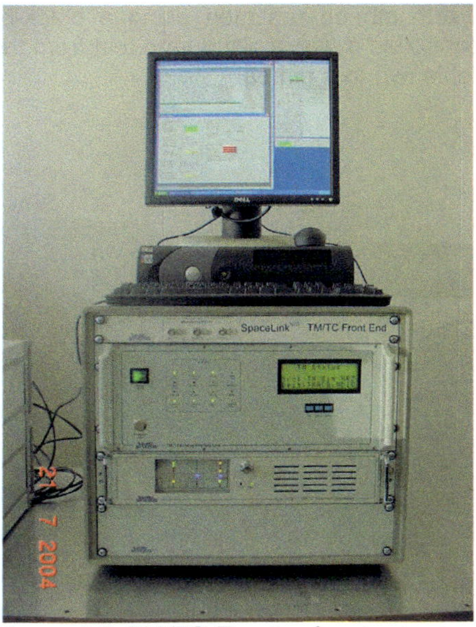

Figure 4.8: TM/TC-Frontend. © Satellite Services B.V.

SCOE Example: Stimulation Equipment for Sensors

Hybrid test benches enable two main types of "Hardware in the Loop" tests which include more equipment than the OBC. On one hand, there are the so-called "open loop" tests. For example, real spacecraft sensors or actuators or payload are connected to the on-board computer via harness. Initial tests only target to evaluate whether

- the equipment can be controlled correctly and,
- the on-board computer receives all acquisition data and status telemetry without errors from the equipment.

Furthermore, it is tested whether the equipment occurrences are connected correctly to the on-board computer ports, for example:

- OBC port A being connected to the solar-array-drive for the +Y solar panel, and
- OBC port B being connected to the solar-array-drive for the -Y panel.

For a subset of critical sensors it might be necessary to stimulate them dynamically and quantitatively, to measure whether the entire system control loop with OBC and sensor in the loop operates as desired. Depending on the different sensor types, various stimulation equipment is required for each.

The following figure shows a detail from the testbed shown in figure 3.27 namely the stimulation system for a satellite Earth sensor. The Earth sensor, (on the left side), is stimulated by infrared Earth albedo radiation using a black-body bolometer. The sensor position can be controlled dynamically during test from the control console via an electrically commandable two axis gimbal.

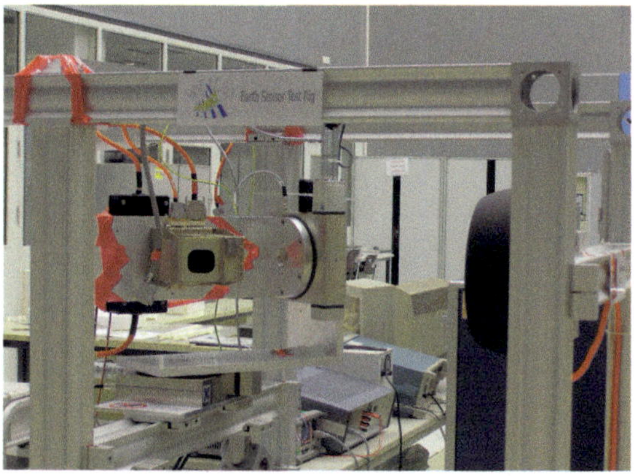

Figure 4.9: Earth sensor test rig. ©Astrium

In a further automated configuration level, the system simulator can be used to calculate attitude and position of the spacecraft. With this information in the test setup depicted above, the Earth position detected by e.g. Earth sensor 1 of the spacecraft can be provided by the simulator and the corresponding control of the test rig can be applied. The OBC then receives the measured signals from the real Earth sensor as if it were in Earth orbit. The OBSW returns them to the attitude control cycle.

The stimulations and / or measurement infrastructures for such closed-loop tests can completely differentiate between the spacecraft system types under test. Their functionalities are individually determined by the definition of tests which are to be run on the installation. The figure on the right shows, as counterpart to the sensor stimulation mentioned above, an actuator test bench for the ATV (see figure 1.6) propulsion system.

Figure 4.10: Propulsion test bench. ©Astrium

4.4 Simulator-Frontend Equipment

The "Simulator-Frontend" equipment serves to connect the on-board hardware, (usually the "Controller-in-the-Loop"), with the still simulated rest of the spacecraft. It is depicted in the figures 3.22 and 3.23 as a layer marked green. The depicted harness lines connect the simulator with the on-board computer. The Simulator-Frontend initially consists of a set of interface cards, which serve to transfer signals from the real spacecraft OBC to the system simulator respectively to return simulated data to the OBC. Since in such a "Hardware in the Loop" configuration the simulator has to respond in real time and thus has to run on a real-time operating system, also the data bus system for interface cards has also to provide real time data transfer features, (e.g. VMEbus).

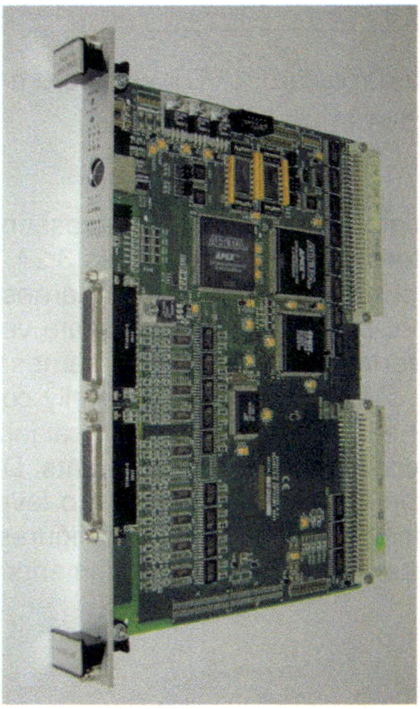

Figure 4.11: Example for a Simulator-Frontend interface card. Photo © Astrium

Figure 4.11 is a photo of an example VMEbus card. The card drivers run as separate tasks on the simulator computer's operating system. The cards are also partially active meaning they have their own control and processing intelligence, usually coded on FPGA chips.

The large number of interfaces and thus interface cards applied in a total system simulation are visualized in Simulator-Frontend rack photos comprised in figures 3.24, (on the left side of the bottom part), 3.25, (the central rack with the large number of incoming cables), and 3.27, (on the right side of the figure).

The following figure 4.12 provides an impression on typical registers, converter chips etc. being part of a simple interface card using the example of an analog-to-digital converter board. The port connected to the on-board computer analog output is depicted on the left side. On the right the simulation is indicated which receives the converted signal now in digital form - which might for example be necessary for the according equipment model in the simulator to calculate its rotational speed and torque. The VMEbus is displayed between the gray area "A/D I/F Card" depicting the components on the card and the "CPU board" representing the computer target hardware of the satellite simulation.

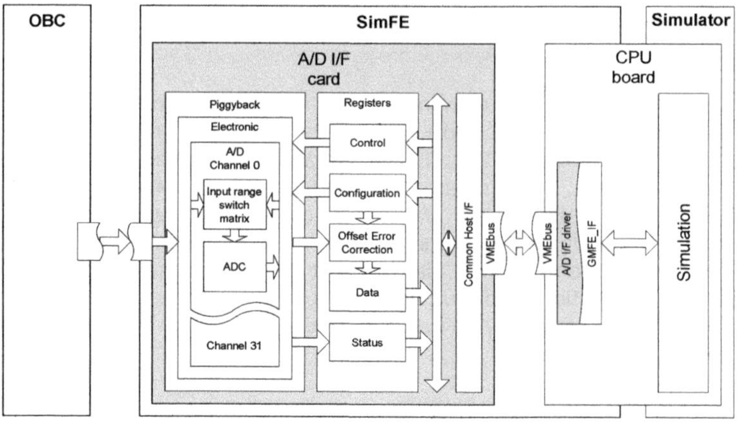

Figure 4.12: Schematic diagram of an interface card, (A/D), with card driver.

© Satellite Services B.V.

The test harness belongs directly to the Simulator-Frontends. It is necessary in order to connect the spacecraft OBC to the Simulator-Frontend. The figures 3.24, 3.25 and 3.27 also give an optical impression of the complexity of these test harnesses. This enables a grasp of the effort for their definition, production and separate verification. Another point not be neglected is that the interface cards, the test harness and the hardware interface on the on-board computer side have to be electrically compatible. This involves substantial effort regarding engineering, design of electrical input/output characteristics down to corresponding signal compatibility measurements. Depending on interface types, the signals may need to be verified with respect to levels, pulse durations, edge angles and protocol sequences. The following two figures, just for illustration, show typical circuit designs taken from design documents of such Simulator-Frontend cards.

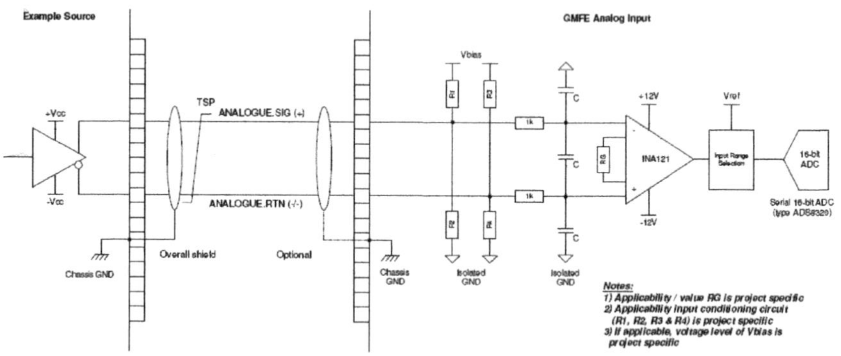

Figure 4.13: Electronic circuit design of an analog frontend-card.

© Satellite Services B.V.

The first one is a conventional electronic circuit for analog signal transmission between signal source (OBC) on the left side, line connection in the middle and card electronics on the right side. The simulator card driver software has to access the analog-to-digital converter, (ADC), of these electronics.

The second circuit diagram shows another typical case: An interface which is used by the OBC to control a power consumer in the real system, (e.g. a pulsed magnetotorquer). However, the simulator does not provide an electrical consumer load. Therefore, a pre-connected so-called "Load Emulator Unit", (LEU), has to generate a voltage output from this current for input to a Simulator-Frontend card. In the figure the LEU is depicted on the left, the conventional Pulse type Simulator-Frontend card is shown in the center.

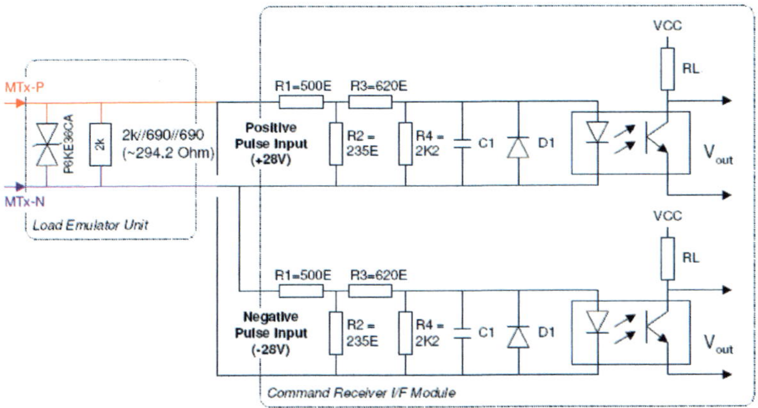

Figure 4.14: Frontend-card with upstream load emulation.
© Satellite Services B.V.

These conversions are still of quite simple nature but there might be other electrical effects to emulate, for example the impedances of electrical coils and similar effects. Already from these simple examples, it can be deduced, that system simulation cannot be limited to the fields of informatics or numerical mathematics, but especially in the case of hybrid test benches, the electrical design of the testbench significantly influences the performance and quality of the simulation. The effort for design and implementation of the Simulator-Frontend defines the critical path of the “Controller / Hardware in the Loop” testbeds in a space project. This results from the complexity of the definition and verification of the Simulator-Frontend. The configuration of the simulator software on the rack defines the availability date of the testbench for application in spacecraft assembly, integration and testing, (AIT).

Another function of the Simulator-Frontend to be discussed is to assure clock synchronicity between the real spacecraft controller, (OBC), and the simulator. If the clocks of simulator and real controller are slowly drifting apart this can induce both numerical problems in longterm tests as well as transmission errors in data protocols. The OBC for example in such cases might read outdated data from the simulator, no data available at the interface or similar effects may occur. To prevent this the

Simulator-Frontend typically is equipped with a pulse generator card which provides an external signal to synchronize the OBC and an internal strobe signal to the simulator. This practice allows very precise synchronization of the two components of the system testbed. The absolute time reference of the testbench may in the case of special "Hardware in the Loop" configurations be provided via network time protocol, (NTP), a GPS-Receiver or other.

4.5 Spacecraft Simulators

The central component of the testbenches of such a development infrastructure is the spacecraft simulator – in the configurations of table 3.2 marked in yellow. The spacecraft simulator is modeling the operational and functional behavior of the satellite. It integrates all technical disciplines such as

- the data exchange between equipment models,
- modeling of system environmental physics - in the case of a satellite including the satellite's body dynamics and the mechanical, electrical and magnetic influences from environment onto the spacecraft,
- and finally modeling the physics inside the system, (in the case of a satellite for example, the satellite's thermal behavior, the electric behavior and so on).

The spacecraft simulator incorporates the models of all the spacecraft's components. These components are modeled in such detail that it is possible to simulate both nominal operating modes as well as failure modes. The failure functionalities inside the equipment models allow a consistent reproduction of failure symptoms of the real spacecraft components for tests of the on-board software. Modeled system components of a satellite are for example:

- The equipment of the attitude and orbit control system, (AOCS)
- The payloads, (normally not covering simulation of the payload science data)
- The power and energy supply system
- The telecommunication equipment

In addition a functional representation of the spacecraft on-board harness is included in the simulation comprising

- power supply harness, and
- signal harness reflecting analog, digital and data bus connections between on-board equipment.

The figure 3.14 depicts a satellite simulator in "Software in the Loop" configuration. It is visible how the topology of the real spacecraft is modeled by equipment models, the simulated harness between them and the harness to the on-board computer. Figure 3.23 shows the satellite simulator in a "Controller in the Loop" configuration. The satellite's simulated equipment is connected by means of the simulated harness to the interface cards and the card drivers of the Simulator-Frontend equipment. The real test harness routing the electronic signals to / from the real on-board computer connects to the in and outputs of the Simulator-Frontend cards.

The simulator itself has to be capable of integrating the system's behavior in the time domain by means of a central numerical solver. In the case of thermal or fluid dynamic systems, the solver additionally has to be capable of handling boundary conditions imposed by the numerical system description. The same applies for the simulation of detailed thermal system behavior and the simulation of power networks. The simulator kernel can be, (if necessary with functional extensions), reused over diverse spacecraft projects since this core set of numerics, model scheduling functions and interface to a control console is project independent. The models of the spacecraft components – for example models of satellite equipment – can also be reused, under the condition that the equipment as well as the operating conditions are the same in both systems.

Since the simulator for hybrid testbench configurations has to be connected to "Hardware in the Loop" and thus has to provide real-time input / output behavior, it is necessary to run the kernel on a real-time operating system with real-time capable data buses and operating systems e.g. on PowerPC cores under VxWorks and VMEbus or on Intel / AMD processors under real-time Linux variant and PCI-X bus. The simulator kernel thus incorporates:

- A numerical Solver for algebraic and differential equation systems
- The interface for simulator commanding via the control console respectively for supplying the control console with simulator telemetry data, (not to be confused with spacecraft telemetry). The data exchange with these testbench components takes place in parallel to the simulator's numerics as separate threads. This enables the sending of commands to the simulator
  - ◇ to command the simulator operation,
  - ◇ to poll or set state variables of simulated spacecraft components,
  - ◇ to poll or set variables of simulated wires,

  while the numeric spacecraft simulation is running.
- A logging interface which allows
  - ◇ the selection of logged simulation parameters,
  - ◇ that are cyclically stored on hard drive with adjustable frequency.
- And finally there normally exists a so called "External Stimulation" interface:
  - ◇ It allows overwriting of the calculated values of a component model with values from a file – e.g to overwrite the values of a startracker in order to achieve a predefined failure profile.
  - ◇ Control settings for frequency of the input data reading, start and end of the overwriting in the course of the simulation etc. are stored and defined in files, (mostly in clear text notation). The overwritten data themselves are stored in ASCII files. The external stimulation is dependent on the occurences of the simulated hardware. It is possible e.g. to,
    - ► simulate startracker 1 and 2 normally while,
    - ► overwrite the values of startracker 3 with failure values.

Furthermore some simulator kernel architectures incorporate the intelligence to distribute the numerical calculation steps between multiple processor nodes. The kernel software of modern simulators today is designed with the use of formal design languages – normally applying the "Unified Modeling Language", (UML). The modeling of the spacecraft equipment is based on the same technique. Details about modeling with UML, the generation of sourcecode from UML and related topics are described in chapter 8.

4.6 Equipment and System Models

The models of spacecraft equipment inside a simulator, the, "equipment models", are emulating the functional and operational behavior of a system element, e.g. the behavior of a sensor of the spacecraft or a power supply device. They incorporate the algorithms to represent the component with the required accuracy and functions to fulfill project specific requirements - e.g. failure models, external stimulations etc. They only model the behavior of the components functionally, but not the internal structure of the real equipment hardware. E.g. for a reaction wheel a functional model computes the wheel's torque, tacho signal, power consumption, wheel rotation rate and the heat dissipation as functions of the supplied motor current. However it does not reflect that there exists a rotor, motor coils, a housing, screws, ball bearings etc.

The algorithms of the component models which are representing the functionality are split into a discrete part, (in most cases modeled as a state machine), which is called in the control cycle and an analog part which is called in both the control cycle and the numeric integration steps.

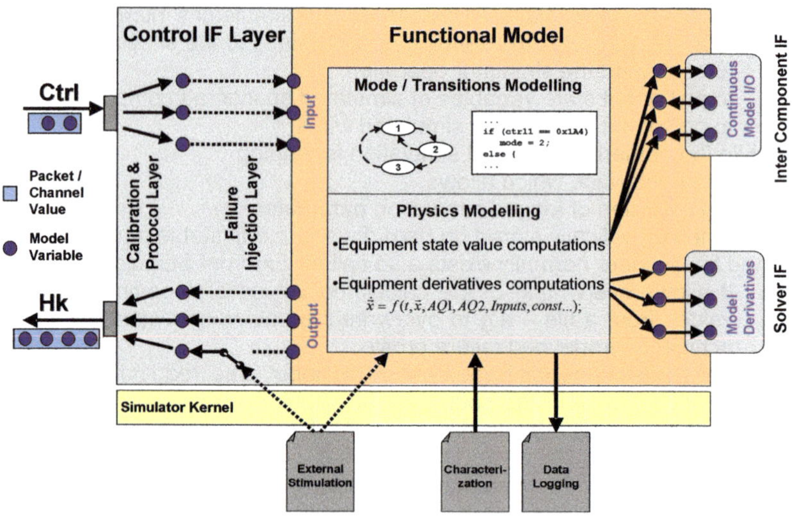

Figure 4.15: Structure of an equipment model.

The component models are equipped with a data interface, ("Ctrl" and "HK" in the figure above), in order to enable commanding via the simulated or "Hardware in the

Loop" OBC. This control interface layer also performs calibration and decalibration of command and housekeeping data between variable's values in engineering units in the functional model and control / housekeeping data protocol packets. Secondly the models provide interfaces to the simulator kernel which are used to

- load the characterization data during initialization,
- to provide data for simulator telemetry to control console,
- and for simulator log data.

In case of an external failure stimulation these interfaces are used to feed the model's numerics or the protocol interface with failure values from files or from control console commands to the simulator. Furthermore the models are equipped with interfaces to the numerical solvers of the simulator. These interfaces are, for example, used to feed the solver with the derivatives of the component's state variables for every integration step. More details on the component numerics are explained in chapter 6. Finally the models have interfaces to each other which are used to transmit physical variables between them - in the figure above labeled as "Continuous Model-I/O". This functionality in many other publications is not described very precisely and therefore shall be elaborated here in a bit more detail.

In diverse publications on simulators it always is distinguished between so called equipment models and system models. Nevertheless by applying a generic description of the simulator technology like it is outlined in chapter 6 this distinction is obsolete. The following explanations make use of the analogy of modeling a satellite again:

If classical equipment models are addressed, the reader immediately thinks of e.g actuators like a reaction wheel. According to the commands from the OBC transmitted over the control interface, the wheel model then feeds back to the "Continuous Model-I/O" a rotational speed, an angular momentum and torque information. Just for now only the torque shall be considered. Another component – for example a magnetic torquer – is producing another torque. The receiver of both torques is a model like all the others, and in this case represents the spacecraft structure. This model of the spacecraft structure - a rigid structure shall be assumed here - simply is not equipped with a control interface to the OBC. The model of the structure receives the torques from the actuator models and eventually from another model representing the space environment. From this input it calculates the overall derivatives which are necessary inputs for the spacecraft attitude integration. The solver has to integrate the angular momentum equations of the spacecraft structure. Based on this example it is visible that the working principle of the actuator, structural and environmental models are in fact the same.

Systems of the type shown in figures 1.12 or 1.10 include more equipment models like pipe tubes, filters, heat exchangers etc. which do not have control interfaces, but a "Continuous Model-I/O" and a solver interface. In the case of such fluid dynamic systems and also in the case of electrical systems, the simulator numerics module not only has to integrate an initial value problem of differential equations but also has to solve in parallel, the complementary boundary value problems as described in chapter 6 in more detail. This fact increases the complexity of the solver interface but does not change the principle of interaction between models and solver.

The effects to be modeled for such space environment models, structural models or equipment models of different kinds are addressed in the following chapter.

Modern system simulators as e.g. described in [16] to [21] or [23] and [120] typically comprise a functionality to load essential characterization data of the simulated system in order to configure the simulation run parameters during initialization. In all systems mentioned above the following settings are loadable:

- The characterization data of the modeled equipment - for example the power consumption depending on the operational mode of the equipment.
- For this configuration sometimes there is a default version of standard settings of the system and separately different versions which store only the configuration deltas to the default values to make the files more clearly arranged.
- In addition there are the files for the configuration of the numerics in which the integration step sizes etc. are defined, the selected numerical solver method and so called scheduling tables – definitions of the sequence in which the models will be called and eventually the timing offsets between equipment model calls.
- Furthermore the configuration of the simulator telemetry and logging usually is loadable from files. In these files, information about simulator telemetry packets, the frequency of telemetry sending and logging are stored.
- For simulators in hybrid testbenches it also is essential that the simulated interfaces between equipment models and Simulator-Frontend cards are configured properly, that the real hardware receives the signals from the models. Because the Simulator-Frontend cards normally are equipped with multiple channels of the same interface type, the configuration information must define which equipment model is mapped to which ports of the card driver - the latter corresponding to certain connectors and pins for the electric I/O-side of the card.
- For these hybrid testbenches, calibration curves have to be loadable for every simulated connection to the card driver as soon as digital / analog converters are used.

All simulator systems today follow the object oriented design paradigm. This means that the code of an equipment model class is only coded once and is only loaded once in the program. For every occurrence of an equipment type in the system, a so called instance, only a copy of the data area of the class is created. This appears to the user as if many equipment components of the same type were loaded.

The simulators are different in respect to the methodology of the system's topology definition. In this case topology means the number of equipment components of a specific type and how they are connected. For some simulators, the number of omponents of a specific type in a system has to be defined already during the design process in the modeling language UML. The sourcecode is then generated from the UML diagrams and compiled. Also the definition of the connections between the models, (which components are connected with which type of interface), have to be specified for these simulators at compile time.

Other systems allow to load the entire system topology, including modeled equipment and equipment interconnections, at simulator initialization time. The student's project *OpenSimKit* [23] consequently follows this configuration design principle. In the case of this simulator, only the model classes of the component types and the connection types are to be coded. The definition of the system topology can be loaded from configuration files at startup - for example whether a system has 2 or 3 sensors of a type.

The configuration files of such kind of simulators today are usually defined via the so-called "Extensible Markup Language", (XML). This means will be treated later in more detail in chapter 8.5.

5 Spacecraft Functionality to be Modeled

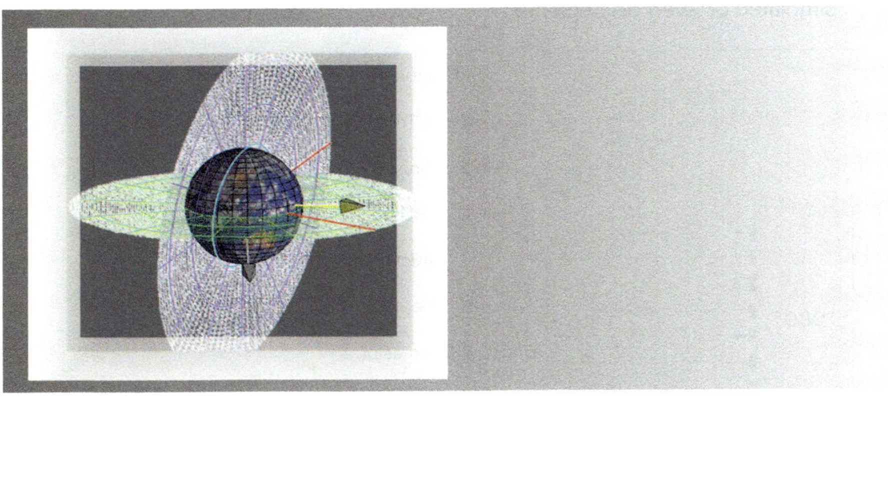

J. Eickhoff, *Simulating Spacecraft Systems*, Springer Aerospace Technology 1,
DOI 10.1007/978-3-642-01276-1_5, © Springer-Verlag Berlin Heidelberg 2009

5.1 Functional Simulation Concept

The term "functional" simulation of spacecraft has already been mentioned. A precise definition is given below. More detailed aspects of the different spacecraft components and the space environment which both have to be modeled in the frame of a functional simulation are treated afterwards. The modeling criteria cited in the following subsections are focused on the satellite domain. However, they can similarly be applied to other spacecraft like shuttles, launchers and transfer vehicles, too.

Basic Concept of Functional Modeling and Simulation:

> Functional modeling implies that just those aspects of a system are simulated which are necessary to supply the item under test, (hardware, on-board software etc.), with realistic data in order to verify it. The models around the item under test do not need to reflect the real overall system topology.

Example - the item under test shall be on-board software of a satellite:

In such a case only those aspects of the satellite are to be simulated, which are necessary to enable the on-board software test.

Negative formulation:

All aspects of the system which are not necessary for test purposes can be simplified or even be neglected.

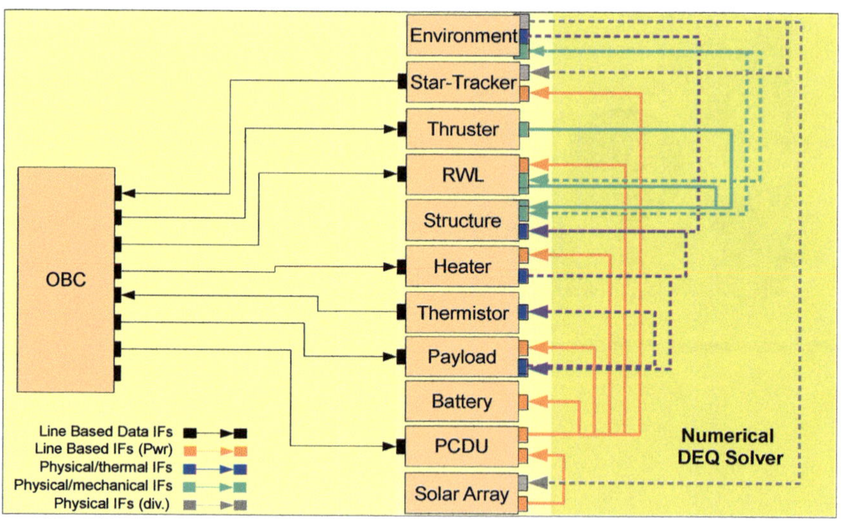

Figure 5.1: Components and interactions to be modeled functionally.

Example - power subsystem:

- For an on-board software test, the data delivered to the on-board computer by a power control and distribution unit are relevant, including measured values of the currents delivered by the solar array, the currents supplied to the power consumers and the charging currents to the batteries.
- Details on the solar array or battery internals, (like the dependency of the cell resistance to the temperature), are of no relevance for such a test.

Example - satellite harness:

- The harness database as used in spacecraft development and required for electrical tests, models all lines including:
  - ◇ Each wire
  - ◇ Shields
  - ◇ Grounding lines etc.
  - ◇ Connector to pin assignments for each wire
- In the simulator only the following functional information is required:
  - ◇ Information on the "functional interface" type - for example whether two spacecraft components communicate via a serial, an analog connection or a data bus etc.
  - ◇ The assignment of "functional interfaces" to "ports" of simulated equipment.
  - ◇ For hybrid testbeds, the assignment of "functional interfaces" to "ports" of front end card drivers.

Example - satellite topology:

- In a functional satellite simulator, the satellite geometry or geometric assembly of equipment does not need to be modeled.
- However, models have to reflect:
  - ◇ The mounting coordinates and orientation of AOCS actuators and sensors using the spacecraft coordinate system.
  - ◇ The representation of the electrical topology w.r.t. equipment redundancy and cross-couplings in the functional interfaces.
  - ◇ The thermal representation of equipment via multiple thermal nodes, if an equipment (e.g. payload) consists of multiple thermally relevant subcomponents.

Example - TC / TM between OBC Simulator and Core EGSE:

The real flow of TC from the Core EGSE packet level to the on-board computer is shown in figure 5.2. The telecommand packets are combined in segments, each of which comprises data for one on-board target address. Afterwards the segments are split up into frames with constant length, which are sent from GCS to the transmitter station. Once again, the frames are broken up there to shorter elements, into so-called Command Link Transfer Units (CLTUs). These units are transmitted partly in parallel to the spacecraft transponder exploiting the bandwidth of the frequency range allocated for the spacecraft. The transponder interface card in the OBC (TTR board) reassembles the received data back to higher layers - for high priority commands at least to frame level, for normal commands up to the TC packet level. The OBSW can access these packets then in the corresponding TTR board registers. A comparable procedure is applied for the inverse direction of spacecraft telemetry transmitted to ground.

In an SVF type testbench the control console sends already CCSDS packets to the simulated OBC. So w.r.t. the OBC input side representing the real OBC's TTR board, only the following points are visible for the OBSW and need to be modeled:

- Read / write registers in the TTR board of the OBC
- Interrupts from the TTR board to the OBSW
- Failure modes of the TTR board

Segmentation and framing usually do not need to be modeled in an SVF type simulation.

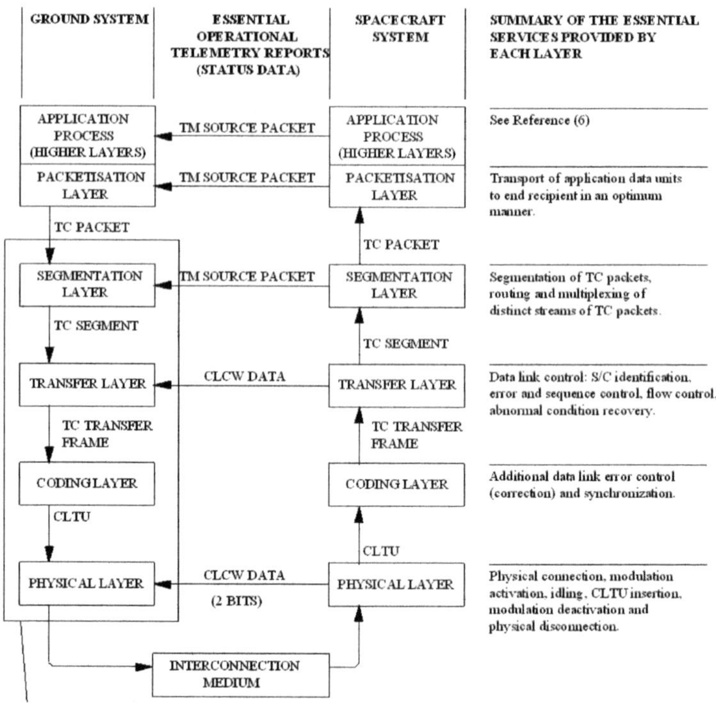

Figure 5.2: Telecommand transfer from ground control system (GCS) to a satellite.
Source: ECSS-E10-71

5.2 Attitude, Orbit and Trajectory Modeling

The orbit propagation calculation as well as partly the attitude calculation for a spacecraft in a simulator have to consider all all those effects, which exert forces or torques onto the spacecraft. With respect to orbital mechanics and attitude dynamics orbit and / or attitude changes have to be integrated over time within the simulator. This has to be done by integration of the impulse equation and angular momentum equations over time, considering the sum of all cited external effects and forces and torques generated by spacecraft actuators (like reaction wheels, thrusters, magnetic torquers etc.).

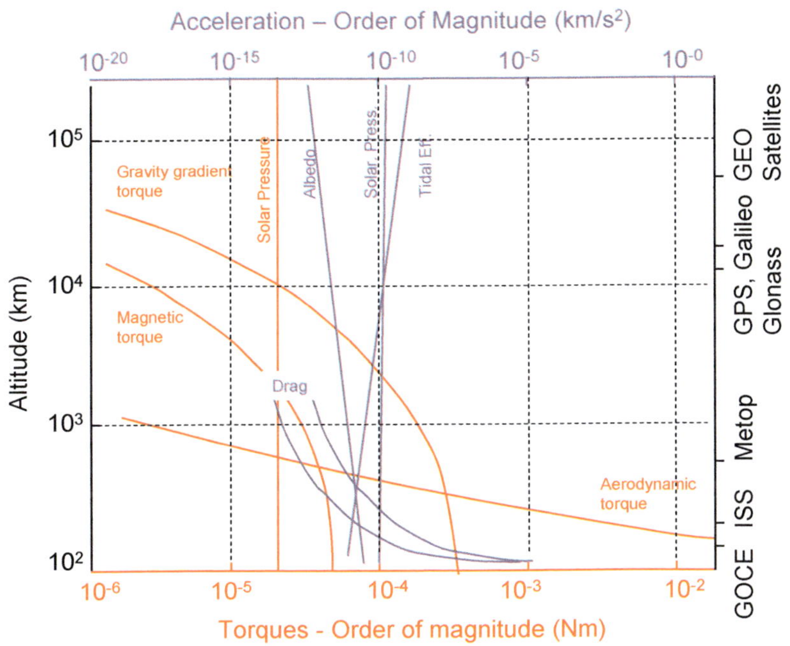

Figure 5.3: Orbit perturbations as a function of the altitude.

Regarding electrical effects, initially the basic information concerning the Sun / eclipse phases, the radiation intensity at the spacecraft position and similar values provided by the environment models are to be considered for simulation of equipment and spacecraft system states. This similarly applies to magnetic field and equipment dipole effects. Therefore, the "space environment model" of a simulator has to reflect all the following external effects:

- Gravitational effects:
  - ◇ Forces and torques impacting the satellite based on anomalies of the Earth's gravitational field

    - ◇ Eventual tidal forces
    - ◇ Forces on the satellite resulting from gravitational effects of other celestial bodies - especially the Moon
- Aerodynamic perturbations based on atmospheric drag
- Perturbations based on solar radiation pressure by the sunlight and by particle flows - only in the Sun phase of the spacecraft orbit
- Forces and torques induced by magnetic interaction between satellite components and the Earth's magnetic field

In the following paragraphs the characteristics of different mission types are listed, including the particular effects, which have to be reflected in an according space environment model.

Low Earth Orbit, LEO:

- Altitude: 250 – 750 km above Earth's WGS Ellipsoid, $v_{S/C} \approx$ 8 km/s.
- Orbit altitude, inclination, eccentricity and Sun synchronicity of the orbit must be modeled.
- Furthermore, perturbations based on atmospheric drag, gravitational anomalies, solar radiation pressure, Moon and planetary effects are to be reflected.
- Corrections by the attitude and orbit control system of the satellite are to be verified by means of simulation.

Medium Earth Orbit, MEO, (e.g. GPS, GLONASS, Galileo):

- Altitude: 10,000-30,000 km above the Earth's surface, $v_{S/C} \approx$ 6-3 km/s.
- Orbit altitude, inclination, eccentricity and Sun synchronicity of the orbit must be modeled.
- Furthermore, perturbations based on gravitational anomalies, solar radiation pressure, Moon and planetary effects are to be reflected.

Geostationary Orbit: GEO:

- Altitude: 35,800 km above the Earth's surface, $v_{S/C} \approx$ 2,6 km/s.
- Orbit altitude, inclination, eccentricity, Sun synchronicity of the orbit and in particular the relative position to the Earth, (North - South, East - West station keeping), must be modeled.
- Perturbations based on gravitational anomalies, Solar radiation pressure, Moon and planetary effects are to be reflected.

Interplanetary Trajectories:

- The orbit propagation to the destination considering all maneuvers, fly-byes, delta-v maneuvers etc. must be modeled.
- The perturbations based on different effects - solar pressure, particle density , magnetic fields of planets etc.
- Complex n-body dynamics problems must be modeled, if necessary.
- The successful orbit / maneuver controls by the AOCS have to be verified.

Lagrangian point missions:

- The orbit propagation to the destination considering all maneuvers and the position control at the destination position with the Lagrangian point specific gravitational sum effects must be modeled.
- The successful orbit / maneuver controls by the AOCS on the flight to the destination have to be verified.
- The same applies for the correct position control at the destination ("station keeping").

Drag-free missions:

- The orbit propagation including drag-free control in the orbit or at the Lagrangian point have to be modeled.
- The successful orbit / maneuver controls by the AOCS on the flight to the destination have to be verified, and the same applies for
- the correct position control at the destination, (drag-free control).

Formation flying missions:

- The orbit propagation has to consider / reflect all orbit and formation correction maneuvers.
- The successful performance of orbit / formation maneuvers and their numeric controls by the AOCS are to be verified.

For all mission types

- the attitude control during operation periods of payloads (fine pointing) has to be modeled.

A more detailed classification of different orbit types can be found in [37]. Substantial modeling examples for orbit dynamics can be found in [38].

5.3 Aspects of Structural Mechanics

Regarding spacecraft structure and its behavior in orbit, the following characteristics have to be modeled in a functional simulator:

- The mass of the spacecraft
- The moments of inertia
- The point of origin of the spacecraft's coordinate system
- The position and alignment matrices of all sensors and actuators relative to the satellite coordinate system
- The flexibility of the structure i.e. if applicable all flexible modes - e.g. of deployed solar panels
- Transient changes if applicable - e.g. during the deployment of solar panels

- Finally fuel sloshing effects have to be considered, if there is a liquid propulsion system on board. This is e.g. relevant for large, geostationary telecommunications satellites designed for long lifetimes, for launcher stages and shuttles.

All these spacecraft characteristics and characteristics variations must be modeled for attitude and orbit integration. The entire motion dynamics of this "structure" under the influence of all orbital perturbation forces and of the controlled actuator forces has to be reflected in the spacecraft simulation model.

5.4 Thermal Aspects

For a functional system simulator like those implemented in SVFs, STBs and similar testbenches for on-board software verification and respectively system tests, it is sufficient to simulate the thermal system aspects in the correct order of magnitude. This implies the following effects have to be detectable by the on-board software from simulated thermistor signals during an orbit simulation:

- Temperatures have to be simulated in the correct order of magnitude.
- Furthermore, the periods in which certain temperatures drop below specified lower limits inducing heaters to be activated by the on-board software have to be in conformity with orbital sequences.
- When the spacecraft leaves the eclipse and enters the Sun phase again the temperatures have to reflect increase, and where exceeding switch levels, OBSW must be able to detect the need for the heaters to be turned off again.
- Furthermore the covered overall temperature simulation range must allow for testing thermal failure detection, isolation and recovery (FDIR) scenarios.

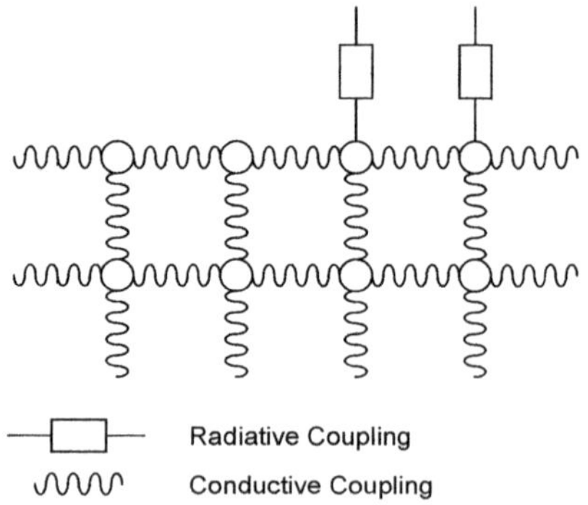

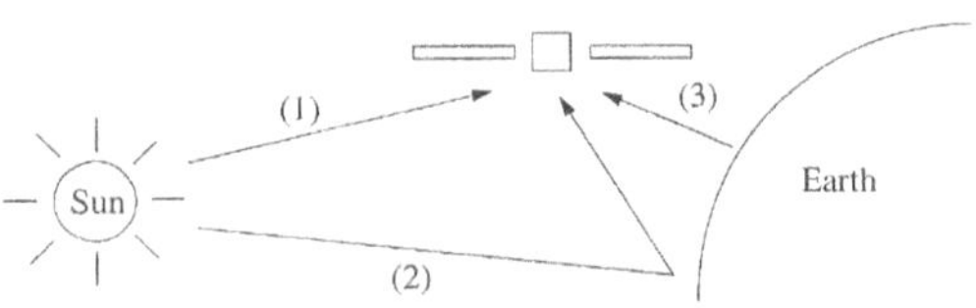

Figure 5.4: Thermal modeling.

Therefore, the following functionalities have to be modeled:

- Solar radiation loads onto the spacecraft outer surfaces - computed by the space environment model)
- Earth albedo radiation loads onto the spacecraft outer surfaces - computed by the space environment model
- Thermal radiation from spacecraft outer surfaces into space
- Coupling of outer surfaces to internal structure via heat conduction and thermal radiation.
- Coupling between internal structure and system components has to be considered, leading off component dissipation especially via heat conduction.

More effective cooling concepts such as heat pipes or stirling coolers etc. might be of relevance for components with high dissipation rates.

Detailed thermal models for spacecraft of a complexity of a medium Earth observation satellite with all its equipment are implemented as complex node models with 3000 to 5000 thermal nodes. The typical number of nodes for a comparative satellite modeled in a functional simulation testbed is reduced to a range of 30 to 50 thermal nodes to provide a qualitatively sufficient precision and in parallel to still allow the simulation to perform in real time on hybrid testbenches such as STBs and EFMs. The following rule of thumb can be assumed:

- One thermal node implemented per occurrence of an equipment which significantly dissipates heat
- One node per installed heater
- One per thermistor
- One per spacecraft outer surface for the modeling of external thermal loads and radiative heat dissipation

5.5 Equipment Modeling

The following pages cover typical spacecraft equipment including their characteristics and the effects which have to be modeled in a functional system simulation. Generally for all equipment types the following characteristics are to be reflected:

- The data interface to the OBC
- The power connections of an equipment component
- The power consumption (dependent on the equipment's mode of operation)
- The thermal dissipation (dependent on the equipment's mode of operation)
- The functionalities of the component electronics regarding:
  - ◇ Its operational modes, commands and telemetry
  - ◇ Its data protocols

Besides standard systems, like the platforms for Earth observation satellites, there are a lot of specific spacecraft types like shuttles, landers, rovers etc. That is why this section only can provide an overview of typical equipment types. For each type the functionalities are cited, which have to be reflected in a functional simulation. The state of the art for the modeling of specific subsystems and equipment can only be determined by literature research.

Control- and Data Processing Equipment

On-board computer

In spacecraft simulators, the on-board computer (OBC) model is definitely the most complex one as it has to model the on-board computer with all of its components to a level of detail which enables to load and to run the real flight software compiled and linked including operating system and BIOS in a "Software in the Loop" setup (SVF) - all compiled for target flight hardware. The following elements therefore typically have to be modeled in the simulation:

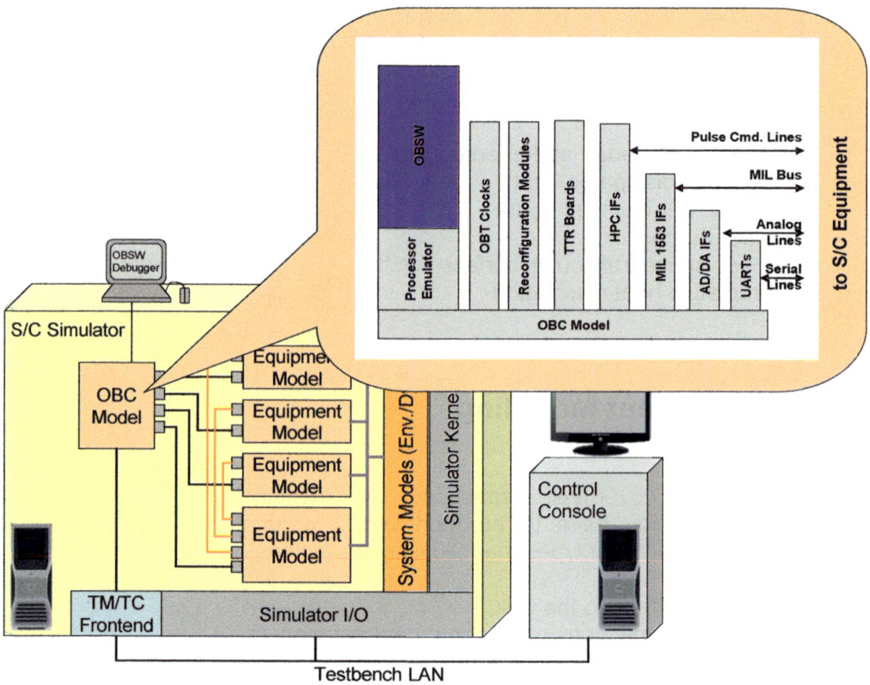

Figure 5.5: OBC model in an SVF (example).

- The central processing unit (CPU) is usually implemented by a processor simulator (for the SPARC[16] based processors, applied in European satellite industry today, these typically are TSIM by Gaisler Research or SimERC by Astrium S.A.S.).
- All OBC subcomponents which do not belong to the processor are modeled in a functional way. This implies that:
  - ◇ Only those functions are implemented which are required to enable the components to correctly respond to commands from the OBSW with correct timing.
  - ◇ The component's I/O-registers have to be modeled (e.g. registers of a data bus controllers).
  - ◇ The ability to reflect and induce component failures by user command must be provided for testing the on-board software's error handling functions.

- The OBC components besides the processor are typically:
  - ◇ Memory of the OBC system (RAM, ROM, PROM, EEPROM)
  - ◇ Memory cards / banks for payload data
  - ◇ Clock module
  - ◇ Reconfiguration logic
  - ◇ Data bus controller (e.g. MIL-STD-1553B bus, SpaceWire)
  - ◇ I/O-subcomponents
  - ◇ Interface modules of all data interfaces to the spacecraft (so-called remote units)
  - ◇ OBC TC decoder, OBC TM encoder, OBC transfer frame generator
  - ◇ Modules for OBC power supply.

- The entire modeling (processor simulation and other components) has to consider
  - ◇ all redundancies,
  - ◇ the I/O-registers and buffers of all components and
  - ◇ the timing for switching / processing processes of all modeled hardware elements.

Figure 5.6 and 5.8 depict such modules including redundancies and cross-couplings in the interconnections of a real OBC. It is evident that, besides the processor model itself which processes the on-board software microcode, there is a big effort in modeling of the remaining OBC infrastructure.

[16] SPARC is a registered trademark of SPARC International.

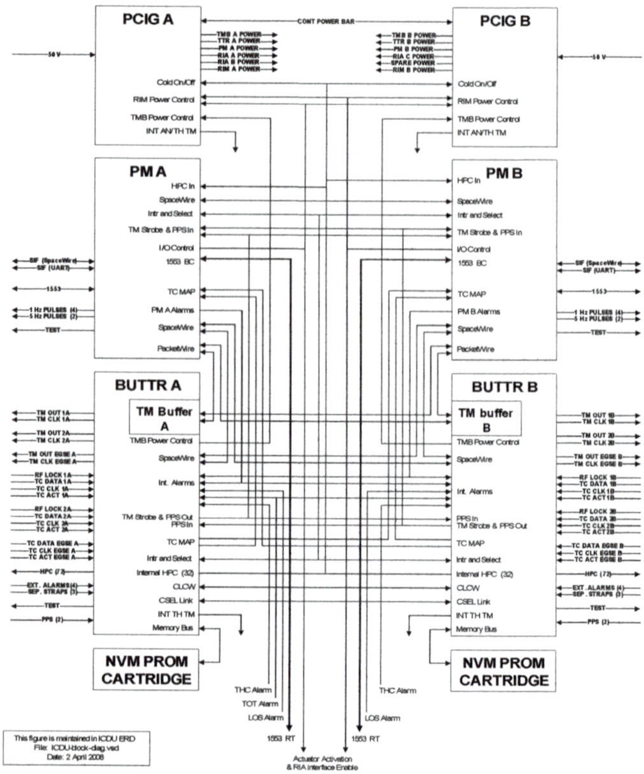

Figure 5.6: Schematic block diagram of an OBC processor module.
© RUAG Aerospace Sweden AB

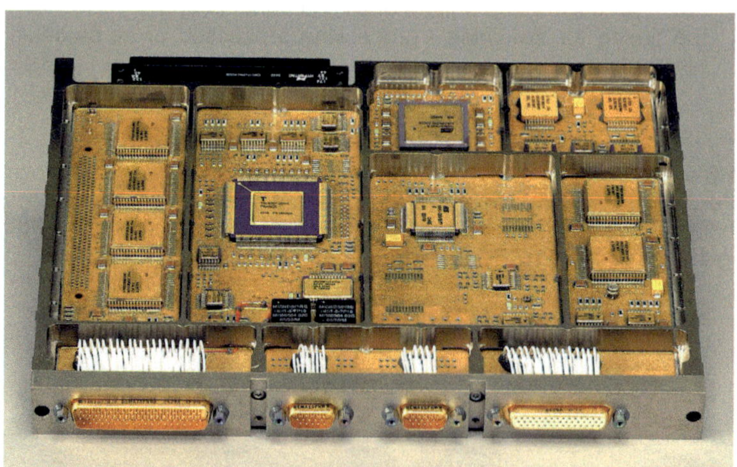

Figure 5.7: OBC processor module. © RUAG Aerospace Sweden AB

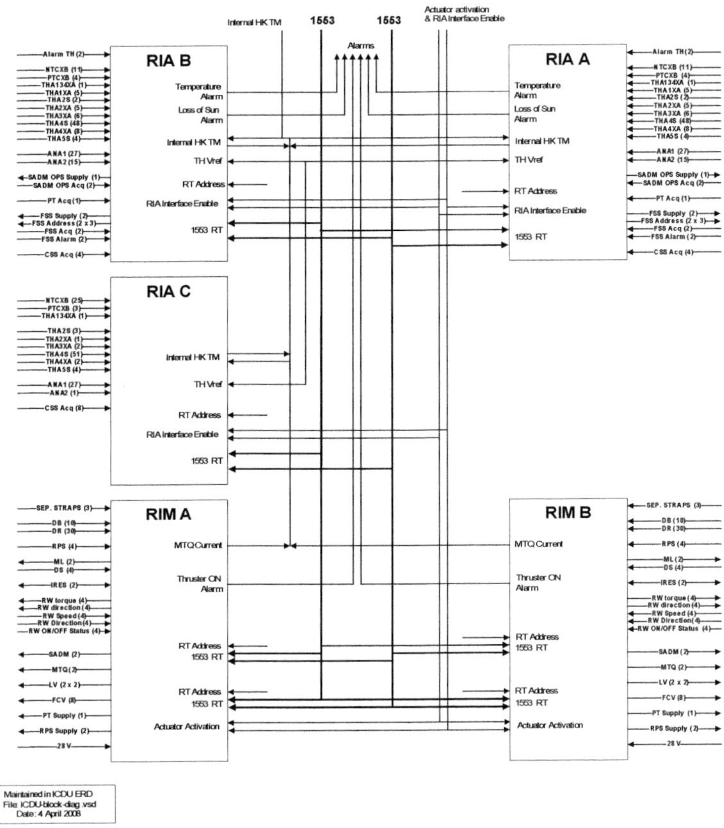

Figure 5.8: Schematic block diagram of an OBC I/O-unit.
© RUAG Aerospace Sweden AB and RUAG Aerospace Austria GmbH

As mentioned before the OBC processor itself in todays projects typically is modeled by using a processor simulator module. This approach results in a pure software model. In the case of an average satellite model equipped with a state of the art on-board computer based on a standard processor like the ERC32, SPARC V7 (single pipeline design) with a typical on-board clock rate of 25Mhz, a 3GHz Pentium PC used as simulator target platform is able to achieve approximate real time performance of the SVF. This means that the simulation of a typical 90 minutes LEO orbit of the satellite also takes approximately 90 minutes computation time on the SVF. Faster simulator target hardware leads to accordingly shorter simulation times per orbit.

The processor simulation of a more powerful multi-pipeline processor (for example the LEON, Sparc V8, providing three pipelines) is much more costly with respect to simulator platform performance. This means that an SVF for a spacecraft equipped with a LEON processor on board, will run accordingly slower. To circumvent this problem a processor emulation can be used. In this case the architecture of the on-board processor is mapped on a field programmable gate-array chip (FPGA) by means of the hardware design language VHDL. The FPGA typically is located on a PCI-card inside the SVF computer. Such a constellation can run 5 to 10 times faster than real-time on a 3GHz Pentium PC for a spacecraft equipped with a 25Mhz on-board LEON processor. This results in a simulation time of 10-15 minutes for the 90 minutes LEO orbit mentioned above. For the two configurations please also refer to the figure below.

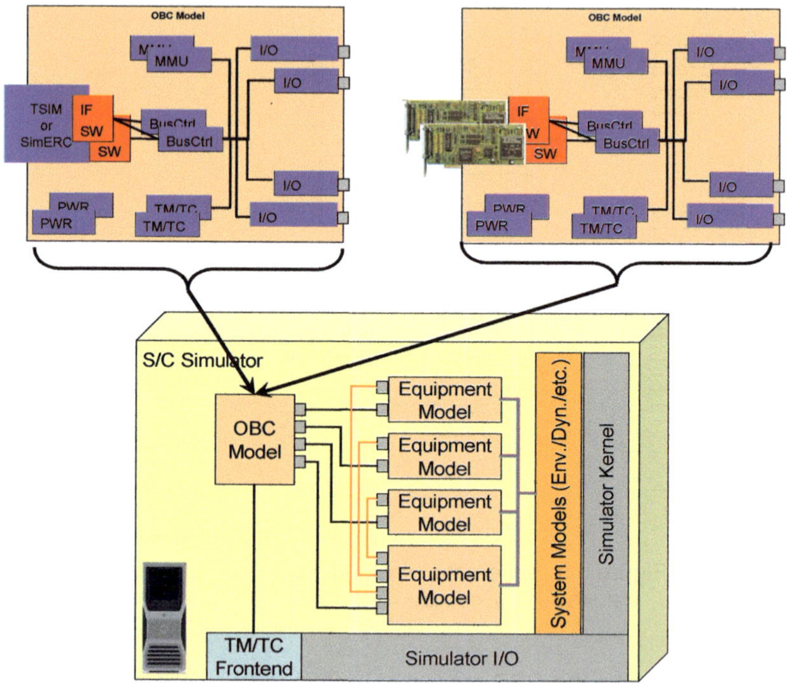

Figure 5.9: Modeling an OBC with processor simulation and respectively emulation.

It is quite evident from figure 5.9 that the modeled I/O-Boards of the simulated on-board computer have to be connected to the simulated harness. This is achieved by the use of a shared memory to which the OBC model writes and from which the harness line model instances read respectively vice versa.

The processor simulation / emulation implicitly represents also the scheduler of the OBC simulation. This central OBC model typically feeds the spacecraft simulator with a synchronizing clock signal. Such a signal also is submitted by the real OBC in a spacecraft to other on-board equipment which requires time synchronization. The synchronization signal typically has a frequency of 1 Hz and therefore is called "Pulse

Per Second" (PPS) signal. A highly precise real-time performance of the simulated processor and as result of the OBC model and finally of the SVF as a whole is not necessary on SVF level because no hardware in the loop is attached to the simulator.

The processor model typically is equipped with an interface to a standardized debugger. This allows to monitor the execution of the OBSW code or to interactively control the execution (stopping, querying variable values etc.) - please also refer to figure 3.14. In the case of a processor emulation by means of an FPGA this functionality however is limited.

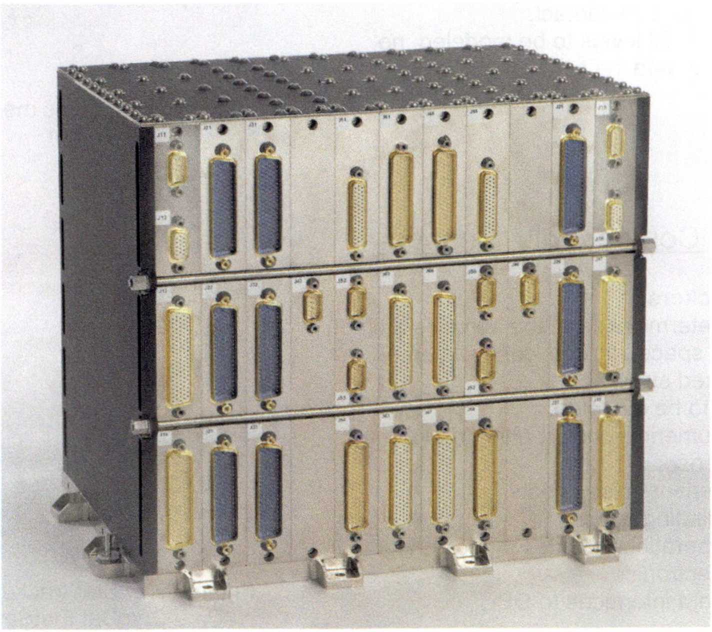

Figure 5.10: Photo of an on-board computer. © RUAG Aerospace Sweden AB

Since spacecraft are equipped with different AOCS, platform and payload components in accordance to their specific mission requirements the OBCs also are equipped with different types of data interfaces and, if necessary, even with different kinds of processors and internal data bus systems. The OBC models have to implement all these characteristics accordingly. The design and implementation of such OBC models typically is carried out by using UML as software design language and C++ as the programming language. Further details on simulator software design and coding techniques are provided in chapter 8.

Mass Memory Systems

- Storage of spacecraft housekeeping and scientific data generated by the payload in the case of satellites as the simulated system.

Aspects to be modeled for functional simulation:

- Storage capacity
- Number of separate storage banks
- For operations simulation (compare chapter 3.2.5):
  - ◇ Data rates provided through housekeeping processes and
  - ◇ payload science data filling memory respectively
  - ◇ memory freed during ground station contact.
- Only fill levels to be modeled, no real data

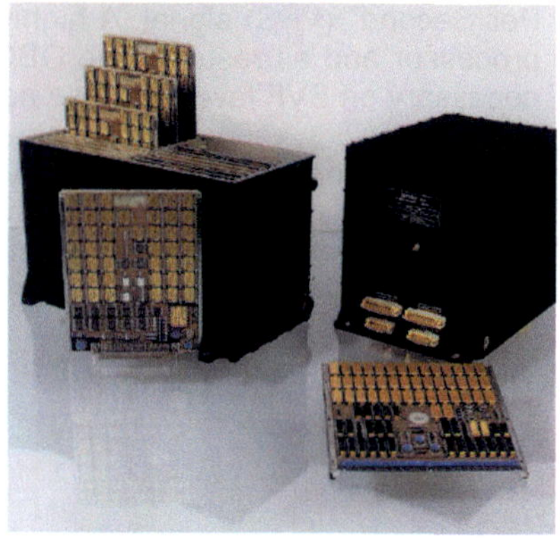

Figure 5.11: Mass memory units.
© Astrium

AOCS Components: Sensors

Star Trackers

- Determination of the sensor's attitude in space through identification of fixed stars.

Aspects to be modeled:

- Numerical correct attitude of the sensor
- Sensor noise effects
- Blinding effects
- Operational modes of the sensor electronics
- Data interfaces to OBC

Figure 5.12: Star trackers in CryoSat FlatSat. © Astrium

Earth Sensors

- Determination of the sensor attitude relative to Earth
- Optical (CCD pixel map) or thermal (slit / thermistor assembly) measurement

Aspects to be modeled:

- Sensor noise, calibration and blinding
- In case of optical and digital sensor:
  - ◇ Operational modes of the electronics
  - ◇ Data interface to OBC

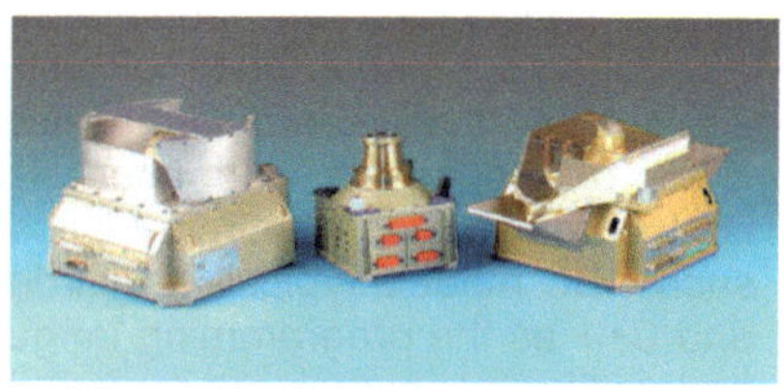

Figure 5.13: Earth sensors.
© Astrium

Sun Sensors

- Determination of the attitude relative to the Sun
- Optical (CCD pixel map) or thermal (slit / thermistor assembly) measurement

Aspects to be modeled:

- Sensor noise, calibration and blinding
- In case of optical and digital sensor:
  - ◇ Operational modes of the electronics
  - ◇ Data interface to OBC

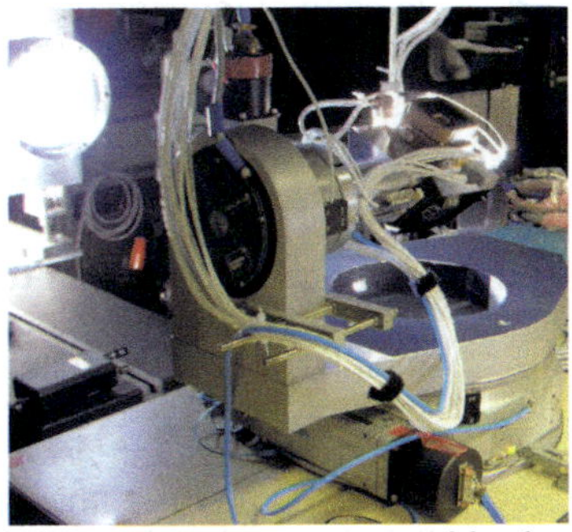

Figure 5.14: Sun sensors on a turntable rig.
© Astrium

Magnetometers

- Measurement of the magnetic flux in mounting orientation.

Aspects to be modeled:

- Sensor noise and calibration
- If necessary permanent distortion by adjacent electrical systems or metal structures

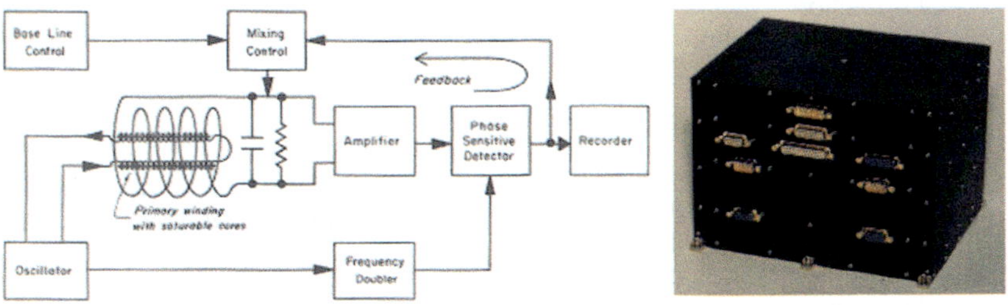

Figure 5.15: Mars Global Surveyor magnetometer. © NASA

Gyroscopes and Gyro-Assemblies

- Measurement of spacecraft rotational rates
- Principle of measurement:
  - ◇ Mechanical: 3 axis mounted gyroscope
  - ◇ Optical: Interference of two laser beams passing an optic fiber coil in opposite direction - "Fiber-Optic Gyro" (FOG)

Aspects to be modeled:

- Gyro electronics, data interfaces to OBC

Figure 5.16: Fiber-optic gyro.
© Astrium

- In case of mechanic gyro: Friction effects
- In case of FOG technology:
  Aging effect on opacity of the fiber as begin / end of life characterization parameters

Accelerometers / Gradiometers

- Measurement of the inertial acceleration mainly to monitor orbital correction maneuvers
- Control spacecraft acceleration for so called "Drag Free" missions

Aspects to be modeled:

- Electronics and its operational modes
- Data interfaces to OBC
- If necessary functionalities caused by the principle of measurement

Figure 5.17: ONERA SuperSTAR accelerometer of GRACE satellite.
© Astrium

GPS/Galileo/GLONASS Receivers

- Measurement of distance between the spacecraft and a GPS satellite in order to determine the spacecraft's position - at least 4 GPS satellites have to be visible by the spacecraft to determine the 3D-position
- Elevation resolution is limited

Aspects to be modeled:

- Numerically correct position of the spacecraft
- Statistical variations within the bounds of the real GPS measurement inaccuracy in order to test OBSW functionalities
- Operational modes of the receiver electronics
- Data interfaces to OBC

Figure 5.18: GNSS receiver. © Astrium

DORIS Receivers (CNES)

- Measurement of distance between the spacecraft and a ground station in order to determine the spacecraft position - at least 4 ground stations have to be visible by the spacecraft to determine the 3D-position
- Special DORIS ground station models needed
- Since the measured distance is used for determination of the orbit position, data are processed on ground and not in the OBSW.

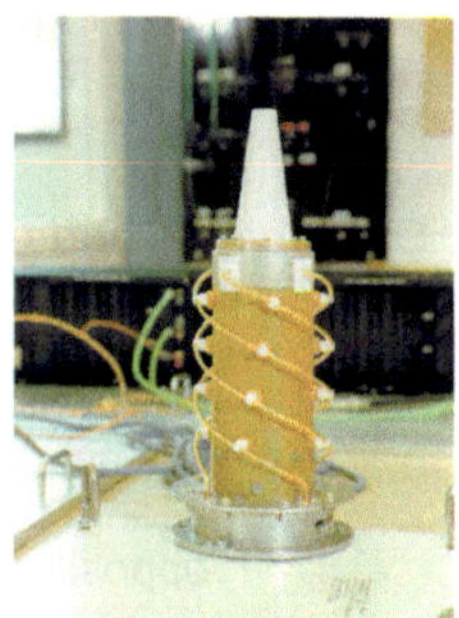

Figure 5.19: DORIS receiver antenna.
© NASA

Aspects to be modeled:
- Operational modes of the receiver electronics
- Data interfaces to OBC
- Dummy measured data packets

AOCS Components: Actuators

Reaction Wheels
- Generation of a momentum along the mounting axis of the wheel through changing the rotational rate of the wheel rotor

Aspects to be modeled:
- Dependency of momentum vs. rotational speed changes
- Dependency of rotational speed changes vs. commanded rate or current
- Sticking and friction effects
- Reaction wheel electronics features and interfaces to OBC
- If necessary special effects during desaturation of the reaction wheels.

Figure 5.20: Reaction wheel setup in a hybrid testbench.
© Astrium

Propulsion Systems for Attitude and Orbit Control
- Cold gas (e.g. nitrogen high pressure systems)
- Mono propellant systems (hydrazine)
- Reignitable bipropellants, mainly used for orbit correction maneuvers

Aspects to be modeled:
- All valves and actuators of the propulsion system that are controlled by the OBC:
  Latch valves, non return valves, flow control valves, pyrovalves, tanks and pressure controllers
- Control of the thruster valves (linear or pulsed)
  and resulting thrust
- Latency of thrust relative to valve opening
- If necessary the electrical system of thruster
  catalytic bed heaters

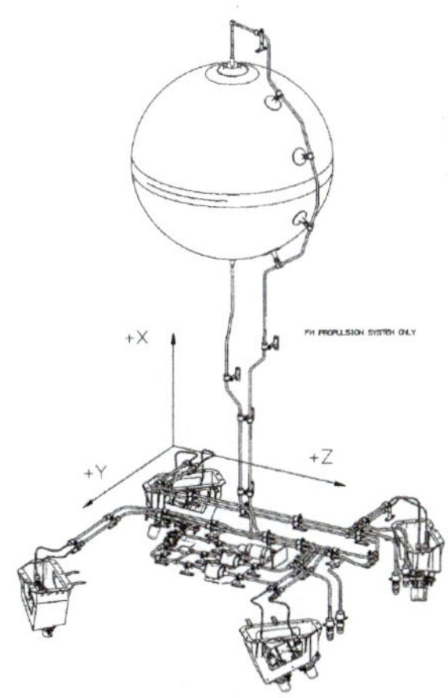

Figure 5.21: Galileo IOV propulsion system.
© Astrium

Magnetotorquers

- Electrical coils generating a magnetic field in order to produce a momentum through interaction with the Earth's magnetic field

Aspects to be modeled:

- Electrical resistance / impedance
- Intensity of the produced magnetic field
- Hysteresis effects

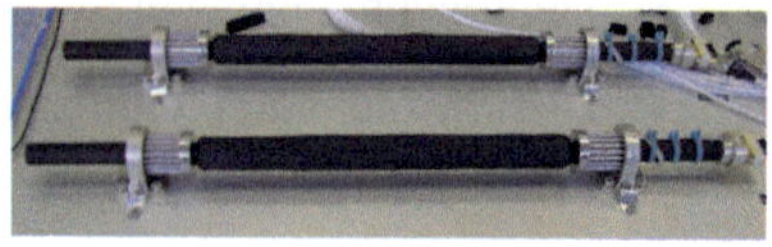

Figure 5.22: Galileo Magnetotorquers.
© Astrium

Ion Propulsion Components

- Applied for fuel efficient but long duration orbit or trajectory maneuvers
- Often used for interplanetary missions

Aspects to be modeled:

- Thruster control versus resulting thrust
- Electrical energy consumption, fuel consumption

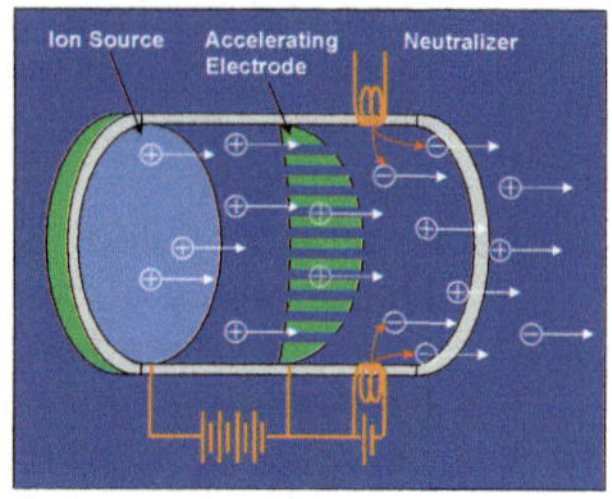

Schematic diagram.
© IRS, Universität Stuttgart

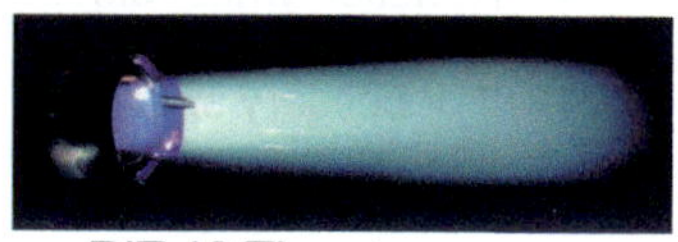

RIT 10 Thruster.
© Astrium

Figure 5.23: Ion thrusters.

Electric µN Engines:
Field Emission Electrical Propulsion, (FEEP)

- Thruster types needed for extremely precise attitude control
- Also used for drag free acceleration control
- And used for position control in formation flight configurations

Aspects to be modeled:

- Thruster control versus resulting thrust
- Electrical energy consumption, fuel consumption
- Functionality and operational modes as well as the electrical characteristics of the typically very complex high voltage electrics which are needed for the heating of the solid propellant and accelerating it.

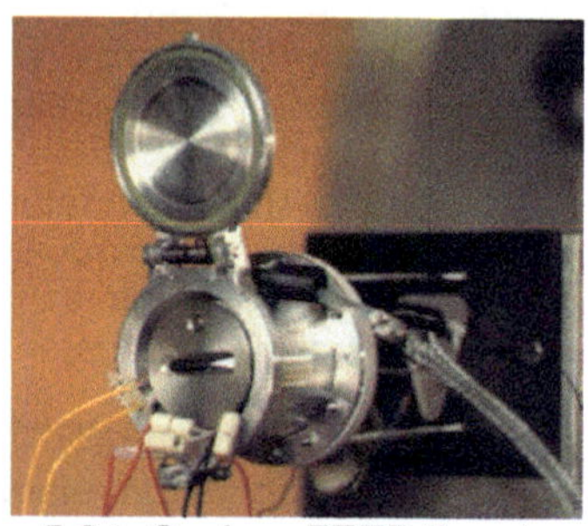

Figure 5.24: Cesium FEEP thruster.
© ESA

Launcher Main and upper Stage Engines

The diversity of launcher and upper stage propulsion systems only can be covered very briefly here touching characteristics of propulsion systems concerning types stem from

- solid propellant boosters to
- liquid propellant systems.
- Furthermore exist cryogenic propellant systems, and
- standard liquid propellant systems.
- The latter again can be broken down according to diverse fuel types.
- Engines furthermore have to be determined between those fed by gas pressured fuel tanks (cf. figure 1.3 and 1.12) and respectively
- those equipped with turbo pump feed systems (compare adjoining figure).
- The engine's thrust characteristics over altitude (atmospheric back-pressure) has to be modeled as well as
- the entire propellant feed system (see e.g. [23]).

Figure 5.25: SNECMA Vulcain II.
Source: Stahlkocher
(GNU Free Documents License)

Power Subsystem Components

Solar Panels
Aspects to be modeled:

- Change of the spacecraft moments of inertia and center of gravity over deployment process
- Blocking during deployment as failure mode
- Mechanical properties like vibration modes
- Aging dependent performance properties (begin / end of life loadable as simulator equipment model start characteristics)
- Redundancies and string switching / cabling
- Failure of cells and strings as failure mode

Figure 5.26: Solar array wing.
© Astrium

Solar Array Drives

Aspects to be modeled:

- Connection to solar panels (Power Lines)
- Connection to power control unit (Power Lines)
- Command lines from OBC to stepper motors
- Stepper motor functions (step, free rotation, hold, active hold etc.)
- Position sensors
- Power consumption of stepper motors
- If necessary internal thermistors
- Failure modes (motor failure, sensor failure, blocking)

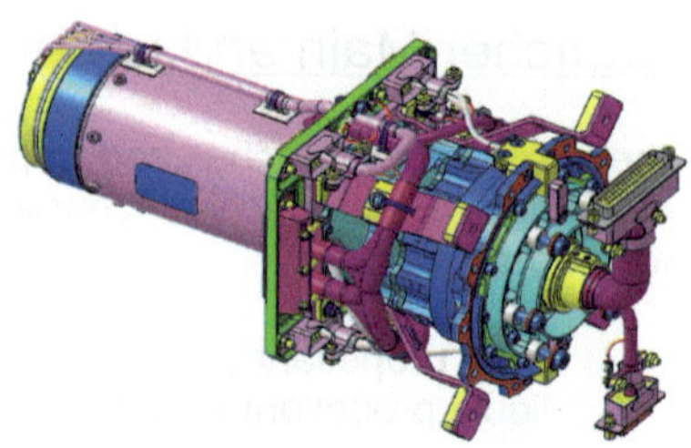

Figure 5.27: Solar array drive.
© Oerlikon Space AG, Zürich

Batteries

Aspects to be modeled:

- Number of battery cells
- Type of battery (Li-Ion, NiMH etc.)
- Charge characteristics
- Current / voltage curves under load
- Self discharge characteristics
- Aging dependent performance properties (begin / end of life loadable as simulator start characteristics)
- Internal circuitry and redundancies
- Failure of battery cells and / or circuitry

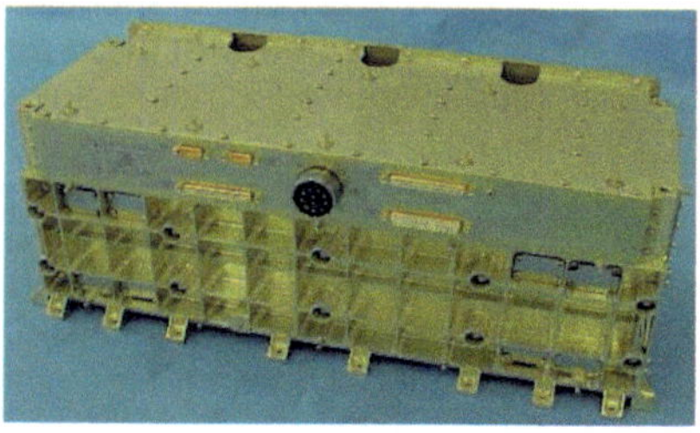

Figure 5.28: Battery stack.
© Astrium

Power Control and Distribution Units

For these units the functionalities for both power control as well as power distribution towards the different consumers are to be modeled. The entire PCDU is commanded via the OBC. Basic power source are the solar arrays. Secondary source (to be charge controlled) is the spacecraft battery.

Aspects to be modeled:

- Connection to OBC and command / control data protocols
- Control functions via high priority command lines
- Connection to the battery
- Connection to the solar panels
- Internal current limitation functionalities
- Internal redundancies
- Relays for user load switching
- Overvoltage monitoring for every channel

Figure 5.29: Power Control and Distribution Unit.
© Astrium

- Protection against foldback currents
- Protection against overcurrents
- Redundancies for every channel
- Power consumption of the PCDU itself
- Heat dissipation of the PCDU

Fuel Cell Subsystems

For modeling of fuel cell systems the reader first is referred back to the figures 1.10 and 1.11. The modeling of such subsystems differs depending on the goal of the simulation. If the goal is the verification of the OBC and the controlling on-board software, then the subsystem can be simplified and can be represented as single unit equipment with according interfaces to the OBC and performance represented by approximations in characteristic curves. If the goal is performance evaluation of the fuel cell system itself, the different elements of the system have to be modeled one by one as shown in figure 1.10. The component types in the system are ranging from the fuel cell stack itself over condensers, membrane separators and pumps to heat exchangers and fans. Fuel cell systems also differ highly in aspects like the type of the electrolyte (mobile / immobile), the type of membrane and whether the system being regenerative or not. A detailed overview on the diverse types and the technical modeling of fuel cells is given in [31] to [33].

Thermonuclear Generators

- Used as energy supply for deep space probes exploring the outer solar system which makes use of solar arrays impossible due to low solar brightness.

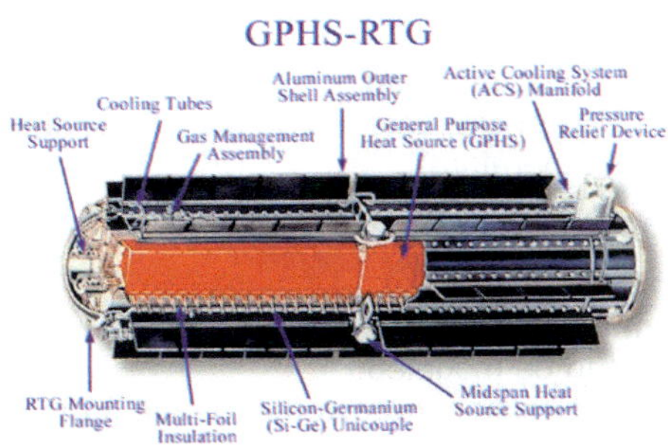

Figure 5.30: Radio-thermonuclear generator of the Cassini space probe. © NASA

Aspects to be modeled:

- Power generation depending on the ratio between ambient temperature and radionuclide temperature due to Seebeck effect
- Current / voltage characterization curve resulting from used materials
- Thermal radiation output depending on the ratio between ambient temperature

and radionuclide temperature
- Electrical connections of the Seebeck elements (including redundancies)
- Type of the radionuclide and thermal characteristic curve depending on operation time (begin / end of mission)
- Cooling system (active cooling system or heat pipe type)

Mechanics / Mechanisms

Solar-Array and Deployment Mechanisms
Aspects to be modeled:
- Duration of deployment process
- Blocking during deployment
- Electrical interfaces
- Change of spacecraft moments of inertia and center of gravity

Antenna Structures and Deployment Mechanisms
Aspects to be modeled:
- Duration of deployment process
- Blocking during deployment
- Electrical interfaces
- Change of spacecraft moments of inertia and center of gravity

Thermal Components

Heaters
Aspects to be modeled:
- Heater characteristic curve (dependency between resistance and temperature)
- Connection between heater and a node of the thermal model

Thermistors
Aspects to be modeled:
- Thermistor characteristic curve (dependency between resistance and temperature)
- Connection between thermistor and a node of the thermal model

Thermostats
Aspects to be modeled:
- Electrical resistance of heater and thermistor element in dependency of temperature
- Switch temperature (eventually hysteresis)
- Connection between thermostat and a node of the thermal model

Satellite / Spaceprobe Payloads

The payloads of a satellite / probe can be diverse, ranging from Earth remote sensing equipment (cameras, radar, etc.) via interplanetary particle or magnetic sensor, spectral cameras or even systems which are payload and AOCS sensor at the same time (atomic clocks, drag free sensors etc.). Therefore also the equipment model characteristics differ largely depending on the type and purpose of simulation testbench, such as:

Figure 5.31: HRSC Mars camera.
© Astrium

- Software verification facility
- Hardware test facility (STB / EFM)
- Mission control simulator
- Formation flight test simulator
- Payload data-processor test simulator.

For a satellite simulator the science data collected by a real payload only have to be modeled with respect to data packet structure and content to a minimum degree so that they can be processed by the satellite's on-board software. The payload's functionalities have to be modeled in higher level of detail to cover the different operational scenarios and to verify:

- Limitations due to thermal effects in the satellite
- Limitations induced by the power supply
- Limitations induced by the amount of available mass storage
- Limitations due to failure modes.

In particular the following aspects have to be modeled for payload units:

- The electronics and their operational modes (normally modeled as state machines)
- The command / control interface to OBC
- Payload science data protocols (science data typically are modeled by dummy packets only with simple counter variables to test the interface from the payload to the OBC and the mass storage devices)
- The thermal behavior and heat dissipation dependent on the payload's operational mode
- The power consumption, also dependent on the payload's operational mode.

Components for Life Support Systems

Due to the complexity and the diversity of life support systems and their variants, and due to the fact that they often are coupled to the spacecraft's power supply or water management systems, life support systems will not be described in detail here. The interested reader can find more details on such equipment in [35] to [36].

Further reading and Internet pages concerning spacecraft equipment modeling are listed in the according subsection of this book's references annex.

The task of the software development team
is to engineer the illusion of simplicity.

Grady Booch

Part II

Simulator Technology

6 Numerical Foundations of System Simulation

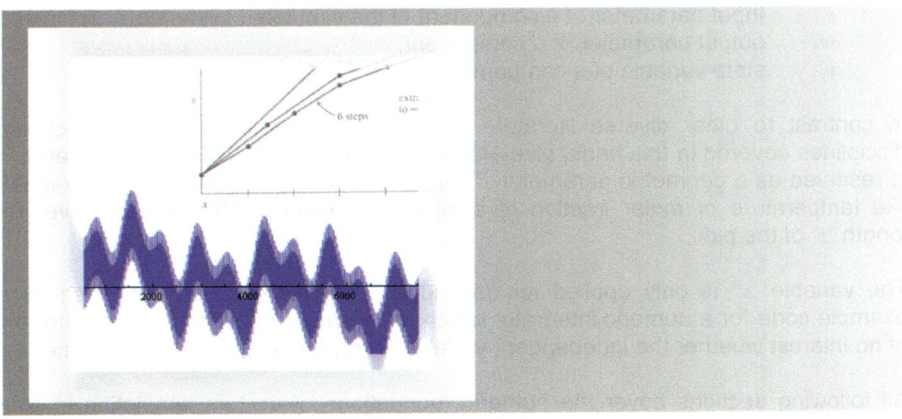

J. Eickhoff, *Simulating Spacecraft Systems*, Springer Aerospace Technology 1,
DOI 10.1007/978-3-642-01276-1_6, © Springer-Verlag Berlin Heidelberg 2009

6.1 Introduction to Numerics

Simulation technology has been applied for years now in computer-based system design in various application fields, from ship building to space technology. It also covers the full scope of problems to be analyzed, ranging from the design of a system component for specific stationary load cases up to overall system simulations for analyses of dynamic system operation.

Diverse commercial or user developed simulation program suites are applied to these tasks in industry. For students who want to familiarize themselves with these techniques it is often not easy to understand the modeling technologies and numeric approaches behind such tools. Despite the fact that for specific simulation application fields, detailed technical literature is available, it is difficult to find an overview on the problem specific appropriate tools and numeric methods. The generation of such a technical and methodical overview is the goal of this chapter, without confusing the reader by diving too deep into specific subtopics, (such as numerical solvers). Further reading is cited wherever the central theme prevents tackling details.

For further elaboration, first the following notation convention shall be defined. Where possible explicit technical variables with standardized variable names are cited, like p for pressure and T for temperature or H for enthalpy, the variables of a simulated component are defined as follows:

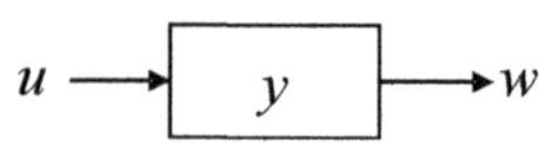

Figure 6.1: Elementary component block.

- u Input parameter of a component of the simulated system,
- w output parameter of a component,
- y state variable of a component.

In contrast to other diverse literature and to keep consistent over all technical disciplines covered in this book, (like AOCS, thermal, power, OBSW), the variable "z" is reserved as a geometric parameter. Taking the example of a pipe, a state variable like temperature or molar fraction of a fluid component can be variable over the length "z" of the pipe.

The variable "x" is only applied for description of generic causalities, such as in example code for a numeric integrator which integrates $y=f(x)$. For the integrator it is of no interest whether the independent variable "x" represents location "z" or time "t".

All following sections cover the numeric foundations of system simulation in state space notation which sums up all input parameters of the component in one vector, all inner state variables in another vector and finally all output parameters too. Thus for more complex components the variables u, v, y, in above figure in reality become vectors. Frequency domain modeling of systems, which is often applied in control engineering for analysis of dynamic responses and control design is only cited at a few points. The transformation from state space domain to frequency domain can be achieved for linearized equation sets by a Laplace transform. At relevant points in the following chapters it will be explained to the reader which goals can be achieved by such computations. The reader however does not need to be familiar with Laplace transformation mathematics oneself.

6.2 Modeling of System Components as Transfer Functions

Functional system modeling has evolved from control engineering and control theory. Today it is supported by very powerful simulation tools like Simulink or Modelica. Especially the tool suite Matlab / Simulink / Stateflow allows for characteristics and performance analysis of very complex systems. Due to a plethora of tool features only a minority of users will wonder about the technologies behind them as well as their strengths and weaknesses.

Function-based approaches adopt their system modeling from transfer functions in mathematics such as:

$$y = \sin(u) \qquad (6.1)$$

In toolkits like Simulink the user can compose entire system models from such functional elements via a graphical user interface. One can of course select not only from simple mathematical transfer functions like the above mentioned sine wave, but from large libraries providing a full spectrum of transfer blocks up to integrators with the possibility to code ones own function blocks.

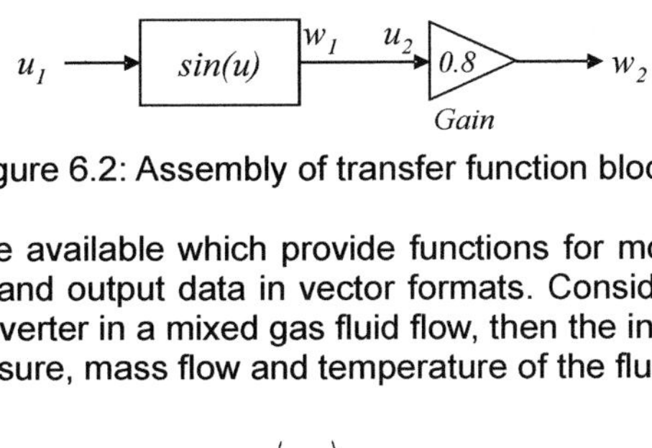

Figure 6.2: Assembly of transfer function blocks.

In addition blocks are available which provide functions for more than one variable, providing their input and output data in vector formats. Considering the simple case of a gas catalytic converter in a mixed gas fluid flow, then the input parameter set can be described by pressure, mass flow and temperature of the fluid.

$$\bar{u} = \begin{pmatrix} p_{in} \\ T_{in} \\ \dot{m}_{in} \end{pmatrix} \qquad (6.2)$$

The output state after the catalytic reaction of the gas mixture can also be described by pressure, mass flow and temperature of the fluid, where a stationary case shall be assumed:

$$\overline{w}=\begin{pmatrix} p_{out} \\ T_{out} \\ \dot{m}_{out} \end{pmatrix}, \text{ with, } \begin{matrix} p_{out}=f_1(p_{in}, \dot{m}_{in}) \\ T_{out}=f_2(T_{in}, \dot{m}_{in}) \\ \dot{m}_{out}=f_3(\dot{m}_{in})=\dot{m}_{in} \end{matrix} \tag{6.3}$$

The change of molar fractions of the gas mixture is considered to be of no interest in this simple example, however it can be reflected by additional entries in the input and output parameter vectors of the component. And here the example already depicts that the diverse output parameters usually are dependent on multiple input parameters which is not directly obvious when using the vector notation which frequently is applied to keep notations compact:

$$\overline{w}=\overline{f}(\overline{u}) \tag{6.4}$$

6.3 Components with Time Response

The next step in functional component modeling is to consider time response of component physics. Most system components in reality do not react purely instantaneously at their output resulting exclusively on input changes via a transfer function. Rather they show integral effects which lead to time delays of output parameter value changes in response to input changes. These effects result from the real component having internal capacitive parameters, the internal state variables $\overline{y}$ cited in chapter 6.1. Thus generally it can be identified that both

- the internal state variables as well as,
- the output parameters,

are dependent on the state variables and the input parameters. More precisely value changes of the internal state variables are directly dependent on the values of input parameters and integratively on their actual state:

$$\begin{matrix} \dot{\overline{y}}=\overline{f}(\overline{y},\overline{u}) \\ \overline{w}=\overline{g}(\overline{y},\overline{u}) \end{matrix} \tag{6.5}$$

Thereby no statement is yet made about the functions $f()$ and $g()$. It can only be identified that modeling systems with time response behavior are to be represented mathematically via differential equations, (DEQs). The system's dynamic behavior can then be computed by integration of these equations over time. The mathematical system description is completed by additional algebraic equations. And moreover the time dependency of the input variables will be considered, which leads to the representation:

$$\begin{matrix} \dot{\overline{y}}(t)=\overline{f}(\overline{y}(t),\overline{u}(t)) \\ \overline{w}(t)=\overline{g}(\overline{y}(t),\overline{u}(t)) \end{matrix} \tag{6.6}$$

Furthermore a component variable, besides being dependent from time and other state variables can also in addition be position dependent, e.g. changing over the running length of a pipe of a propulsion system. This leads to the even more generic equation set

$$\begin{aligned}\dot{\overline{y}}(z,t)&=\overline{f}(\overline{y}(z,t),z,\overline{u}(t))\\ \overline{w}(t)&=\overline{g}(\overline{y}(l,t),\overline{u}(t))\end{aligned} \tag{6.7}$$

where in this example l represents the overall length of the pipe. In the previous equations always component input, internal state and output have been considered. As soon as the component is integrated into a system, mathematical modeling in addition has to reflect that diverse output parameters directly or indirectly, entirely or partly may be fed back to component input. The treatment of such coupling matrices shall be postponed until chapter 6.9.

For an analysis of resulting equation types, the internal state of a component shall be considered as non location dependent. The resulting equation types, equation systems and their numeric solutions are presented step by step.

To be simulated, systems either are either

- time-**continuous**:

$$\dot{\overline{y}}=\overline{f}(\overline{y}(t),\overline{u}(t)) \tag{6.8}$$

- or time-**discrete:**

$$\overline{y}_{n+1}-\overline{y_n}=f\left(\overline{y_n},\overline{u_n}\right) \tag{6.9}$$

Since time-discrete systems, (like state machines), in most cases do not impose significant problems in their numerical representation in simulators, the following elaborations shall concentrate on time-continuous systems.

The following differential equation types are common to appear during system numeric description:

- Continuous **non-linear** - the following equation cannot be transformed directly into the form of equation (6.6):

$$\dot{\overline{y}}\,\overline{y}-\overline{y^2}=f\left(\overline{u},t\right) \tag{6.10}$$

- Continuous **linear** - the derivatives of $\overline{x}$ only are multiplied by algebraic factors and only are linked with each other by + / - operators:

$$a_2\ddot{\overline{y}}+a_1\dot{\overline{y}}+a_0\overline{y}=f\left(\overline{u},t\right) \tag{6.11}$$

- For these equations it can in addition be determined between continuous linear **time-variant** equations

$$a_1(t)\dot{\bar{y}}+a_0(t)\bar{y}=f(\bar{u},t) \tag{6.12}$$

- and continuous linear **time-invariant** ones:

$$a_1\dot{\bar{y}}+a_0\bar{y}=f(\bar{u},t);\;\; a_0, a_1=const. \tag{6.13}$$

Resulting equation systems only consisting of this equation type are called "continuous, linear, time invariant systems“, (CLTI systems).

In practice it mostly turns out that the balance equations which are worked out to represent a component's physical behavior are not yet differential equations in the forms depicted above. Instead the balance equations in physics mostly are partial differential equations,(PDEs). In these the state variables, that are to be integrated over time, are not only time dependent but also dependent on other parameters such as location. How such balance equations look like and how they can be transformed to ordinary differential equations (ODEs), is treated in the following sections using various examples.

6.4 Balance Equations

Before an equation set similar to those cited in those four a.m. types (6.10) to (6.13) is available, first the physics of the system or component to be simulated shall be formally described and afterwards in addition some mathematical conversions will be applied. As the next chapters will show, the description of physical processes in system components via balance equations in most cases leads to a set of partial differential equations together with additional algebraic equations.

6.4.1 Equation Set for Fluid Systems

Fluid systems in spacecraft can be found e.g. in:

- Propulsion and engine control systems
- Tank pressurization systems
- Complex power systems such as fuel cell systems
- Environmental conditioning and life support systems
- Active cooling / heating systems
- Laboratory rack systems

The balance equations set up for each individual system component are balancing amounts for which the changes can exclusively be described by the following three essential processes:

- Storage
- Transport, flow
- Transformation, conversion

The control volume in most cases is a fixed volume element, as far as possible identical to the spacecraft component to be modeled for simulation. For the balanced amount enclosed in the control volume the following equation applies,

$$\frac{d}{dt}\int_V M \; dV = -\oint_A \overline{\varphi} \, d\overline{A} + \int_V S \; dV \tag{6.14}$$

where:

M	is the volume specific amount to be balanced,
V	is the control volume,
A	is the surface of the control volume,
φ	represents the amount passing the system borders, and
S	is the volume specific conversion rate inside the control volume.

Applying Gauss' law the surface integral can be transformed to a volume integral:

$$\frac{d}{dt}\int_V M \; dV = -\oint_V \nabla \overline{\varphi} \, dV + \int_V S \; dV \tag{6.15}$$

Or in differential form:

$$\frac{\partial M}{\partial t} = -\nabla \overline{\varphi} + S \tag{6.16}$$

In this notation now all necessary balance equations can be represented for system component simulation.

In the following sections, the balance equations for thermal and fluid dynamic systems are used as examples since they are specially well suited to demonstrate certain numeric characteristics. Equation systems for attitude dynamics on orbit etc. are treated later. The balance equations for thermal dynamic systems usually are:

- The equation of continuity in form of a mass balance
- Compound balances for compounds in material mixes
- The momentum equation
- The Energy equation

Equation of Continuity

Considering the equation of continuity, the balanced amount M corresponds to the mass per volume, i.e. the density ρ of the fluid. The source / sink term q in this case is equal to zero since mass neither can be generated nor eliminated. Chemical conversions inside the component from one chemical matter to another are not a topic of the equation of continuity but of compound balance equations. So the equation of continuity can be written as:

$$\frac{\partial \rho}{\partial t} = -\nabla(\rho \bar{v}) \tag{6.17}$$

For simple system components which can be represented with a unidimensional flow through, this simplifies to:

$$\frac{\partial \rho}{\partial t} = (\rho v)_{ein} - (\rho v)_{aus} \tag{6.18}$$

Compound Balance

Compound transport inside the considered control volume results from flow, chemical conversion and convection, the latter resulting from concentration gradients. In such compound balance equations, for one compound in a mixture, the balanced amount is the volume fraction c of the considered compound. For a mix of i substances, for each of them such a compound balance can be established,

$$\frac{\partial c_i}{\partial t} = -\nabla(c_i \bar{v}) - \nabla(-D \nabla c_i) + \frac{\dot{m}_i}{\rho_i V_{ges}} \tag{6.19}$$

, where the last term covers the chemical conversion of the substance, (source / sink term). In unidimensional representation this results in:

$$\frac{\partial c_i}{\partial t} = -\frac{\partial (c_i v)}{\partial z} - \frac{\partial}{\partial z}\left(-D_{i,j}\frac{\partial c_i}{\partial z}\right) + \frac{\dot{m}_i}{\rho_i V_{ges}} \tag{6.20}$$

Respectively applying constant diffusion coefficients it simplifies to:

$$\frac{\partial c_i}{\partial t} = -\frac{\partial (c_i v)}{\partial z} + D_{i,j}\frac{\partial^2 c_i}{\partial z^2} + \frac{\dot{m}_i}{\rho_i V_{ges}} \tag{6.21}$$

Principle of Linear Momentum

For the linear momentum equation, the balanced amount is the specific linear momentum flow. The equation reads as follows:

$$\frac{\partial(\rho\bar{v})}{\partial t}=-\nabla(\rho\bar{v}\bar{v}^T)-\nabla(\bar{\bar{\tau}}+\rho\bar{\bar{\delta}})+\bar{f} \tag{6.22}$$

Here, $\bar{f}$ represents the resulting force flow onto the fluid inside the control volume, e.g. resulting from gravitation, mechanical linear or centripetal forces.

- For Newtonian fluids, the viscosity stress tensor is linearly dependent on the velocity gradient.
- For isotropic fluids with the viscosity stress tensor following Stoke's hypothesis
  - ◇ and in component representation,
  - ◇ and under the assumption that the only resulting force impacting the fluid,

  the following equation results,

$$\frac{\partial}{\partial t}(\rho v_i)=-\sum_j\frac{\partial}{\partial z_j}\rho v_i v_j-\frac{\partial p}{\partial z_i}-\sum_j\frac{\partial}{\partial z_j}\tau_{ij}+\rho g_i \tag{6.23}$$

with $i=x,y,z$ and $j=x,y,z$

Energy Balance

For the energy balance equation, usually the intrinsic energy of the component is used as the balanced amount, in volume specific representation. The flux parameters represent the net energy flow into the control volume consisting of:

- The net inflow of intrinsic energy
- Technical work on the control volume surface
- Heat flow into control volume due to heat conduction
- Friction work of the flux in control volume and technical work done
- Net increase of potential energy

The energy equation thus reads as follows:

$$\frac{\partial}{\partial t}\left(\rho u+\frac{1}{2}\rho\bar{v}^T\bar{v}+\rho g h\right)=$$
$$-\nabla(\rho u\bar{v})-\nabla(\rho g h v)-\nabla\left(\left(\frac{1}{2}\rho\bar{v}^T\bar{v}\right)\bar{v}^T\right)-\nabla(p\bar{v})+\nabla(\lambda\nabla T)-\nabla(\bar{\bar{\tau}}\bar{v})$$
$$+\dot{Q}_{internal}/V_{cumul}+\dot{W}_{t,internal}/V_{cumul} \tag{6.24}$$

In some cases the energy balance also is established using the overall energy density, (not only intrinsic energy), or the overall enthalpy density.

Examples for Algebraic Coupling Equations

Algebraic coupling equations define fixed interdependencies between the diverse parameters of those balance equations. For thermal / fluid dynamic systems these typically are the thermodynamic state equations as shown in the examples below. For other systems these also can be force or momentum equilibrium equations or other principles.

$$\rho = \rho(p, u) \tag{6.25}$$

$$T = T(p, u) \tag{6.26}$$

$$\frac{p}{\rho} = R\,T \tag{6.27}$$

$$\partial u = \rho c_v \partial T \tag{6.28}$$

6.4.2 Equation Set for Spacecraft Dynamics

The equation system for attitude dynamics of a satellite - rigid body system - shall serve as example for the equation types appearing in this domain. The variation of the position vector r applies as:

$$\begin{aligned} \dot{r}_x &= v_x \\ \dot{r}_y &= v_y \\ \dot{r}_z &= v_z \end{aligned} \tag{6.29}$$

The variations of velocities can be described by

$$\begin{aligned} \dot{v}_x &= f_1(F_x) \\ \dot{v}_y &= f_2(F_y) \\ \dot{v}_z &= f_3(F_z) \end{aligned} \tag{6.30}$$

with F being the sum of all forces interacting with the spacecraft body. For the gradients of rotation rates, the following non-linear differential equation system applies:

$$\overline{\overline{\Theta}} \cdot \dot{\overline{\omega}} = \overline{N} - \dot{\overline{H}} - \overline{\omega} \times (\overline{\overline{\Theta}} \cdot \overline{\omega} + \overline{H}) \tag{6.31}$$

with:

$\overline{\overline{\Theta}}$	Moments of inertia
$\overline{\omega}$	Rotational rate
$\overline{N}$	Sum of all torques
$\overline{H}$	Internal angular momentum - e.g. through reaction wheels

And finally for attitude variations, described through quaternions[17] instead of trigonometric equations, the following system of first order ordinary differential equations can be set up:

$$\begin{aligned} \dot{q}_1 &= f_1(q_1, \dot{\omega}_x, \dot{\omega}_y, \dot{\omega}_z) \\ \dot{q}_2 &= f_2(q_2, \dot{\omega}_x, \dot{\omega}_y, \dot{\omega}_z) \\ \dot{q}_3 &= f_3(q_3, \dot{\omega}_x, \dot{\omega}_y, \dot{\omega}_z) \\ \dot{q}_4 &= f_4(q_4, \dot{\omega}_x, \dot{\omega}_y, \dot{\omega}_z) \end{aligned} \qquad (6.32)$$

Equation (6.31) already represents a non-linear differential equation system of first order. In practice this becomes more complicated through the elaboration of the matrix $\overline{N}$.

Nonlinearities in these equations of dynamics especially result from forces / torques imposed on the spacecraft through gravitation, solar pressure, magnetic fields etc. The sufficiently precise modeling of all these effects make the equations complex in real application.

Detailed literature concerning attitude and orbit control of spacecraft are listed in the according subsection of this book's references annex.

6.4.3 Equation Set for Spacecraft Electrics

The physics of spacecraft power electrics, namely of:

- Solar cells / solar panels
- Batteries / battery stacks
- Power controllers, and
- Power distribution units

for the purpose of system simulation can be formulated as a system of first order linear differential equations - partly with non constant coefficients resulting from characteristic curves of solar cells, battery cells etc.

Further reading on mathematical modeling of satellite power electrics is listed in the according subsection of this book's references annex.

[17] Concerning definition and deduction of quaternions please refer to the relevant literature on attitude and orbit control of spacecraft, such as [40], chapter 12.1.

6.5 Classification of Partial Differential Equations

The balance equations and other physical equations from chapter 6.4 in some cases are ordinary differential equations, (ODEs), in other cases they are partial differential equations, (PDEs). For the numeric integration of system behavior over time these PDEs have to be converted into ODEs beforehand. The first step towards this is a closer analysis of the type of occurring PDEs.

In analogy to the example balance equations cited in chapter 6.4 and their characteristics, a second order partial differential equation of the following type shall be considered:

$$a\frac{\partial^2 y}{\partial t^2}+2b\frac{\partial^2 y}{\partial t\,\partial z}+c\frac{\partial^2 y}{\partial z^2}+d\frac{\partial y}{\partial t}+e\frac{\partial y}{\partial z}+fy-g=O \qquad (6.33)$$

The nature of the solution of such PDEs is dominated by the main term which comprises the first three equation elements. Three basic types of PDEs can be determined:

$$\begin{array}{ll} \text{hyperbolic} & \\ \text{parabolic} & \text{, with the discriminant } b^2-ac \\ \text{elliptic} & \end{array} \quad \left\{ \begin{array}{l} >0 \\ =0 \\ <0 \end{array} \right.$$

None of the previously treated balance equations comprise higher derivatives w.r.t. time:

=> $a=0$

Also mixed derivatives w.r.t. time *t* and location z do not appear:

=> $b=0$

In the balance equation for compounds for mixed materials and in the energy balance equation higher order derivatives w.r.t. location z appear:

=> $c\neq 0$

Thus the equation sets treated in chapter 6.4 can be identified as being parabolic type DEQs.

6.6 Transformation of PDEs into Systems of ODEs

For solving a system of PDEs with derivatives concerning time and location, first a discretization of the balanced control volume w.r.t. location has to be undertaken. This can be achieved by discretization into nodes. E.g. a simple pipe or a reactor can be discretized into slices, a 3-dimensional spacecraft geometric structure can be discretized into thermal grid-nodes. The subsequent sections describe the dependencies of state variables of different nodes from each other. For demonstration, again a balance equation of the most complex type treated so far is applied as an example - the balance of a fluid compound, i:

$$\frac{\partial c_i}{\partial t} = -\frac{\partial (c_i v)}{\partial z} + D_{i,j}\frac{\partial^2 c_i}{\partial z^2} + S_R \tag{6.34}$$

Whereby S_R represents the production rate of the compound i, which is the reaction rate. The equation represents a nonlinear PDE type - of so-called diffusion / convection type. Reducing the focus for a moment to one single flow component (without component index i) the equation reads:

$$\frac{\partial c}{\partial t} = -\frac{\partial (c v)}{\partial z} + D\frac{\partial^2 c}{\partial z^2} + S_R \tag{6.35}$$

For local discretization it shall be assumed that the component can be discretized into L equidistant elements of the length Δx , indexed as $[1...L]$. Then the local derivative at position l in the pipe can be discretized as

$$\left.\frac{\partial c}{\partial z}\right|_l \approx \frac{c_{l+1} - c_{l-1}}{2\Delta z}$$

which leads to:

$$\left.\frac{\partial^2 c}{\partial z^2}\right|_l = \left.\frac{\partial\left(\frac{\partial c}{\partial z}\right)}{\partial z}\right|_l \approx \frac{1}{\Delta z}\left[\frac{c_{l+1} - c_l}{\Delta z} - \frac{c_l - c_{l-1}}{\Delta z}\right] = \frac{c_{l+1} - 2c_l + c_{l-1}}{\Delta z^2} \tag{6.36}$$

The partial differential equation, $\frac{\partial c}{\partial t} = -\frac{\partial (c v)}{\partial z} + D\frac{\partial^2 c}{\partial z^2} + S_R$, thus can be transformed into a system of L ordinary differential equations of the form,

$$\frac{d c_l}{d t} = \alpha c_{l-1} + \beta c_l + \gamma c_{l+1} + S_R \tag{6.37}$$

whereas the following subterms are defined as:

$$\alpha=\frac{D}{\Delta z^2}+\frac{v}{2\Delta z} \qquad \beta=\frac{-2\ D}{\Delta z^2} \qquad \gamma=\frac{D}{\Delta z^2}-\frac{v}{2\Delta z}$$

Since the compound concentrations are dependent on each other, the differential equation system must be solved simultaneously.

Each differential equation only describes the time behavior of one compound concentration at one location over time. A solution of the DEQ system represents the computation of time behavior of one compound concentration along one line, $x{=}const$, over time, so in one pipe or reactor segment. Thus this approach commonly is called the "Method of Lines".

$$\frac{d c_l}{d t}=\alpha c_{l-1}+\beta c_l+\gamma c_{l+1}+S_R$$

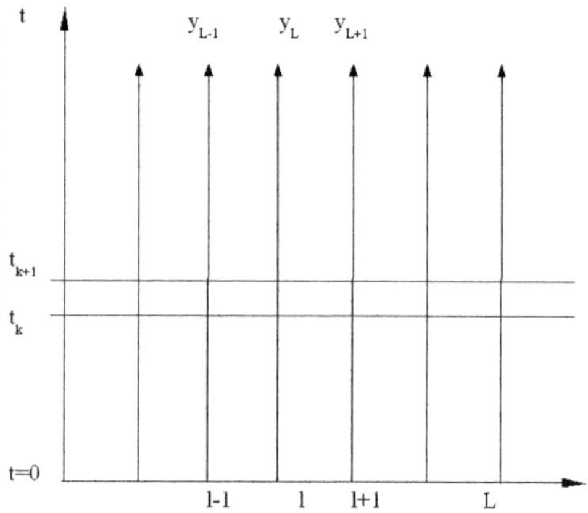

Figure 6.3: Method of lines.

For system states to be described by multiple parameters, the following equation represents the same approach assuming that the system state can be completely described by pressure P, temperature T and compound concentration c:

$$\begin{aligned}
\frac{d c_l}{d t}&=f_{cl}(c_{l-1},c_l,c_{l+1},T_{l-1},T_l,T_{l+1},p_{l-1},p_l,p_{l+1},x,t)\\
\frac{d T_l}{d t}&=f_{Tl}(c_{l-1},c_l,c_{l+1},T_{l-1},T_l,T_{l+1},p_{l-1},p_l,p_{l+1},x,t)\\
\frac{d P_l}{d t}&=f_{pl}(c_{l-1},c_l,c_{l+1},T_{l-1},T_l,T_{l+1},p_{l-1},p_l,p_{l+1},x,t)
\end{aligned} \tag{6.38}$$

For a generic parameter set, not only compound concentrations, the equation in vector format reads as follows:

$$\overline{\overline{B}}\frac{d\overline{y}}{dt}=\overline{f}(\overline{y,x,t}) \tag{6.39}$$

With: $\overline{\overline{B}}=\begin{bmatrix}1 & & & & \\ & \ddots & & & \\ & & 1 & & \\ & & & \ddots & \\ & & & & 1\end{bmatrix} \qquad \overline{y}=\begin{bmatrix}y_1\\ \vdots\\ y_l\\ \vdots\\ y_L\end{bmatrix} \qquad \overline{f}=\begin{bmatrix}f_1\\ \vdots\\ f_l\\ \vdots\\ f_L\end{bmatrix}$

In both cases this leads to a linear system of ordinary differential equations with non-constant coefficients as depicted before in equation 6.12. The physical behavior of such systems thus can be described by a system of, eventually partial, differential equations. By application of local discretization methods this PDE system can be mathematically transformed to a system of coupled ordinary differential equations, extended by the already discussed algebraic coupling equations. Such an equation system is called a differential algebra system or DA-system for short. Differential equations of higher order can be converted to DEQ systems of 1st order.

6.7 Numerical Integration Methods

System simulation requires solving the integration of a differential equation system as treated in the previous sections over time, starting from a consistent set of initial conditions. Mathematically in a first simplified formulation this reduces the task to the solution of an

- initial value problem of
- a differential-algebra system
- of first order.

For stability of the solution and for numerically precise reasons it is essential to

- apply numerical solving techniques suited for the equation types, and to,
- assure the consistency of the initial state of the integration for all variables.

The type of ordinary DEQs appearing in the DA-system was elaborated previously:

$$\overline{\overline{B}}\frac{d\overline{y}}{dt}=\overline{f}\left(\overline{y,x,t}\right) \tag{6.40}$$

Algebraic equations appear as material state equations, (e.g. gas equation), or when modeling system components as discrete state machines. Numerically the algebraic coupling equations are treated as DEQs without derivative terms and are solved together in the overall DEQ system. Before treating specifics of this approach, the most common mathematical integration methods for solving such initial value problems shall be treated.

Euler Method

The most simple approach is the explicit Euler method. This method integrates over the entire interval Δx in one step and with one value for the derivative.

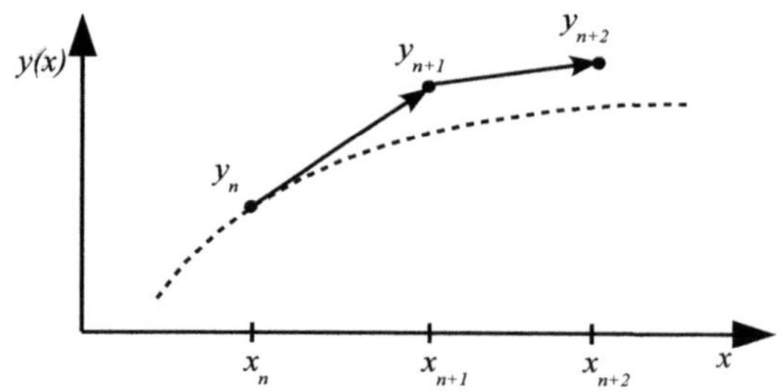

$$y'=f(x,y) \quad y(x_0)=y_0$$
$$h=\Delta x$$
$$y_{n+1}=y_n+hf(x_n,y_n) \qquad (6.41)$$

Figure 6.4: Explicit Euler method.

The explicit Euler method numerically only converges for system behavior parameter values with negative gradient. An example where this simple method can be adequate is the state integration of the gas thermal dynamics in the high pressure bottles of the system depicted in [23]. Here mass contained, pressure and temperature all have a negative gradient over the entire system operation time. Semi-implicit Euler methods which are also stable for DEQ systems with positive derivatives will be treated later in chapter 6.12.

Runge-Kutta Method

The most widespread integration methods are variants of the Runge-Kutta type. They exist in multiple variants with different order. The most commonly used are:

2nd order – Midpoint method
4th order – classic Runge-Kutta method
6th order – Runge-Kutta-Fehlberg method

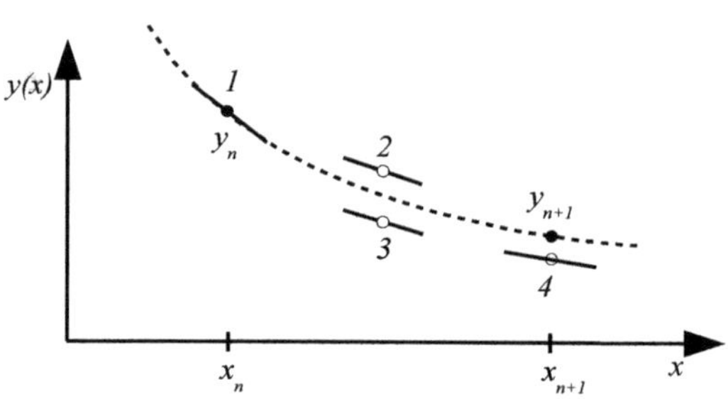

Figure 6.5: 4th order Runge-Kutta method.

$$h=\Delta x$$
$$k_1=hf(x_n,y_n)$$
$$k_2=hf\left(x_n+\frac{h}{2},y_n+\frac{k_1}{2}\right)$$
$$k_3=hf\left(x_n+\frac{h}{2},y_n+\frac{k_2}{2}\right)$$
$$k_4=hf(x_n+h,y_n+k_3)$$
$$y_{n+1}=y_n+\frac{k_1}{6}+\frac{k_2}{3}+\frac{k_3}{3}+\frac{k_4}{6}$$

(6.42)

These methods do not integrate in one step over the entire interval, but instead, intermediate "test computations" are made as can be seen in the above formulae for the Runge-Kutta 4th order. At each of the test points the local derivative is computed which is used for the next test step. At the end the final result is computed via an embedding formula. For details on the (very good) stability and the achievable precision, the reader is pointed to the corresponding mathematics literature. All Runge-Kutta methods cited here are of the explicit type.

Richardson Extrapolation

A completely different technique is followed by the Richardson extrapolation method. They first compute the integration over the interval with few intermediate steps - two in the figure below. Thereafter they increase the number of intermediate steps by even numbers and successively achieve a more and more precise result. Finally via the evolution of the value of the intermediately computed test results, the final function value which would result from infinitely small step size can be estimated. A popular subtype of these algorithms is the so-called Gragg-Bulirsch-Stoer method.

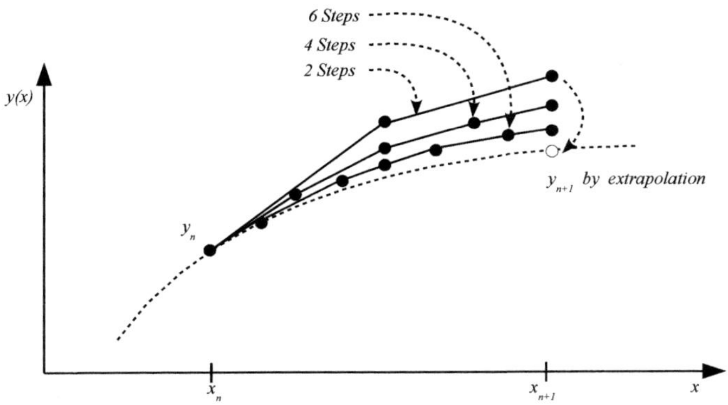

Figure 6.6: Richardson extrapolation.

The following Java code example computes the initial value problem for

$$y' = 2 \sin x + (\tan x)\, y \quad \text{with} \quad y(0) = 1 \tag{6.43}$$

in the interval [0,1] from 2nd to 12th order.

```
public class Main {
    /* a = interval - initial value: */
    static int a = 0;
    /* b = interval - terminal value: */
    static int b = 1;
    /* exact solution */
    static double lsg = 2 / Math.cos(b) - Math.cos(b);

    /**
     * @param args the command line arguments
     */
    public static void main(String[] args) {
        start();
    }
```

```
    /* function to integrate */
    public static double f(double x, double y) {
        double z = 2 * Math.sin(x) + y * Math.tan(x);
        return z;
    }

public static void start() {
    /* Set the order here: */
    // The order needs to be divedable by two (2,4,6...)
    double ordnung = 12;
    double maxk = Math.ceil(ordnung / 2);

    double nk = 0;
    double hk = 0;


    /* Declaration of a field y[] for the function values
       at the supporting points */
    double[] y = new double[(int) Math.pow(2, maxk) + 1];

    /* initial value y(a)= */
    y[0] = 1;

    /* Declaration of a 2-dimensional field T[][] for the interim values
*/
    double[][] T = new double[(int) maxk + 1][(int) maxk + 1];


    /* Part 1: T[k][1] */
    for (int k = 1; k <= maxk; k++) {
        /* nk = Number of bands */
        nk = Math.pow(2, k);
        /* hk = band size */
        hk = (b - a) / nk;
        /* Anlaufrechnung nach Euler-Cauchy */
        y[1] = y[0] + hk * f(a, y[0]);
        /* Calculation of function values
           at the sampling points according to
           the tangent trapezoidal formula */

        for (int i = 2; i <= nk; i++) {
            y[i] = y[i - 2] + 2 * hk * f(a + (i - 1) * hk, y[i - 1]);
        }

        /* Calculation of T[k][1] from the function values
           at the sampling points */
        T[k][1] = (y[(int) (nk - 1)] + y[(int) nk]
                  + hk * f(a + nk * hk, y[(int) nk])) / 2;
    }


    /* Part 2: Extrapolation of the interim values */
    for (int o = 2; o <= maxk; o++) {
        for (int k = o; k <= maxk; k++) {
            T[k][o] = T[k][o - 1] +  (T[k][o - 1] - T[k - 1][o - 1])
                    / (Math.pow(2, 2 * o - 2) - 1);
        }
    }
```

```
    /* Part 3: Print-out of the solution into a table. */
    System.out.println("Order | k| nk| Tk[x][x] | Exact solution |
                        Rel. err.");

    for (int i = 1; i <= maxk; i++) {
        String myOrdnung   = Double.toString(i * 2);
        String myK         = Double.toString(i);
        String myNk        = Double.toString(Math.pow(2, i));
        String mySolution  = Double.toString(lsg);
        String myRelError  = Double.toString(Math.abs((T[i][i] - lsg) /
                                                      lsg));
        System.out.println(myOrdnung + " | " + myK + " | " + myNk
                    + " | " + T[i][i] + " | " + mySolution + " | "
                    + myRelError);

    }
  }
}

Output:
--------------------------------------------------------------------------------
Order| k   |   nk | Tk[x][x]           | Exact solution    | Rel. err.
  2.0| 1.0|   2.0| 3.1486986551235145 | 3.161329129493711 | 0.0039953050925195435
  4.0| 2.0|   4.0| 3.172940317313528  | 3.161329129493711 | 0.003672881672297243
  6.0| 3.0|   8.0| 3.1616979644949303 | 3.161329129493711 | 1.1667086409262056E-4
  8.0| 4.0|  16.0| 3.161338582359264  | 3.161329129493711 | 2.9901554585989617E-6
 10.0| 5.0|  32.0| 3.161329227808318  | 3.161329129493711 | 3.109913667664177E-8
 12.0| 6.0|  64.0| 3.161329129896201  | 3.161329129493711 | 1.273167154728482E-10
```

Further Methods

Beyond these multi-step methods exist approaches which integrate into their extrapolation formula one or more historic sample point values, e.g. the Adams-Bashforth methods (cf. [47]) not further described in this volume but worth considering.

Constant Step Sizes vs. adaptive Step Size Control

For most of the discussed methods, in addition variants exist which provide adaptive stepsize control. Varying integration-step size is an efficient technique for numerical effort optimization in conventional simulations. However for real-time systems this is not a suitable technique since interval sizes Δx are prescribed by data exchange intervals to hardware in the loop.

Consistent Initial Conditions

In system simulation the consistency of the initial conditions of all state variables at integration initiation is essential for the numeric stability of the applied method. In process engineering of complex plants therefore simulators exist which prior to dynamic behavior simulation compute a static system state solution - in other words solving the DEQ set for static system state. For a satellite the system state at launcher separation can be taken as a consistent initial data set for simulation start. For this point usually all state parameters such as attitude, velocity vector, subsystem states etc. are known.

6.8 Integration Methods Applied on System Level

For treating system modeling and simulation basics on **system** level two exemplary applications will be discussed in parallel in the subsequent sections,

- a propulsion system for rocket stages, and
- a satellite attitude and orbit control system (AOCS).

Please also refer to the two following figures. Hereby simulation of the system's time behavior, can be formulated as initial value problem to be integrated over time starting from a known system state at $t=0$.

Propulsion System for Rocket Upper Stages

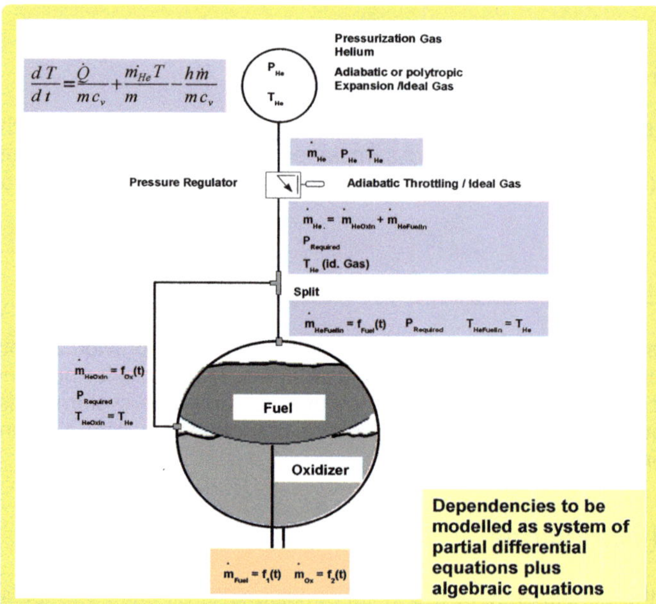

Figure 6.7: Propulsion system for rocket stages.

Satellite Attitude Control System

Below again the satellite model as already presented in chapter 5:

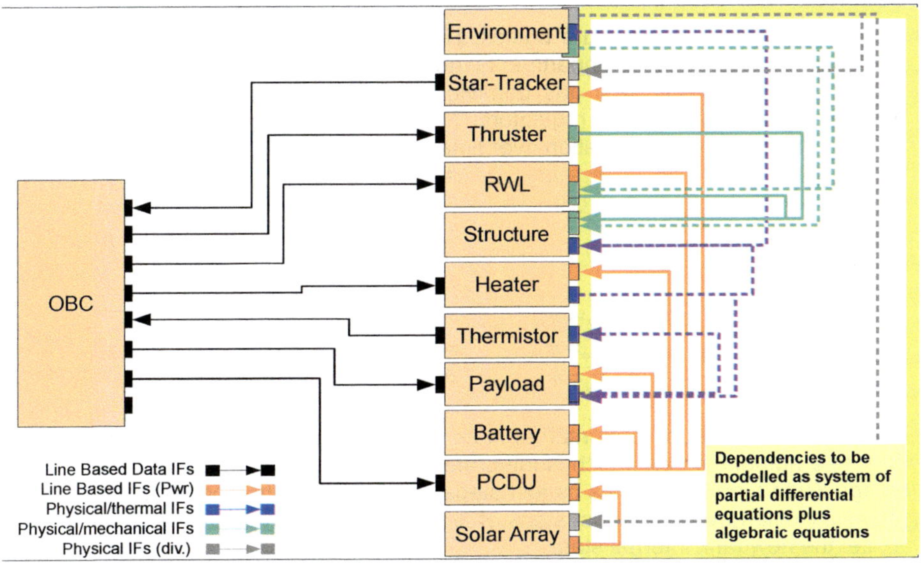

Figure 6.8: Satellite model - components and interconnections.

In a modular simulator, it is not the goal to reformulate the DEQ set for the entire system and to program a fixed solution for each system type or variant of the AOCS or the propulsion system. The ideal solution would be to assemble the system level model from a library of equipment models which can be taken from a library and can be "clicked" together in a graphical editor according to the real system's functional layout. The simulator then should be able to load the configuration and to run the simulation - which however is extremely complex w.r.t. software implementation.

For the implementation of such initial value problem solvers, two basic approaches can be followed. The first one requires compromises with respect to achievable solution precision, the second one is in principle exact. Both approaches are discussed further, using the a.m. spacecraft systems as application examples.

Simplified Integration on System Level – Example 1

Considering the model of the tank pressurization system (please refer to figure 6.7) the mass flows of oxidizer and fuel towards the engine are regulated according to required ΔV-of the launcher over time. Flow control either is performed via valves, or - if the stage is accordingly equipped - by controlled turbo-pumps. The cited mass flows of oxidizer and fuel extracted from the tank compartments represent the external boundary conditions:

$$\dot{m}_{Ox} = f_1(t) \quad \text{und} \quad \dot{m}_{Fuel} = f_2(t)$$

From the Helium high pressure bottles, Helium expands accordingly and flows into the fuel and oxidizer tank compartments, hereby modeling all thermal dynamic effects like temperature drops, real gas behavior, condensation of oxidizer and pressure drops in pipes and pressure regulators. The derivatives of state variables modeling the tank inner state for integration over time compute as follows:

$$\begin{aligned}
\dot{y}_a &= \dot{m}_{OxCondens} &&= f_a(\bar{y}, t) \\
&\vdots && \vdots \\
\dot{y}_l &= \dot{m}_{FuelCondens} &&= f_l(\bar{y}, t) \\
\dot{y}_m &= \dot{m}_{HeOxIn} &&= f_m(\bar{y}, T_{HeOxIn}, t) \\
\dot{y}_n &= \dot{m}_{HeFuelIn} &&= f_n(\bar{y}, T_{HeFuelIn}, t)
\end{aligned} \tag{6.44}$$

Only the derivatives of tank internal state variables thus are computed and handed over to a numeric solver which is integrated into the tank model. The solver - e.g. of Runge-Kutta type - computes its test steps under the assumption of constant Helium inflow condition over all its test steps of Δt. Therefore it computes the solution for $t + \Delta t$, i.e. the new tank internal state.

The states further upstream in the system then compute the Helium pressure gas mass flows as average values over the entire time step Δt and thus from the resulting state changes in the Helium high pressure bottles as:

$$\begin{aligned}
\dot{m}_{He} &= \dot{m}_{HeOxInAverage} + \dot{m}_{HeFuelInAverage} \\
\dot{T}_{He} &= f_p(\bar{y}_{Bottle}, \dot{m}_{He}, t) \\
&\vdots
\end{aligned} \tag{6.45}$$

The model of the Helium bottle e.g. computes the temperature and pressure of outflowing gas under consideration of real gas effects and pressure drops. Therefore via pipes, filter, regulators (left out here), the new inlet state conditions for the tank compartments result for time equal to $t + \Delta t$, i.e. for integration of the next time step.

Implemented DEQ solution method on system level:

This approach corresponds to

- a Runge-Kutta method on tank level, but,
- an Euler method on system level.

Advantages:

- The DEQ subsystems are decoupled and their integration can be coded directly inside the component model.

- For each type of component an individually best suited numerical method can be selected.
- No overall equation system is generated which might eventually show up stiffness effects etc. (Topics which will be treated later in chapter 6.12).

Disadvantages:

- Since the temperature, pressure and mass content constantly decrease inside the high pressure bottles, the Euler method implemented is still converging. Thus for this specific system layout no disadvantages appear. But this is an exceptional case.

For a system simulator which shall provide the feature to "click" together the system layout or to define it simply in an input file, such a program concept is pretty straightforward since each component implicitly comprises "its" numerics. An overall system wide consistency (especially concerning hydraulic boundary conditions) must be achieved by additional means. An open source system simulator following this concept can be downloaded from [23].

Simplified Integration on System Level – Example 2

A similar approach as in example 1 is applied in the following simple attitude control application:

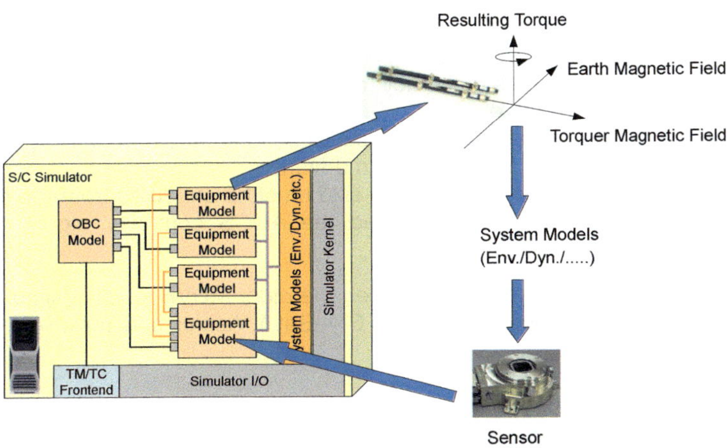

Figure 6.9: Satellite attitude control loop.

In this system the attitude change is actuated by means of a magnetotorquer which produces a magnetic field interacting with the Earth's magnetosphere and thus producing a torque onto the spacecraft. For computation the resulting torque is assumed to be constant over the time interval Δt. This already is a simplification, since the torque as result from interaction of torquer field and Earth magnetosphere can be computed exactly for time t. But already after the first Runge-Kutta test step the state of the spacecraft - i.e. its attitude - has changed minimally and thus the torquer axis direction has changed w.r.t. Earth magnetic field which implicates a change in resulting momentum onto the spacecraft.

With this simplifying assumption of a constant torque over the entire interval Δt in this case the attitude integration up to $t+\Delta t$ is performed. Internally the dynamics integrator can compute with e.g. a Runge-Kutta method of 4th order under consideration of additional environmental disturbance forces etc.

After attitude integration the sensor measures the new attitude parameters and forwards the measurements to the OBC.

<u>Implemented DEQ solution method on system level:</u>

- The approach represents a Runge-Kutta method on the level of attitude integration (including all disturbances like solar pressure etc.), but,
- it represents an Euler method on system level.

Advantages:

- The DEQ subsystems are decoupled which results in a simple computation sequence on system level.
- For a system simulator which shall provide the feature to "click" together the system layout or to define it simply in an input file, such a program concept is pretty straightforward since each model computes its equation system locally.

Disadvantages:

- Reduced numeric accuracy: This modeling approach is not suited for testbenches with high accuracy requirements nor for longterm runs since errors add up over time.
- The mathematic stability is only assured up to a certain - relatively small - step size, which however is not known in advance.
- The maximum allowable time step size must be extracted from the data of the controller designers which defined the AOCS controller algorithms in the OBSW.

Systems following this approach have been the older MDVE satellite simulators from Astrium GmbH for the space progammes CryoSat 1, Aeolus and TerraSAR (cf. [14]).

Exact Integration on System Level – Example 3

The following examples shall treat exact integration on system level. For the first example again the tank pressurization system from figure 6.7 is considered, however this time with slightly modified modeling. All derivatives of the entire pressurization system including all components now are integrated together by a common DEQ solver. The equation system looks as follows:

$$
\begin{aligned}
\dot{y}_a &= \dot{m}_{OxCondens} &&= f_a(\overline{y}, t) \\
&\;\;\vdots && \\
\dot{y}_l &= \dot{m}_{FuelCondens} &&= f_l(\overline{y}, t) \\
\dot{y}_m &= \dot{m}_{HeOxIn} &&= f_m(\overline{y}, T_{HeOxIn}, t) \\
\dot{y}_n &= \dot{m}_{HeFuelIn} &&= f_n(\overline{y}, T_{HeFuelIn}, t) \\
&\;\;\vdots && \\
\dot{y}_o &= \dot{m}_{He} &&= f_o(\overline{y}, t) \quad = \dot{m}_{HeOxIn} + \dot{m}_{HeFuelIn} \\
\dot{y}_p &= \dot{T}_{HeBottle} &&= f_p(\overline{y}_{Bottle}, \dot{m}_{He}, t) \\
&\;\;\vdots &&
\end{aligned}
\qquad (6.46)
$$

All derivatives are registered as common data structure with the solver. The solver computes the test steps (4 in case of classic Runge-Kutta) and computes the final state results for the overall system at $t + \Delta t$. In contrast to equation 6.45 here specially has to be pointed out that:

- The overall mass flow of Helium is no longer computed as average out of the 4 Runge-Kutta steps of the tank solution only, but in each test step here it is recomputed on system level.
- Furthermore now the state equations for the Helium bottles also are solved via the same Runge-Kutta method on system level.

Implemented DEQ solution method on system level:

- The approach corresponds to a Runge-Kutta method on system level.

Advantages:

- The approach is mathematically exact which results in
- maximum accuracy and stability.

Disadvantages:

- The stiffness of the DEQ system is not considered in advance (cf. Chapter 6.12) which however for this simple system does not yet impose problems.
- For a system simulator which shall provide the feature to "click" together the system layout or to define it simply in an input file, such a program concept is very ambitious since
  - ◇ each equipment model has to register its derivatives with the solver, and
  - ◇ because depending on the user defined system layout (one high pressure bottle or two) the solver has to integrate a different number or even different types of DEQs.

Exact Integration on System Level – Example 4

The last considered system represents an exact integrated variant of example 2 - please also refer to the figures 6.10 and 6.11. The first figure shows

- in similarity to figure 6.8, and
- reduced to mechanical effects in the AOCS (green dotted lines),

how the equipment models and environment model first exchange variables (often called "Algebraic Quantities") according to the actual spacecraft state. E.g. the thruster models, reaction wheel models and space environment model "report" forces and torques towards the spacecraft structure model.

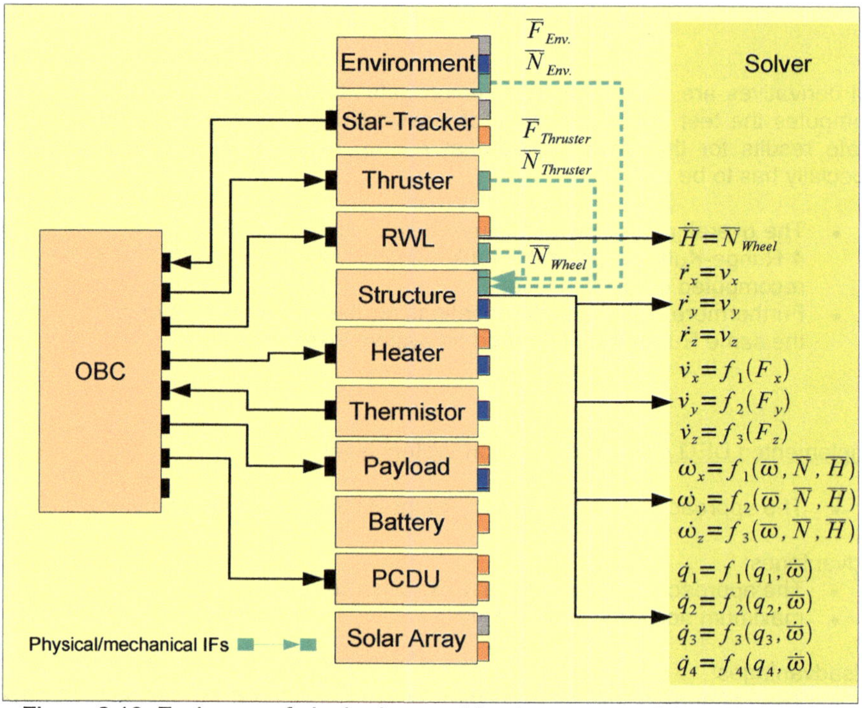

Figure 6.10: Exchange of algebraic quantities between models and reporting derivatives to the solver.

From there the models compute the state variable derivatives - in the figure below the reaction wheel and the structure model - and register them with the solver. The integrator now performs one test step integration. Figure 6.11 then depicts returning the new state variables - namely rotational momentum, position and rotational rate - to the various equipment, depending on which model requires which state variables. Thereafter the same cycle starts again for the next Runge-Kutta test step and so forth.

The implementation of such an approach implies an adequate model architecture for the equipment models as shown in figure 4.15. Especially refer here to the model interfaces for handing over the derivatives to the solver, for read-back of the integrated state variables and for exchanged algebraic parameters (see "Continuous Model IF" in figure 4.15).

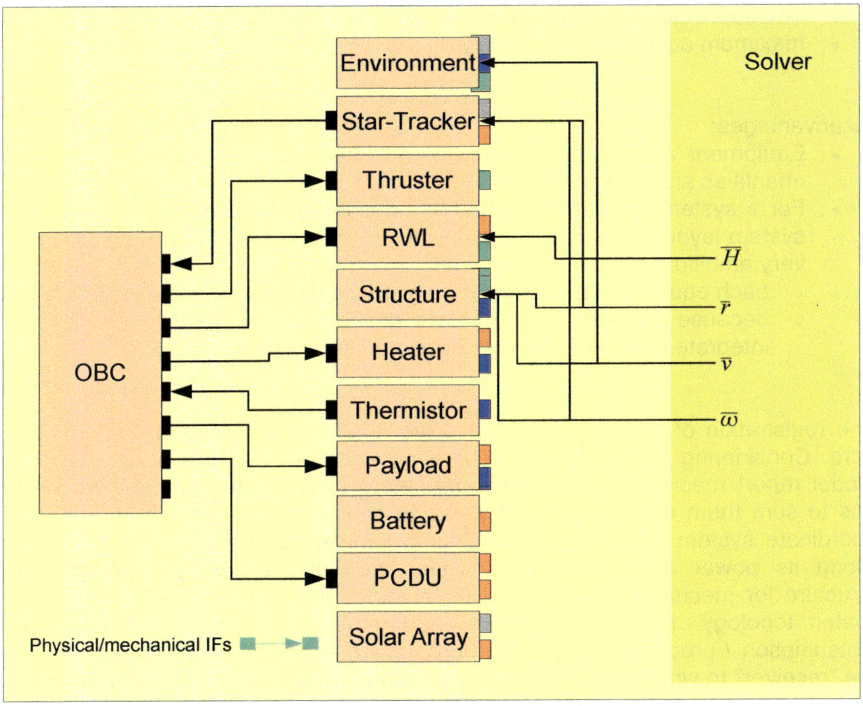

Figure 6.11: Feedback of integrated state variables towards the equipment models.

The computation flow for such a system according to figures 6.10 and 6.11 is:

- Reading of actuator control values from OBC
- Call of the Model-Interface-Layer for reading and decalibration of data from protocol
- Computation of discrete model parts (state machine)
- Integration cycle of the central solver:
  - ◇ Computation of the continuous parts of the equipment model and handover of derivatives to solver
  - ◇ Integration step
  - ◇ Computation of integrated final values from test steps and hand back of state variables to the models
- Call of Model-Interface-Layer for calibration of output via protocol
- Writing output to line models towards OBC

Figure 6.12: Computation flow sequence.

Implemented DEQ solution method on system level:

- The approach corresponds to a Runge-Kutta method on system level.

Advantages:
- The approach is mathematically exact which results in
- maximum accuracy and stability.

Disadvantages:
- Equipment and system models need interfaces for exchange of algebraic quantities and derivatives.
- For a system simulator which shall provide the feature to "click" together the system layout or to define it simply in an input file, such a program concept is very ambitious since:
  - ◇ each equipment model has to register its derivatives with the solver, and
  - ◇ because depending on the user defined system layout the solver has to integrate a different number or even different types of DEQs.

The registration of derivatives with a solver shall be analyzed in a bit more detail here. Considering first figure 6.10, here all actuators and the space environment model report mechanical influence parameters towards the structure model which has to sum them up and eventually has to convert them into a central spacecraft coordinate system. In an electrical system similarly all electrical equipment would report its power consumptions. However therefore a "receiver" model like the structure for mechanics or a PCDU for electrics must exist, which - depending on system topology - has to handle a varying number of input forces, torques or power consumption / production rates. The force / torque or power "suppliers" must know the "receiver" to whom they have to report right from program start and must register themselves as input "suppliers" there. A reaction wheel e.g. only registers a torque, the space environment model has to register both forces and torques - depending on attitude and position of the spacecraft.

These features for registration of parameters with a "receiver" can be supported by so-called service oriented architectures (SOA) of the simulator kernel. More details on this very advanced approach are discussed in chapter 8.

A similar mechanism has to be applied in the second step. Here first all equipment and system models (e.g. space environment model) register the derivatives they compute with the solver (see figure 6.10) and they register, which state variables they later need back for their local computations (see figure 6.11). These two variable sets however can differ. E.g. the startracker model in figure 6.11 does not compute any position vectors, rotation rates etc., however it requires attitude angles and position vector of the spacecraft for computation of its quaternion output for the OBC.

Here again such a service oriented architecture on simulator kernel level is required. The numerical solver must provide the according functionality and the models must perform the registration steps at simulator initialization time. A technical solution for this problem are entirely dynamic simulator kernel architectures such as *OpenSimKit* introduced in V4.0 which still is in research phase. Such systems are highly dynamic and allow in one case to e.g. load a spacecraft configuration with 3 reaction wheels and in the next run a configuration with 4 wheels. For changing the design only a different simulation input file is required. All the internal mode registering with a solver is "black box" technology and the user is not burdened with these features.

An alternative approach can be a "hard coding" of the model / solver interlinking in the software architecture UML model at spacecraft simulator design time. From the UML design, the fixed code for one spacecraft topology and a fixed networking of models and solver is then generated. Eventual design changes in the real spacecraft need a design adaptation of the simulator on UML level and a regeneration of code. Infrastructures in satellite engineering applying this concept are the newer MDVE simulators from Astrium applied in GOCE, LISA-Pathfinder and Galileo-IOV as well as the latest Astrium "Simulator 3rd Generation" infrastructure implementations for the ESA projects Bepi-Colombo, Sentinel 2 and EarthCARE (cf. [16], [17] and [19]).

6.9 Boundary Value Problems in System Modeling

The integration of differential equation systems was described in the previous chapters 6.7 and 6.8. It must be performed over the entire equation system in parallel as far as the equations are coupled via state variables or derivatives. This however only covers the integration of a pure initial value problem of a DEQ set. The cited example of an actuated AOCS and the integration of spacecraft attitude change over time is an application where such type of initial value problem solving is sufficient.

This technique is no longer adequate for a system with additional boundary conditions. The mathematical task then is to perform a state integration over time for a combined initial value / boundary value problem. The following simple life support system depicts such a type of system.

This depicted system - from an early design phase of the Columbus laboratory - basically focuses on correct temperature control of the cabin and the "standoff" with the experimental racks. Temperature control is achieved through a variation of the airflow via the controlled fan, a controllable flow distribution between cabin and standoff as well as by control of the airflow passing a heat exchanger for cooling. Air cleaning, humidity control and CO_2 absorption as well as oxygen supply are performed via the "Inter Module Venting" interface to the ISS core modules.

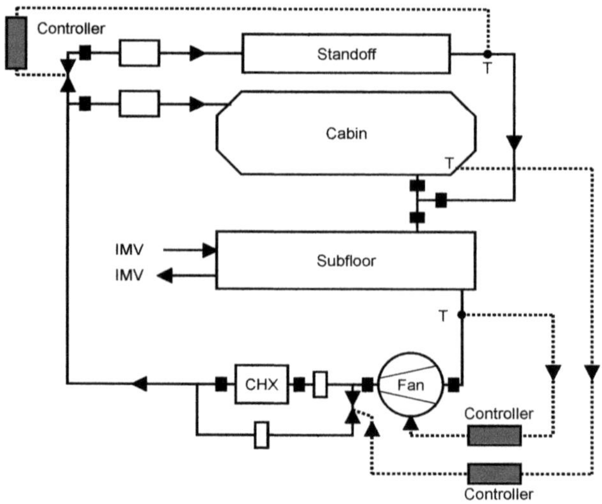

Figure 6.13: Early phase design of the Columbus life support system - similar in [129].

In such system designs, bifurcations - called "split" here - and junctions occur, and the self adjusting distribution of airflows isn't calculable anymore via a pure forward integration from a known state. The situation shall be explained in more detail at hand of a more simplified model of the life support system - please refer to the figure below where all valves, controllers and flanges are removed.

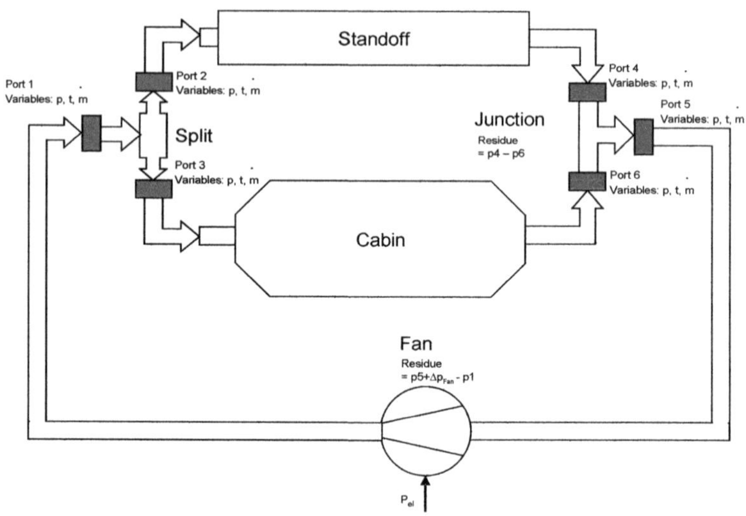

Figure 6.14: Abstraction of the life support system.

During operation of such a system e.g. the cabin temperature can change over time due to dissipated heat from the astronauts. This results in a change of air density and in the volume flow which drives the pressure drop. However at the point of junction, the boundary condition of pressure equality of the flows from cabin and standoff has to be assured. Therefore the changing mass flows between cabin and standoff have to be recalculated after each integrated time step.

A similar effect occurs if due to experimental parameters in the standoff racks, the airflow resistance increases and as a result the pressure drops in standoff. Also in this case a real variation of the mass flows between cabin and standoff results.

In reality such variations of the mass flows occur gradually over time. In simulation however the time increase is discretized, so that the integrative states in cabin and standoff (e.g. temperature) vary between two time steps. After each integration step the boundary condition of pressure equality would be infinitely violated, assuming same mass flow distribution as in the previous step. A so-called residue for this pressure equality occurs which has to be eliminated by calculation of adapted mass flow distribution at the split.

A further similar effect occurs if the overall flow resistance of the entire system increases. Under the assumption of a constant pumping power, the pressure increase and / or mass flow at the fan will change in such a case. Also this residue between pressure increase at the fan and pressure drop over the system has to be eliminated before a new time step integration can be performed - since otherwise the mass flow rates do not represent real world conditions.

Both effects, the variation of mass flow distribution between standoff and cabin as well as change in overall mass flow also can occur together (which in practice happens in most cases). The challenge is to design the simulator software in such a way, that it

- in the first place can identify the points of residues, and
- that on the other hand it can identify which parameters are to be readjusted, how to eliminate the residues between two time step integrations and thus to compute a physically valid state.

Since the equipment output variables are depending on their input data (directly or indirectly), the simulator has to identify relevant components and their adequate variables to be tuned for residue elimination. Here it is to be considered that e.g. for simple electric system components (e.g. resistors or impedances) the output parameter values linearly depend on the input values. Similar dependencies apply for thermal systems. For fluid systems, the dependency of output from input parameters mostly can be identified directly, however the dependencies are nonlinear. The elimination of residues can only be achieved by application of numerical root finding methods. For the overall elimination of all residues in the system, an equation system of the following form has to be solved:

$$\overline{R} = \overline{\varepsilon} \qquad \text{with} \qquad \overline{\varepsilon} \rightarrow \overline{0} \qquad (6.47)$$

For numerical root finding an entire pool of mathematical methods is available which are treated later. In the system of figure 6.14 the following residue conditions are to be fulfilled:

- At the junction, the pressures of port 4 and 6 must be equal, and
- for the fan pressure increase must be equal to the pressure drop over the entire system back to fan.

The degrees of freedom are the mass flow distribution at the split and the resulting mass flow over the fan. Thus for residue elimination an equation system of two equations for two residues is to be solved where two degrees of freedom can be varied. The equation system thus is well defined for this case (overdetermined and undetermined residue equation systems are tackled later to keep this example straightforward). Intuitively understandable residue variables and variables for the degrees of freedom are used in this example, such as pressures and mass flows. In more complex systems like fuel cells or chemical plants, very less intuitive variables can be sources of residues. E.g. in a chemical distillation splitter, the sum of molar outflows of liquid and gas phase has to be equal to 1.0. Any deviations therefrom become a residue.

To eliminate residues the simulator can be implemented in a way that the user has to manually identify the subsystem parts to be iterated for elimination of each residuum, so-called "meshes". This principle is applied for example by the simulator programs SIMTAS *Object* [130], and *OpenSimKit* [23]. In both cases the meshes are to be defined in according input file sections of the software.

For the system in the above example, one mesh would be defined which leads from split via cabin and standoff to the junction and which would iterate the junction residue value pressure difference until resulting in a value below a selected limit. A second mesh thereafter would vary the overall mass flow in the system until the residue in the fan (pressure increase minus pressure drop) is sufficiently low. In case the mass flow over the entire system would change significantly during this outer mesh iteration, the assumed mass flow at split for previous inner mesh iteration would cause invalidation. In such a case the inner mesh has to be reiterated and so forth until all residues are below permitted limits.

The technique of letting the user define the meshes in particular is not a nice approach since one has to determine how to place the meshes. This task is acceptable for a simulation expert but definitely makes the use of the software unnecessarily complex for a beginner. Therefore the goal must be to find an algorithm which automatically can achieve this residue elimination after each time step integration and which can identify at handover of the system topology the according residue variables and the degrees of freedom.

For the above system such an algorithm simplified could look as follows, (cf. figure 6.14):

- Origin of the algorithm is a residue, assumed e.g. for the junction in figure 6.14.

- As key parameters in model programming the two parameters determining the residue are taken - here the two input pressures at the in-ports of the junction. The occurrence of the junction model registers these variables in a table, i.e. $p4$ and $p6$.
- Since the residue-relevant parameters $p4$ and $p6$ are input parameters of the junction occurrence, the algorithm continues to find the according degree of freedom upstream from the junction.
- First it searches whether any component directly upstream from the junction input ports offers a pressure variable as degree of freedom at its outlet.
- Cabin or standoff however cannot offer this, however their output pressures depend (non linearly) on their state variables and input variables.
- Therefore from the side of the junction, the upstream components cabin and standoff (upstream of $p4$ and of $p6$) are queried for a qualitative dependency table, which output value towards junction depends on which input value (independent of linearity). Cabin and standoff provide such a table in the form:

$$\begin{aligned} p_{out} &= f1(p_{in}, \dot{m}_{in}) \\ \dot{m}_{out} &= f2(\dot{m}_{in}, T_{in}) \\ T_{out} &= f3(\dot{m}_{in}, T_{in}) \end{aligned} \tag{6.48}$$

- The occurrence Cabin 1 thus can complement the information of the variables table cited above and can identify that $p4$ depends on $p2$ and $\dot{m}_2$.
- The occurrence Standoff1 can complement the information of the variables table cited above and can identify that $p6$ depends on $p3$ and $\dot{m}_3$.
- To manipulate $p4$ for junction residue elimination thus a component has to be found which offers a degree of freedom for either $p2$ and / or $\dot{m}_2$ and / or a component upstream the standoff has to be found which offers a degree of freedom for either $p3$ and / or $\dot{m}_3$.
- The component connected to the input ports 2 and 3 of cabin and standoff is the split. The split contains again a similar dependency table of output and input parameters:

$$\begin{aligned} \dot{m}_{out,A} &= free \\ \dot{m}_{out,B} &= free \\ p_{out,A} &= p_{in} \\ p_{out,B} &= p_{in} \\ T_{out,A} &= T_{in} \\ T_{out,B} &= T_{in} \end{aligned} \tag{6.49}$$

Of course in the physical implementation of the split model it is considered that $\dot{m}_{out,A} + \dot{m}_{out,B} = \dot{m}_{in}$ has to apply, i.e. that not both mass flows are freely selectable due to conservation of mass. However the algorithm here can automatically identify the split as component which allows variation of mass flow distribution in case the residue of the junction is the above limit.

In a similar approach the algorithm proceeds for the outer residue 1, the pressure increase at the fan and the drop over the system. The latter is computed by $p1$-$p5$ so directly concerning port variables of port 1 and port 5 of the fan. The fan itself is considered to be isothermal and has registered in its component dependency table:

$$\begin{aligned} T_{out,B} &= T_{in} \\ p_{out} &= f(p_{in}) \\ \dot{m}_{out} &= free \end{aligned} \tag{6.50}$$

Here the algorithm can identify that for elimination of the fan's residue, the mass flow passing the fan itself has to be varied.

Following this approach, only the equipment model programmer for a new model has to implement the according functionalities for registering the parameter dependencies according to equation 6.48 in the system table. Since the model programmers usually have detailed knowledge about the physical interdependencies, this will not impose a problem for them. The simulator software thereafter can find itself the residue variables, it can identify other components providing degrees of freedom for residue elimination and it can automatically iterate these meshes found. The simulator user only needs to assemble his system from precoded equipment models and no longer has to care about boundary value problems, meshes and so forth. This technique is very challenging concerning implementation but has clear advantages.

6.10 Root Finding Methods for Boundary Value Problems

As the example from figure 6.14 demonstrated, the residue values at the split and fan can be regarded as functions of the degree of freedom variables (mass flow and mass flow distribution) at the fan and split respectively. For both residues the value has to be minimized - ideally to zero. Thus a mathematical root finding has to be performed. Before treating the mathematics in more detail it shall be reminded that

- such a numerical root finding process has already to be performed once after simulation initialization to achieve a physically valid state over the entire system, and
- that all root finding algorithms only converge within a certain interval around the root value.

Thus in any case at initialization time of the simulator for all system state variables (here pressures, mass flows, temperatures) values of technically realistic magnitude have to be provided as initial values. The first residue iteration after simulation initialization serves for the following purpose:

The system engineer who designed the system and who wants to simulate dynamic load cases for design verification, dynamic behavior analysis etc., usually knows the design parameters of the system equipment (fan power, heat dissipation of experiments in standoff etc.). In addition he knows the magnitude of the operational system parameters for a certain load case - for the discussed system these are

- magnitude of the cycling overall mass flow at a specified fan electric power, and
- magnitude of distribution of mass flows between cabin and standoff.

With this it is usually assured that the initial values specified for simulation initialization are sufficiently close to real values of the system in operation, so that the root finding process for fine tuning of the initially assumed parameter values converges. At simulator start the exact values are not yet known, thus first a root finding computation varying the degree of freedom parameters accordingly has to be performed. In the system discussed above this eliminates the residues at junction and fan. The state achieved thereafter is the computation of a quasi-stationarity system state for the life support system. Thereafter a time step integration can be performed to integrate the initial value problem for internal state variables of the equipment models (gas concentrations in cabin, wall temperatures, gas temperature etc.). After each such integration step of the initial value problem, again a residue elimination step is performed. Result logging and reporting by the simulator always are performed after the residue elimination step, because only at that point a physically valid system state is computed.

The method of root finding itself shall be described applying the example of the rather intuitive Newton method. The following figure depicts - for a one dimensional case - the stepwise approach to find the root from a starting point x_1 over multiple intermediate steps via intersections of the curve's tangents with the x-axis.

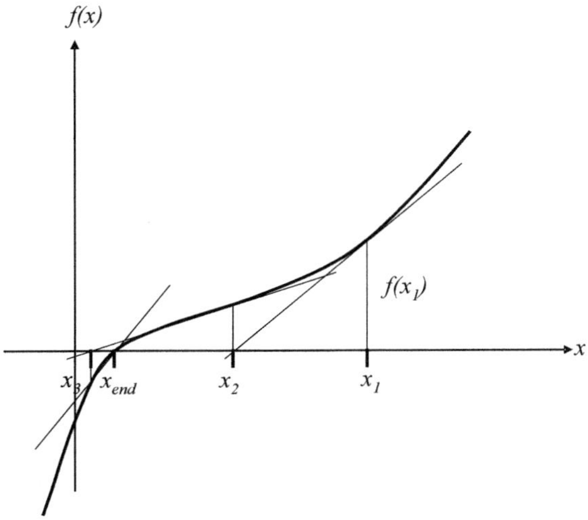

Figure 6.15: Stepwise approximation of a function's root.

Generally the following formula can be derived and which is to be applied long enough until the root is approached with sufficient precision:

$$x_{n+1}=x_n - \frac{f(x_n)}{df(x_n) \ / \ dx} \tag{6.51}$$

This was for the single dimensional case and pretty intuitive. If multiple residues are to be solved and which are dependent on multiple degree of freedom variables as in the example of figure 6.14, the approach can be extended for multiple dimensions. Accordingly equation 6.51 then modifies to

$$\overline{x}_{n+1}=\overline{x}-\overline{\overline{J}}(\overline{x})^{-1}\overline{f}(\overline{u}) \tag{6.52}$$

whereby $\overline{\overline{J}}$ is the so-called Jacobi matrix, the tensor with the partial derivatives of the function $\overline{f}$:

$$\overline{\overline{J}}=\frac{\partial \overline{f}}{\partial \overline{x}}=\begin{bmatrix} \partial f_1/\partial x_1 & \partial f_1/\partial x_2 & \cdots & \partial f_1/\partial x_n \\ \partial f_2/\partial x_1 & \partial f_2/\partial x_2 & \cdots & \partial f_2/\partial x_n \\ \vdots & \vdots & \ddots & \vdots \\ \partial f_n/\partial x_1 & \partial f_n/\partial x_2 & \cdots & \partial f_n/\partial x_n \end{bmatrix} \tag{6.53}$$

Since solving equation 6.52 via the computation of the inverse of the Jacobi matrix is numerically costly, instead the following linear equation system is solved:

$$\overline{\overline{J}}(\overline{x_n})\Delta \overline{x_n} = -\overline{f}(\overline{x_n}) \tag{6.54}$$

The degree of freedom variables $\overline{x}_{n+1}$ then result from:

$$\overline{x}_{n+1}=\overline{x}_n+\Delta \overline{x} \tag{6.55}$$

Basically here for both the single as the multidimensional case the problem exists that the functional dependency between residue and degree of freedom variables cannot be formulated by algebraic equations.

If between degree of freedom component (e.g. split in a.m. example) and residue component (e.g. junction in a.m. example) only equipment is located with algebraic transfer functions, then the derivatives of the residue dependent on the degree of freedom variables are directly computable for the Jacobi matrix. If however between split and junction, components are located which comprise a behavior that e.g. has to be approximated via a performance chart, then the derivatives of the residue functions from them have to be numerically interpolated. Usually this is done from difference to the system residue iteration of the time step before. An example (following the Newton method) for a two dimensional case can be found on the Internet in [50].

Here the residue elimination was explained applying the relatively intuitive and easy to understand Newton method. Convergence performance and stability of the root finding as dependent from deviation of initial value at start of the process shall not be treated further here. These are topics of the according mathematically specialist literature. For such root finding problems today very high performance methods are available, e.g. the "Conjugate Gradient Methods" or "CG-Methods" (cf. [52]).

In the fields of chemical plant or power plant construction mainly the performance at the stationary optimum operation point is of interest, whereby for such systems the complexity is by far higher than for typical space systems like fuel cell systems, life support systems or propulsion systems. In plant construction and engineering for these tasks specific tools for modeling the stationary operational case of the plant exist, so-called "Flow-Sheeting" tools. As already explained, the root finding after a time step integration in a system corresponds to finding a quasi-stationary system solution. The mentioned Flow-Sheeting tools comprise functions for also handling over and under determined residue equation systems not treated here yet. For this topic of handling residue elimination generically in complex systems, the reader is pointed to the specific literature on system flow sheeting techniques, e.g. to [51].

6.11 Numerical Functionalities for Control Engineering

6.11.1 Mathematical Building Blocks and their Transformation to RPN

Especially for tools applied in control engineering like Simulink or Modelica, it has proven good practice to support assembly of the control cycle via graphical function blocks and to parameterize these blocks thereafter. This approach is sufficiently known, see e.g. figure 3.9.

However how can it be achieved in software to allow startup of a simulation run immediately after having clicked together the system layout? For answering the concept, first a purely algebraic function chain shall be considered. One of the functional elements - here without internal state for simplification - could look as follows:

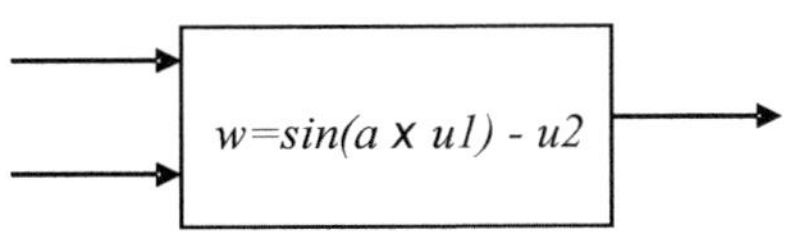

Figure 6.16: Block element with defined function.

The block result w to be computed from the inlet parameters u shall be computed according to the mathematical formula:

$$w = \sin(a \times u1) - u2 \tag{6.56}$$

Assuming availability of a module editor allowing to type in directly the mathematic formula of equation 6.56 into the above block, or to load the formula from input file, this would be a handy approach for building a simulation. In a Simulink diagram alternatively also the implementation could be based on low level blocks as visualized in the layout below.

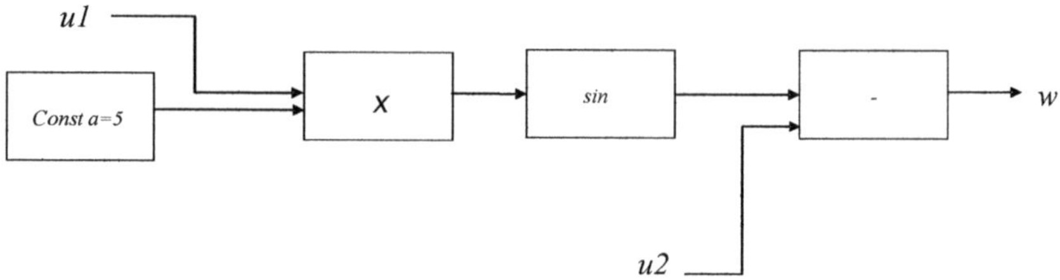

Figure 6.17: Assembly of elementary function blocks.

If it is desired however to directly specify the formula instead of assembling the function chain from such ultra low level blocks for + , - , x , / , sin , exp and so on, then this would result in having the formula text of equation 6.56 in the function block which is not compiled and thus is not executable for the computer. Either the formula has to be written in a real programming language and has to be compiled - which shall be avoided - or a parser is needed which "reads" the function text and generates an executable function stack in "Reverse Polish Notation"[18] (RPN) - also known as postfix notation. This parsing requires two loops. In the first pass the formula is transformed into an RPN stack still only containing text blocks of the formula. This algorithm works as follows:

1. Find first argument (here of sine function): Result: $a \times u1$
2. Subloop: Eliminate brackets:
   - 2.1. Search argument or operand: Result: a →put on stack
   - 2.2. Search argument or operand: Result: "x"
     Operand requires two arguments →memorize operand
   - 2.3. Search argument or operand: Result: *u1* →onto stack
   - 2.4. Operand can be applied →"x" onto stack

a
u1
x

   Intermediate result on stack:
   End of subloop reached.

3. Search argument or operand: Result: *sin*
   Operand requires one argument
4. Operand can be applied →"sin" onto stack

[18] Concerning RPN e.g. refer to http://en.wikipedia.org/wiki/Postfix_notation

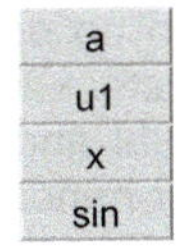

Intermediate result on stack:

5. Search argument or operand: Result: "-"
   Operand requires two arguments
   (one argument is provided by the stack) →memorize operand
6. Search argument or operand: Result: u2 →onto stack
7. Operand can be applied →"-" onto stack

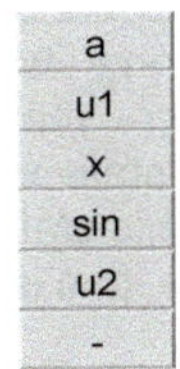

Intermediate result on stack:

8. End of formula text parsing

After completion of this first parsing loop, a text stack is available as a recipe in RPN as to compute sequentially. Now in a second loop this has to be transformed to a function stack with pointers to the component's input variables, pointers to computation functions for x , - , sin and with conversion functions for design parameter values (a). After this second loop the following interlinking of the RPN stack and elementary computations is accomplished.

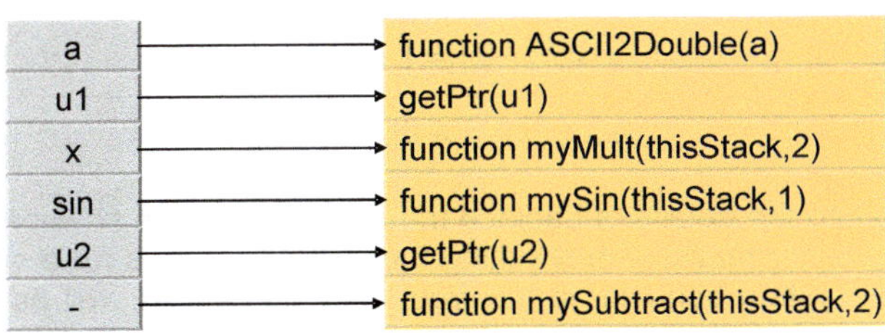

Figure 6.18: RPN text stack and function stack.

This function stack of the numerical block model is now directly executable. The result can be computed.

Also important is that the stack only has to be generated once via the parser, when reading the formula of the block model at simulation initialization time. During cyclic simulation computation for each step only new values for the input variables *u1* and *u2* will be handed over - the pointers to their memory and the pointers to functions of the computation stack remain unchanged. Thus by new evaluation of the same function stack, the result of *w* for the next step can be computed again directly.

By this means for such block models a functional evaluation is possible without sourcecode compilation. The stack representation works without brackets and parentheses. Also no operator precedence rules are necessary to be implemented. Besides this the stack representation in RPN opens further algebraic functionalities which are essential and will be treated in chapter 6.11.3. In addition this technique also is suitable for handling hierarchical modeling like depicted e.g. in figure 3.11.

6.11.2 Linearization of System State Equations

The previous explanations on system modeling and computation of the time response all were based on system description in state space which is applied in control engineering when analyzing system behavior over time. All system parameters which may change over time during system operation, can be aggregated to a state vector. The state vector moves along a curve - a trajectory - over time within this state space. For e.g. integration of an orbit position of a rocket or satellite over time, this is the appropriate numeric formulation.

However in automation and control engineering, often not the exact initial state change over time is of interest, but e.g. the oscillation frequency of a system as a function of the excitation function, the system damping, precision of control, eigenfrequencies etc. For a system composed of a mass and a spring e.g. for computation of the oscillation frequency it is of no matter whether at the start of movement, the mass was deflected up or down, and in which direction it starts moving. The analysis of a system w.r.t. its oscillation frequencies is not performed in state space but in the frequency domain, i.e. the Laplace transformed domain of the system function. The definition of the Laplace transformation is as follows:

$$F(s)=L\{f(t)\}=\int_0^\infty e^{st} f(t)\,dt \qquad s=\sigma+i\omega;\quad \sigma>0;\quad t\geq 0 \tag{6.57}$$

Without diving into the details of Laplace transformation, here shall be explained only that after this transformation the resulting function $F(s)$ allows for a very comfortable analysis of the system behavior in the frequency domain as well as the analysis of system response to impulse or frequency excitations. Precondition for the transformation from $f(t)$ into $F(s)$ however is linearity of $f(t)$.

In the descriptions concerning initial value problem integration for system simulation and during the descriptions on residue iterations for boundary problem solving, so far no constraints occurred concerning system equations linearity. For those computations in control engineering which require a Laplace transformation to the frequency domain, now this additional linearity constraint for the differential equation arises.

Linear systems are to be described by linear differential equations of n-th order, please refer to equation 6.11. In case a nonlinear equipment equation has to be converted to linear form, it is linearized around the operating point. Again applying the example of a mass-spring system this is e.g. the zero position. For a complex nonlinear equipment with complex time behavior, eventually this linearization has to be repeated for each time-step. The functions f and g in equation 6.5 are approximated by a Taylor series around the operating point (terminated after the first term). Thus results

$$\begin{aligned}\dot{\bar{y}} &= \bar{\bar{A}}\,\bar{y} + \bar{\bar{B}}\,\bar{u}\\ \bar{w} &= \bar{\bar{C}}\,\bar{y} + \bar{\bar{D}}\,\bar{u}\end{aligned} \qquad (6.58)$$

with the Jacobi matrices

$$\bar{\bar{A}} = \frac{\partial \bar{f}(\bar{y},\bar{u})}{\partial \bar{y}} = \begin{bmatrix} \partial f_1/\partial y_1 & \partial f_1/\partial y_2 & \cdots & \partial f_1/\partial y_n \\ \partial f_2/\partial y_1 & \partial f_2/\partial y_2 & \cdots & \partial f_2/\partial y_n \\ \vdots & \vdots & \ddots & \vdots \\ \partial f_n/\partial y_1 & \partial f_n/\partial y_2 & \cdots & \partial f_n/\partial y_n \end{bmatrix} \qquad (6.59)$$

and

$$\bar{\bar{B}} = \frac{\partial \bar{f}(\bar{y},\bar{u})}{\partial \bar{u}} = \begin{bmatrix} \partial f_1/\partial u_1 & \partial f_1/\partial u_2 & \cdots & \partial f_1/\partial u_m \\ \partial f_2/\partial u_1 & \partial f_2/\partial u_2 & \cdots & \partial f_2/\partial u_m \\ \vdots & \vdots & \ddots & \vdots \\ \partial f_n/\partial u_1 & \partial f_n/\partial u_2 & \cdots & \partial f_n/\partial u_m \end{bmatrix} \qquad (6.60)$$

The numerics of an equipment model thus are transformed to a "subsystem" with following flow diagram in state space[19]:

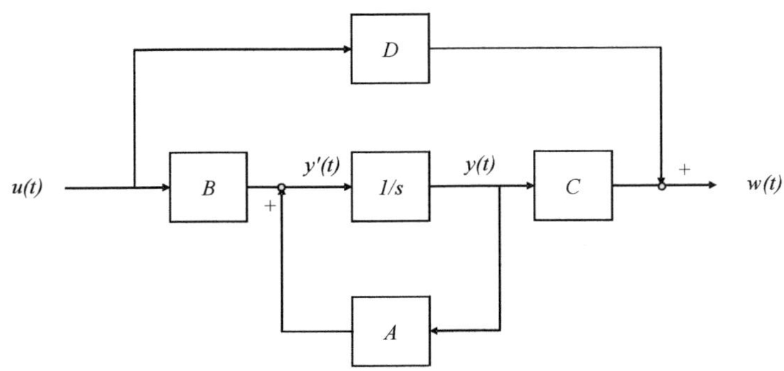

Figure 6.19: Equipment numerics as state space flow diagram.

So the linearization of system non linear functions or characteristic curves can be interpreted as their approximation by tangents at the operating point. The latter can be a stationary operating point, a position of rest or a quasi stationary operating point reached via initial value problem integration.

Similar as in the step for residue elimination in chapter 6.10 also here problems arise concerning the effort for computation of the Jacobi matrices. However for block

[19] 1/s is the representation of an integral term in the Laplace transformed frequency domain.

models which allow transforming their function into RPN stack notation, here an elegant algorithmic solution exists as explained in the following chapter.

6.11.3 Linearization by Algorithmic Differentiation

For computation of the Jacobi matrices $\bar{\bar{A}}$ and $\bar{\bar{B}}$ to linearize the functions f and g the function code for the partial derivatives to $\bar{x}$ and $\bar{u}$ is generated by means of an algorithm. This algorithm works similarly to the generation of an executable RPN function stack as shown in chapter 6.11.1. For its work it needs a table with the analytic derivatives of all elementary mathematical functions and the rules for how to differentiate a function. The mathematical function in the block model - e.g. of equation 6.56 - is treated by this algorithm as long as only elementary differential quotients exist in the form dE/dx whereby E is an elementary operand, i.e. a variable or a constant. For $E=x$ the derivative is equal to one. In case E is a constant the derivative is 0. For this algorithm again the RPN stack is the adequate output representation.

For example for the function of equation 6.56 the partial derivative to $u1$ - which is

$$\frac{\partial v}{\partial u1} = a \times \cos(a \times u1) - 0 \qquad (6.61)$$

can be computed by applying this technique. So from the RPN stack representation of the original function, the algorithm can generate the RPN stack of the partial derivative and from there the stack with pointers to variables and computation functions can be generated for numerical computation.

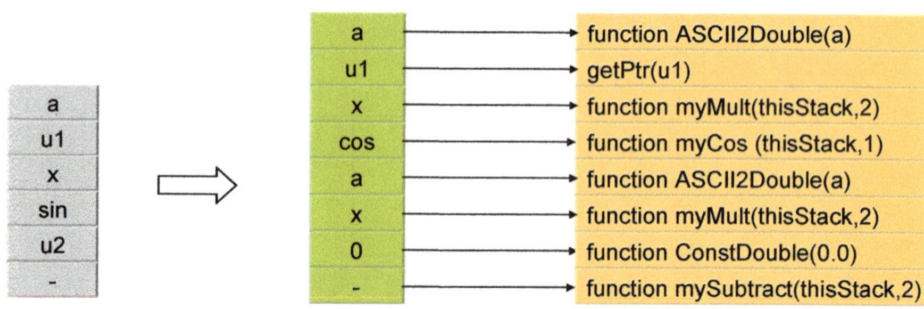

Figure 6.20: From RPN function stack to derivative function stack.

With this method system model equations can be entered directly into the simulator in state space representation, and they can be algorithmically differentiated. This avoids necessity for programming such block models in the classic sourcecode approach. For large systems however, the memory requirements for the stacks of all partial derivatives of all functions to all variables for all component models are significantly high. In addition the effort for preparing the differentiation rules for all elementary mathematic functions is a big challenge for development.

Furthermore this method needs the model's mathematics to be algorithmically formulated as in equation 6.1 and it does not allow for embedding components which work with characteristic diagram interpolations. Therefore this technique mainly is adopted for simulations in control and automation engineering for design of complex system controllers and their algorithms.

6.12 Semi-Implicit Methods for Stiff DEQ Systems

The considered explicit initial value problem solving methods so far are suited for most types of not too complex systems without problems. However one effect of the occurring differential equation systems thus far was neglected - the so-called "stiffness" of the DEQ system. The stiffness especially has to be considered for systems which for their simulation generate a large number of different types of equations and which include a significant amount of feedback in system topology. Even though there exists no single formal definition of stiffness this chapter is intended to impart a basic idea for readers from the engineering domain and shall explain how stiffness limits initial value problem solving by means of explicit methods.

Differential equation systems are stiff, if their overall solution is largely driven by one element only, while other solution elements have only negligible influence.

Example: Chemical Reactor Chain:

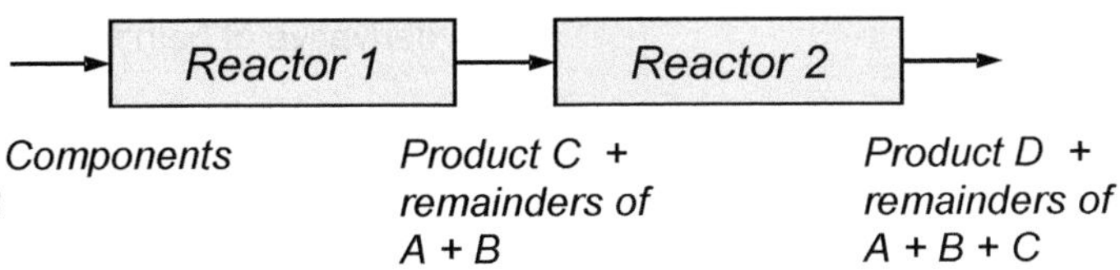

Figure 6.21: Example system - chemical reactor column.

This example shall provide an intuitive explanation. The following facts shall apply:

- The reaction A+B shall be slow, i.e. at the outflow of reactor R1 is still significant remainders of A and B can be found.
- The reaction C to D shall be almost instantaneous.

With this the amount of the product D is mainly driven by reaction 1. If now the differential equation system for the entire column is set up, with high probability this will result in a stiff DEQ system.

Example: Purely Numerical Demonstration

The following DEQ system shall be given for purely numerical demonstration:

$$\frac{du}{dt} = 998u + 1998v$$
$$\frac{dv}{dt} = -999u - 1999v \tag{6.62}$$

With the initial conditions:

$$u(0) = 1$$
$$v(0) = 0 \tag{6.63}$$

The exact solution (which can be determined analytically) reads:

$$u = 2e^{-t} - e^{-1000t}$$
$$v = -e^{-t} + e^{-1000t} \tag{6.64}$$

Both equations in the system above in principle, converge fast towards zero. However for values close to $t=0$ both equations are largely dominated by their second term. The function gradients are extreme in the range of $t=0.005$ to $t=0.015$ and in addition thereafter the sign of the second order derivative of both functions changes.

To integrate this system numerically, for reasons of stability (not for precision!), a time step size of $\Delta t < 1/1000$ has to be chosen. Otherwise the solution will diverge - as indicated in the second part of the figure, which is a zoom-in of $v(t)$ for values of t close to zero.

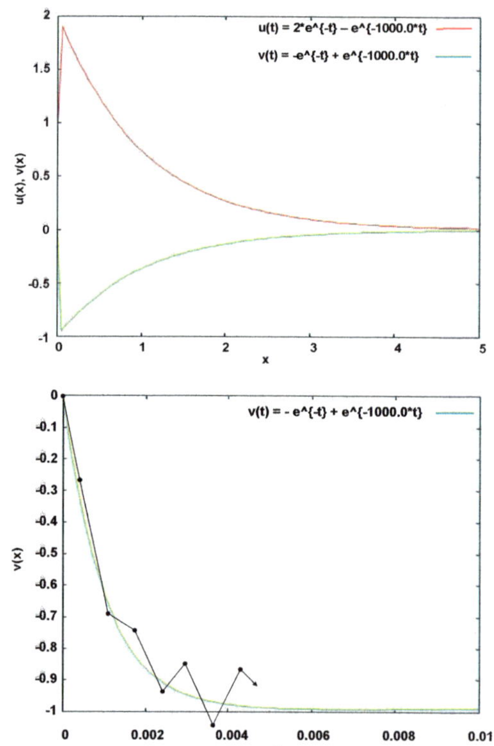

Figure 6.22: Stability problem in numeric initial value problem solving.

Implicit Algorithms

To bypass this problem so-called implicit algorithms can be applied. These are algorithms which compute the function derivatives, not only at known and historic function sample points, but also for points in the area up to $t+dt$. As example the implicit Euler method is discussed here. Since it is a method of Euler type, for the variables y at time $t+dt$ the equations apply:

$$\begin{aligned} &\overline{\overline{B}}\frac{d\overline{y}}{dt}=\overline{f}(\overline{y},\overline{x},t) \\ &\overline{y}_{k+1}=\ \overline{y}_k\ +\ dt\,\overline{f}(\overline{y}_{k+1},\overline{x},t) \end{aligned} \tag{6.65}$$

For the solution of the differential equation system next the time coordinate has to be discretized (time index k, step size Δt). Thus applies:

$$\overline{y}_{k+1}=\ \overline{y}_k\ +\ \Delta t\,\overline{f}(\overline{y}_{k+1}) \tag{6.66}$$

This equation however cannot be used before the term $\overline{f}(\overline{y}_{k+1})$ is formulated as expression depending exclusively on parameters at sample point k. Therefore around the old solution (=> "semi" implicit Euler method) an according Taylor series is developed which is terminated after the 2nd term:

$$\overline{f}(\overline{y}_{k+1})\ \approx\overline{f}(\overline{y_k})+\left.\frac{\partial\overline{f}}{\partial\overline{y}}\right|_k\,d\overline{y}\ =\ \overline{f}(\overline{y_k})+\left.\frac{\partial\overline{f}}{\partial\overline{y}}\right|_k(\overline{y}_{k+1}-\overline{y_k}) \tag{6.67}$$

Introducing this into the above equation leads to the linear equation system:

$$\overline{y}_{k+1}\ =\ \overline{y_k}+\Delta t\left[\overline{\overline{B}}-\Delta t\left.\frac{\partial\overline{f}}{\partial\overline{y}}\right|_k\right]^{-1}\cdot\overline{f}(\overline{y_k})\quad\text{mit}\quad \overline{\overline{B}}=\begin{bmatrix}1 & & & & \\ & \ddots & & & \\ & & 1 & & \\ & & & \ddots & \\ & & & & 1\end{bmatrix} \tag{6.68}$$

Thus all terms at time sample point $k+1$ are only dependent on parameter values at sample point k. If the step size h is made sufficiently small, it is adequate to evaluate once per integration step the term

$$\left[\overline{\overline{B}}-\Delta t\left.\frac{\partial\overline{f}}{\partial\overline{y}}\right|_k\right]^{-1}$$

which requires an inversion of this matrix. The structure of the matrix however mainly is dominated again by a Jacobi matrix $\partial\overline{f}/\partial\overline{y}$.

The inversion of the matrix, which is required for each time step integration of the equation system, induces significant numerical effort. To minimize this the equation system is somewhat rearranged, so that the result is:

$$\underbrace{\left(\frac{\bar{\bar{B}}}{\Delta t}-\frac{\partial \bar{f}}{\partial \bar{y}}\bigg|_k\right)}_{\bar{\bar{A}}}\bar{y}_{k+1}=\underbrace{\left(\frac{\bar{\bar{B}}}{\Delta t}-\frac{\partial \bar{f}}{\partial \bar{y}}\bigg|_k\right)\bar{y}_k+\bar{f}\left(\bar{y}_k\right)}_{\bar{b}} \tag{6.69}$$

For computation of a new system state thus a linear equation system has to be solved, which has the following structure:

$$\bar{\bar{A}}\bar{y}=\bar{b} \tag{6.70}$$

It is important here that in this equation system only partial derivatives $\partial \bar{f}/\partial \bar{y}$ at the sample point k occur, which can directly be determined. The effort for the solution of equation 6.69 depends on the structure of the matrix $\bar{\bar{A}}$. Since the matrix $\bar{\bar{B}}$ essentially is a diagonal matrix populated by ones, the structure of $\bar{\bar{A}}$ mainly is driven by the Jacobi matrix:

$$\frac{\partial \bar{f}}{\partial \bar{y}}=\begin{bmatrix} \partial f_1/\partial c_1 & \partial f_1/\partial c_2 & & & & & \\ \partial f_2/\partial c_1 & \partial f_2/\partial c_2 & \partial f_2/\partial c_3 & & & & \\ & \partial f_3/\partial c_2 & \partial f_3/\partial c_3 & \ddots & & & \\ & & \ddots & \ddots & \ddots & & \\ & & & \ddots & \ddots & \ddots & \\ & & & & \ddots & \ddots & \ddots \\ & & & & & \ddots & \ddots \end{bmatrix} \tag{6.71}$$

If the simulated system is equipped with feedback loops, the Jacobi matrix includes additional coupling terms. In the case of a fluid system with simple output to input feedback, the Jacobi matrix contains one additional coupling term at position *(1,L)*. Element 1 of the component thus is dependent on results from the last element, but not vice versa (assuming no diffusion against flow direction).

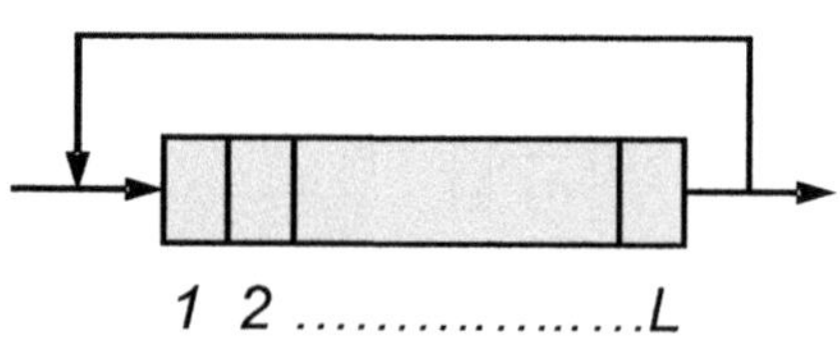

Figure 6.23: Fluid system with feedback.

$$\frac{\partial \overline{f}}{\partial \overline{y}} = \begin{bmatrix} \partial f_1/\partial c_1 & \partial f_1/\partial c_2 & & & & & & \partial f_1/\partial c_L \\ \partial f_2/\partial c_1 & \partial f_2/\partial c_2 & \partial f_2/\partial c_3 & & & & & \\ & \partial f_3/\partial c_2 & \partial f_3/\partial c_3 & \ddots & & & & \\ & & \ddots & \ddots & \ddots & & & \\ & & & \ddots & \ddots & \ddots & & \\ & & & & \ddots & \ddots & \ddots & \\ & & & & & \ddots & \ddots & \ddots \\ & & & & & & \ddots & \ddots \end{bmatrix} \qquad (6.72)$$

Integration over Time - Multi Parameter System

For the example with multiple parameters, the functions *f* always are dependent on all parameters, e.g. from compound concentrations, temperatures and pressures of all three elements to be considered, which are *L-1*, *L* end *L+1*. As result the Jacobi matrix becomes more complex. More precisely formulated, for the example of a one dimensionally discretized component without feedbacks, it becomes block-tri-diagonal:

$$\frac{\partial \overline{f}}{\partial \overline{y}} = \begin{bmatrix} \begin{bmatrix} \partial f_{c1}/\partial c_1 & \partial f_{c1}/\partial T_1 & \partial f_{c1}/\partial p_1 \\ \partial f_{T1}/\partial c_1 & \partial f_{T1}/\partial T_1 & \partial f_{T1}/\partial p_1 \\ \partial f_{p1}/\partial c_1 & \partial f_{p1}/\partial T_1 & \partial f_{p1}/\partial p_1 \end{bmatrix} & \begin{bmatrix} \partial f_{c1}/\partial c_2 & \partial f_{c1}/\partial T_2 & \partial f_{c1}/\partial p_2 \\ \partial f_{T1}/\partial c_2 & \partial f_{T1}/\partial T_2 & \partial f_{T1}/\partial p_2 \\ \partial f_{p1}/\partial c_2 & \partial f_{p1}/\partial T_2 & \partial f_{p1}/\partial p_2 \end{bmatrix} & [\,] & [\,] \\ \begin{bmatrix} \partial f_{c2}/\partial c_1 & \partial f_{c2}/\partial T_1 & \partial f_{c2}/\partial p_1 \\ \partial f_{T2}/\partial c_1 & \partial f_{T2}/\partial T_1 & \partial f_{T2}/\partial p_1 \\ \partial f_{p2}/\partial c_1 & \partial f_{p2}/\partial T_1 & \partial f_{p2}/\partial p_1 \end{bmatrix} & \begin{bmatrix} \partial f_{c2}/\partial c_2 & \partial f_{c2}/\partial T_2 & \partial f_{c2}/\partial p_2 \\ \partial f_{T2}/\partial c_2 & \partial f_{T2}/\partial T_2 & \partial f_{T2}/\partial p_2 \\ \partial f_{p2}/\partial c_2 & \partial f_{p2}/\partial T_2 & \partial f_{p2}/\partial p_2 \end{bmatrix} & [\cdots] & [\,] \\ & & \ddots & \\ [\,] & [\,] & [\cdots] & \begin{bmatrix} \partial f_{cL}/\partial c_L & \partial f_{cL}/\partial T_L & \partial f_{cL}/\partial p_L \\ \partial f_{TL}/\partial c_L & \partial f_{TL}/\partial T_L & \partial f_{TL}/\partial p_L \\ \partial f_{pL}/\partial c_L & \partial f_{pL}/\partial T_L & \partial f_{pL}/\partial p_L \end{bmatrix} \end{bmatrix}$$

(6.73)

Implicit Algorithms - Summary

In contrast to the explicit Euler method, the semi implicit method is sufficiently stable and also converges for positive derivatives. The selection of appropriately small step sizes to achieve desired numeric accuracy has to be considered separately. Absolutely essential for the integration of stiff DEQ systems and for the stability of the solution is a normalizing of the parameters to achieve that all integrated parameters lie in the same value range.

Here only the semi implicit Euler method was considered for application for stiff systems. In analogy also implicit, respectively semi implicit methods of Runge-Kutta type or Richardson extrapolation methods exist (the so-called Rosenbrok methods and also implicit Gragg-Bulirsch-Stoer methods). For these the reader is directed to the according specialist literature, e.g. [53], [54], [55], [56]. Furthermore - also in analogy to the explicit ones - implicit predictor-corrector methods exist. Also here the author points to the according literature such as e.g. [56], [57], [58].

Further reading and Internet pages concerning numeric methods are listed in the according subsection of this book's references annex.

7 Aspects of Real-time Simulation

CryoSat © Astrium

J. Eickhoff, *Simulating Spacecraft Systems*, Springer Aerospace Technology 1,
DOI 10.1007/978-3-642-01276-1_7, © Springer-Verlag Berlin Heidelberg 2009

7.1 Time Definitions

During simulation runs, telemetry data packets are generated by the on-board software and also simulator telemetry data packets from the simulator side, both being routed to the control console. Spacecraft telecommands and respectively simulator commands are sent from control console to spacecraft on-board software and respectively to the simulator core. All these packets are stamped with time signatures to be able to track the ongoing activities chronologically. However different types of time information have to be distinguished to avoid misinterpretations. Therefore the most important time definitions are covered again below.

Simulator Session Time: Actual wall clock time, called “Zulu-Time” or “Local Time”, mostly given in “Universal Time Code”, (UTC), in notations like year:day:hr:min:s:ms. This time is usually applied for time-stamping of TC/TM packets between spacecraft respectively simulator and the control console.

Simulation Runtime (SRT): Time in ms or s, which counts up as soon as the simulator is computing - i.e. is not in "suspend" mode. This sort of time counter is needed for the numerical solvers inside the simulator for integrating the time dependent state variables. This time often is called "SRT" or “tSim”.

Simulated Mission Time (SMT): This is the time of the simulated mission situation in space which in most cases lies in the future since at time of simulation the spacecraft is still in build phase and is not yet launched. SMT is mostly represented as "Modified Julian Date" time. It is relevant for the correct representation of celestial body positions relative to the spacecraft in the simulation. Furthermore it is relevant for navigation equipment models of on-board GPS receivers and the like. SMT is usually computed in UTC or "Global Positioning System", (GPS), Time.

On-board Time (OBT): OBT typically in the first place is a counter in the on-board software which starts counting up when the hardware or simulated on-board computer is booting. Usually the OBT can be set via telecommand to an absolute value, the SMT, since also the on-board software needs absolute time information. Also here in most cases UTC or GPS Time is used.

The time systems are not treated here in detail, however the reader should be aware that the baseline for the time systems are different astrometric or historic references. Thus dependent on the mission it must be decided in which time reference frame data has to be processed.

$$\text{Local Time} \neq \text{GPS Time} \neq \text{TAI (Atomic Time)} \neq \text{UTC} \neq \text{UT1} \neq \text{ET / TDT}$$

In chapter 5 of [38] detailed definitions of the cited time references are explained. UTC and GPS Time only differ by leap seconds introduced into UTC.

For the time clock strobes, eventually multiple sources exist on board the spacecraft. Firstly the on-board computer in any case is equipped with internal quartz clocks for each of its redundant sides. Furthermore high precision time reference generators might be on board in the form of e.g. GPS or Galileo receivers, and in case of GPS and Galileo system satellites themselves even extremely accurate atomic clocks are available on board the real spacecraft as time reference. In the area of functional system simulation it usually is not necessary to reflect such timing precisions and differences between the time scales. However a spacecraft simulator must provide the means for testing in flight resynchronization of the on-board software e.g. from OBC inner quartz clock to GPS receiver signal and vice versa.

A further problem is to guarantee for a system simulation, that the numeric system state integration over time propagates synchronized to the time progress "seen" by the on-board software in the OBC in the loop or in a simulated OBC. This topic of time synchronization is the topic of the next chapter.

7.2 Time Synchronization

A central problem in system simulation occurs especially for hybrid testbenches. It concerns the time synchronization between the on-board software in the real OBC and the spacecraft simulator, a topic which directly affects the transfer of signals between real OBC and simulated spacecraft equipment. However before treating the topic of time synchronization, the scheduling of signal data input / output of the on-board computer itself shall be sketched out, which of course is controlled by the on-board software.

A so-called "Polling Sequence Table", (PST), in the on-board software controls when which spacecraft equipment (and which of its interfaces) is commanded, respectively when result data are polled by the OBC. This sequence table in some spacecraft (e.g. for Satellites) can still be adapted in orbit. Such a PST can comprise a fixed simple sequence (such as in case of the Galileo-IOV satellites) where all OBC I/O's are handled with the same cycle frequency. Or, such a PST can comprise multiple OBSW cycles, because not all equipment or all interfaces of them are handled with the same frequency. The length of the PST is the time interval, corresponding to the lowest equipment control frequency. Below some examples are given, again at hand of satellite projects:

CryoSat 1:

- Overall PST length: 4s.
- PST comprises 4 AOCS control cycles.
- The AOCS control cycle frequency is 1Hz, which means that within 1 second all sensor data are polled by the OBSW and all control signals are submitted to the actuators.
- The control of all other units, such as payloads, power subsystem and thermal subsystem only occurs once within the PST, i.e. only every 4 seconds. The access to these units from OBC is distributed over the PST to average OBC CPU load and data bus loads.

GOCE:

- AOCS control cycle frequency 10Hz which induces much higher constraints for the simulator hardware in hybrid testbenches. PST length 1s.

Galileo-IOV:

- PST cycle frequency for all equipment (avionics, platform, payload) 5 Hz, i.e. PST length: 0.2 s.

Compliant to these equipment acquisition and control cycles of the OBSW, the system simulator must accept actuator data and provide result data / sensor data via the simulated OBC / equipment interfaces. Now also on board a real spacecraft for simplification of data handshake synchronization, the OBC distributes a cyclic clock strobe signal (in most cases 1 pulse per second and therefore called PPS signal) via dedicated pulse lines.

- For most units, especially those which are only distributing data when being explicitly polled by the OBSW, this is only a sync strobe to keep their internal electronic clocks aligned with the OBC.
- However there also exists equipment which actively send data to the OBC without prior acquisition and asynchronously to the PPS signal, e.g. certain types of startrackers.
- And thirdly there exists equipment which is polled by the OBSW in event driven mode, not aligned with the PPS signal, e.g. certain GPS receivers.

Figures 5.6 and 5.7 depict an example OBC core module. Here visible are the processor modules "PM A" and "PM B" and outbound from them to the left respectively, right border of the figure, the pulse lines for 1 Hz and 5 Hz strobes are drawn. Figure 7.1 sketches out the time signal generation in a simulated OBC of an SVF testbench.

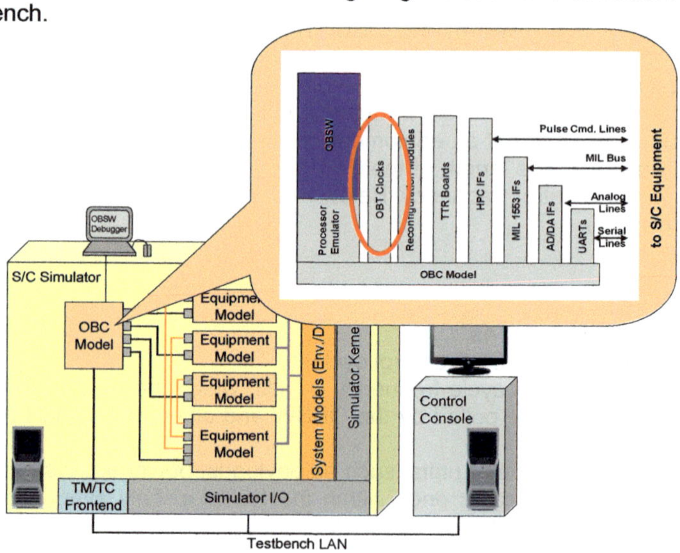

Figure 7.1: Time reference and pulse generation in simulated on-board computer.

The time synchronization in an STB with OBC as "Hardware in the Loop" can be achieved by two means: Firstly the PPS output connector of the real OBC can be coupled to an appropriate Simulator-Frontend card for routing the signal to the simulator and e.g. triggering according cycle scheduling events of the simulator kernel. The simulator kernel then has to react to these events or interrupts and has to synchronize its time propagation accordingly.

Another method uses the already cited possibility that modern OBCs also allow themselves to synchronize to external clocks, e.g. to the accurate GPS receivers or, in case of navigation satellites, themselves even to the extremely accurate atomic clocks aboard. In such a case a Simulator-Frontend card with internal signal generator can be applied which is cabled to distribute its strobe both to the OBC hardware as well as routing it to the system simulator. This case is depicted in the figure below.

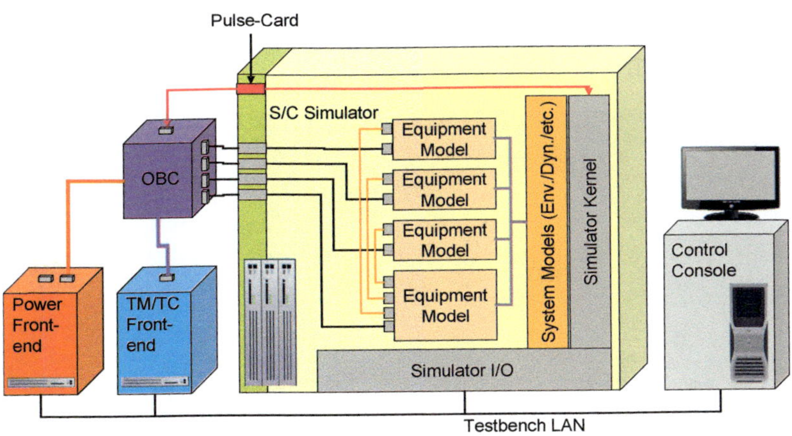

Figure 7.2: Time synchronization in an STB.

7.3 Modeling Time in a Simulator

For the correct interaction of on-board software data handling and simulated spacecraft equipment, it must be assured that if ever the OBSW acquires data from the simulated equipment, results of sufficiently actual state are available. The approach still can be simplified in purely simulation based setups like SVF configurations. Here the strategy could be to alternatively schedule the OBC model and the rest of all spacecraft models and to enable a data handover in between. This however no longer is adequate for hybrid testbenches since the simulation cannot be synchronized to the low level data bus protocols. The functionality here must represent the real in-orbit situation where the OBSW directly interacts with spacecraft equipment.

To principally guarantee data availability for the OBC (hardware or simulated) when acquiring data from simulated spacecraft equipment, the following approach is taken: At the OBC side of the simulated OBC's interconnection lines corresponding data buffers have to be established. These are implemented into the I/O-ports of the OBC model where the simulated lines are connected to. For a hybrid testbench these memory buffers are implemented inside the Simulator-Frontend card electronics and are fed from simulator side by the models via the mapped simulated lines. The data flow via hardware cabling from card to OBC as soon as the OBSW acquires the data from the card. The writing of OBC command data via I/O-port buffers of an OBC model or via Simulator-Frontend card buffers towards simulated equipment is just handled by analogy. Please also refer to figure 3.23 and the explanations on Simulator-Frontend cards in chapter 4.

By this means, time integration of simulated physical states inside the simulator can be performed by fixed time step numerical solvers whilst data acquisition from and commands to simulated equipment by the OBC are handled according to the settings in the OBSW's PST.

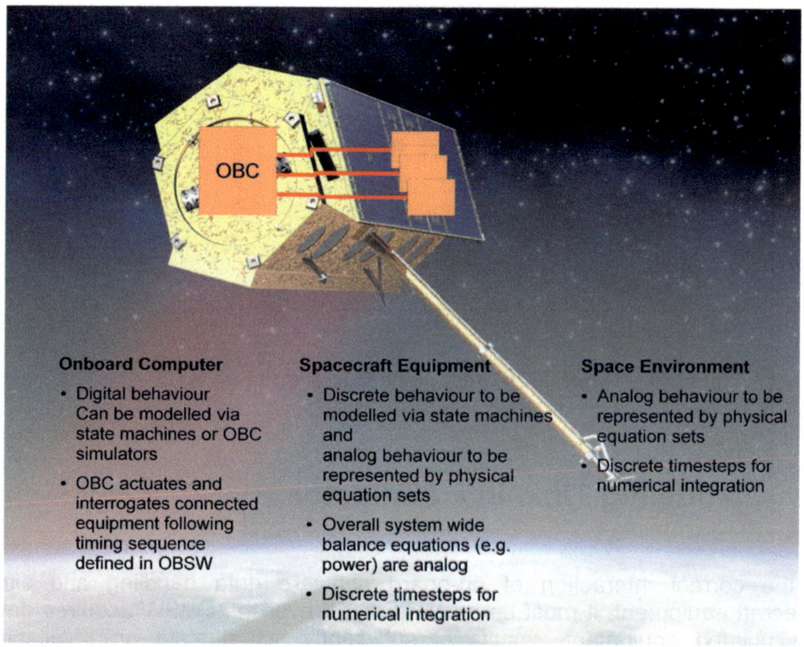

Figure 7.3: Elements of a spacecraft simulation and their characteristics.

A further effect concerning data exchange between OBC and simulated spacecraft has to be tackled. First of all it has to be avoided, that both OBC and simulator access the exchange memory of a data line at the same time. Since both sides are not aware of the counterpart's activities this only can be achieved by implementing all these cited data buffers as toggle buffers, where the card electronics gives access to either buffer from simulator side respectively outer OBC side, switching over after each access. Furthermore to assure stability of the overall control, the simulator has to compute with finer time step size than the one prescribed by the OBSW PST frequency. More precisely the simulator has to compute at least with two times the frequency of the PST to comply to the Nyquist-Shannon criterion of control stability. Anyway, already due to limits concerning integration step sizes of the numerical solvers, the simulator frequencies will be higher. Please also here refer to the explanations on numerics given in chapter 6. Therefore explicitly the difference shall be emphasized between

- the data exchange of OBC and the simulated spacecraft equipment, and
- the pre-computation and provision of result data in the toggle buffers with higher frequency for control stability reasons and reasons of numerical accuracy.

The integration step frequency of the simulator numerics thus has to be an integer multiple of the OBSW PST. This fine strobe for numerics control the simulator kernel usually generates internally as configurable multiple of the PPS signal it receives. In the following figure the PPS pulses are marked in orange color. The simulator computes with 10 Hz and the fine strobes are visualized in gray.

Figure 7.4: PPS strobe (in orange) and fine strobe (in gray).

What exactly is computed in each fine strobe cycle interval, depends on the integration method on system level as they have been discussed in chapter 6.

Simplified Integration on System Level

In the following figure and paragraphs an example is cited which represents an Euler method on system level. Before analyzing this scheduling table it first has to be explained that the slots marked with "OBC" only are slots where the simulator reads data from the OBC outgoing lines' exchange memory and makes them available for access by the spacecraft equipment models, as well as it writes back result data to the OBC input lines' exchange memory areas. Whether the OBC really feeds / reads the data at these points in time is of no interest in the first place since the data are intermediately buffered. As explained, usually the simulator internal frequency is significantly higher than the OBC's PST frequency.

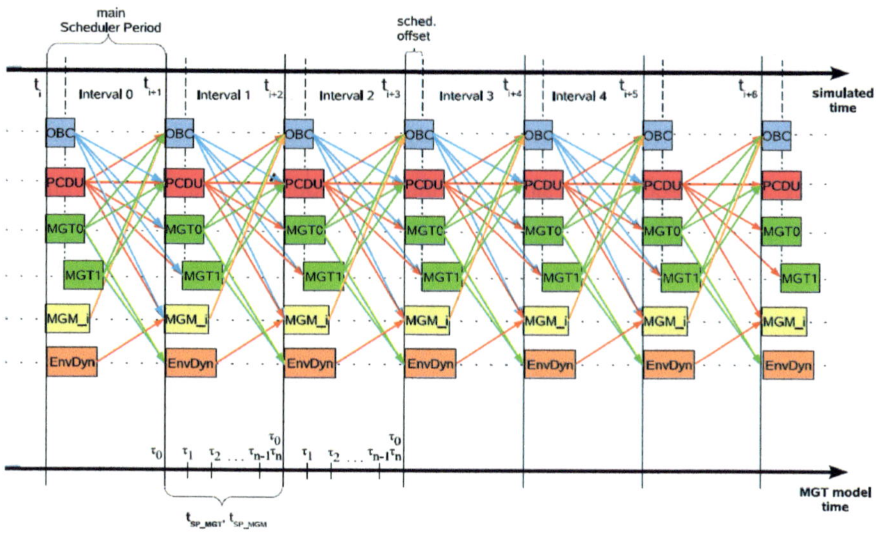

Figure 7.5: Scheduling table of a system simulation.

© IRS, Universität Stuttgart

Furthermore in this figure it is visible that within one time interval, model computations can be sorted according to a predefined sequence. Here magnetotorquer 1 (MGT1) computes after magnetotorquer 0 (MGT0). Within a time interval all models "see" the same time value.

What however is also visible here, is that the input data for the magnetotorquer are read at a point in time, the torquer computes its results in the next interval and in a further one these are then considered by the environment / dynamics model (EnvDyn). The computed changed spacecraft attitude then is made available to the sensor model (magnetometer here, MGM) and it takes again a further step until the sensor measurement data are available in the output lines' exchange memory to the OBC.

Although this process is permanently running in parallel so that in each interval results are made available for the OBC, they however are based on input which is several simulator cycles "outdated". In case the OBC has updated the actuator values in between the results are no longer entirely precise. This approach corresponds to an Euler method on system level as explained in chapter 6.8 example 2.

Exact Integration on System Level

For a highly precise integration on system level, the simulator has to provide the functionalities to

- read actuator data from OBC in each fine strobe interval,
- to then perform the entire computation cycle as depicted in the figures 6.10 and 6.11 and as it was summed up in figure 6.12., and finally
- at the end of the interval it has to provide output results to the OBC.

Specific system models which consume significant CPU resources but towards which the precision requirements are relatively moderate, optionally also can be computed only in each n-th simulator cycle. The same applies for simulator data logging to file and provision of simulator telemetry to the control console.

Example of staggered scheduling:

- Thermal model (computing in intervals 1,3,5...)
- Logging to disk (computing in intervals 2,4,6...)

When the simulator receives the start command for computation from console, it may initiate real computation not before receiving the main strobe from the external clock reference (PPS). When simulation stop is initiated the simulator still has to continue work until the end of the main cycle. This guarantees a restart of the computation synchronized to the OBSW PST after simulator suspend is canceled.

When simulations are applied on multicore computer platforms, multiple schedulers in the simulator kernel can be active in parallel. All however have to be synchronized to the same main strobe. This topic is further treated in the next chapter.

7.4 Real-time Parallel Processing

Especially in hybrid testbenches, the real-time response behavior of the simulator is a key issue. It has to assure in any case, in time availability of I/O-data for access by the OBC. In no circumstance (also not when in parallel data logging to disk or user commands build up) it may happen that the OBC misses computed data.

Therefore during design phase of such a hybrid testbench, the required numeric CPU performance and Simulator-Frontend I/O-performance have to be analyzed. For a spacecraft of the complexity of an average Earth observation satellite, the real-time simulation today can be implemented on a single simulator CPU core.

In case of spacecraft with extreme requirements (e.g. satellites or space probes with very advanced pointing requirements or transfer vehicles with docking functionalities) very short AOCS control cycle times result. Thus the need for application of multiple simulator cores might arise to suit all real-time constraints - in case the simulator kernel architecture supports such techniques.

Concerning parallelizing of numerical computations it first has to be explained that such functional simulation cannot be parallelized with the same techniques as applied e.g. for finite element problems in structural mechanics or as thermal analyses based on lumped parameter node models. The latter types of simulations can be well distributed on massively parallel supercomputers since for each element the same equation set has to be solved and residues are to be eliminated after each computation step. However in functional system simulation, each model class reflects physical behavior of different equipment and thus different equation types and equation sets.

Parallelization therefore has to be achieved by distributing the computations to multiple threads which then are allocated to different simulator CPU cores. Very simplified it could assume that one thread is generated per model instance and each CPU core computes one or more model threads. The treads are allocated to the CPUs according to their complexity so that all results are available in time for OBC data access also under worst case conditions.

Simplified Integration on System Level

For an Euler method based integration on system level (see chapter 6.8 example 2) one could assume for simplification that according to figure 7.5 data acquisition from OBC by the simulator and write back into the exchange memory is handled by one simulator thread. Computation of the demanding power control and distribution unit (PCDU) model is handled by a second thread. Magnetometers are handled by a third one, space environment and dynamics model by a fourth and magnetotorquers by a fifth one. Please therefore also refer to the following figure.

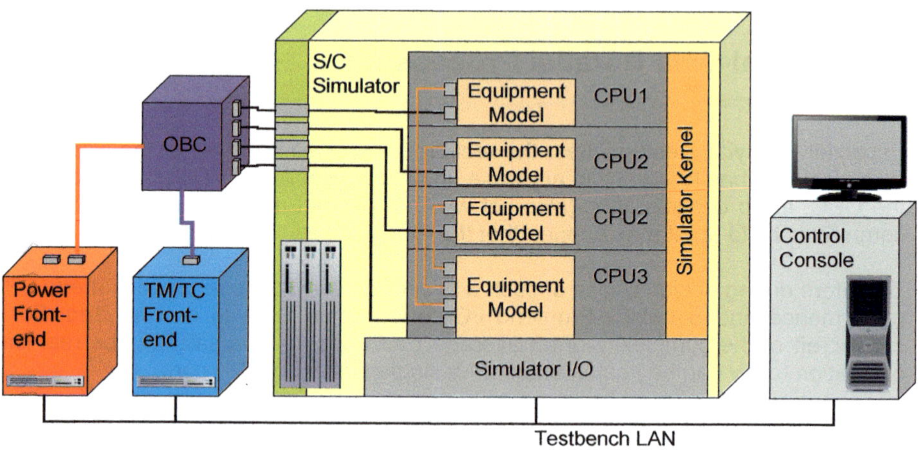

Figure 7.6: Computation of entire equipment models distributed to CPU cores.

Exact Integration on System Level

The situation again becomes significantly more complex as soon as exact central DEQ solving techniques are to be applied on system level.

The computation of the derivatives $\dot{\bar{y}}$ of its state variables can be performed individually by each model occurrence - see also equation 6.5 and figure 6.10. Thus these steps can be distributed to multiple threads and CPU cores.

The computation of the test step state variable results in e.g. a Runge-Kutta method k_1 to k_4 (see equation 6.42) as well as the computation of the final state vector $\bar{y}_{n+1}$ for the entire DEQ system however can only be performed by one central algorithm on one CPU node.

Combined initial value / boundary value problems even complicate the simulation parallelization further. The residue elimination for diverse boundary conditions in the system, as explained in chapter 6.10, is extremely complex w.r.t. parallelization. No fully generic means for this problem exists. For further particulars on these problems, for approaches and selected solutions the reader is referred to the according specialists literature in process engineering, e.g. to [51].

8 Object Oriented Architecture of Simulators and System Models

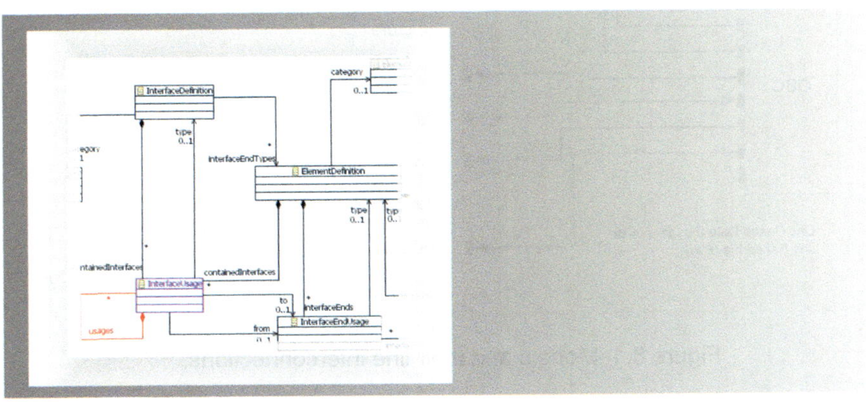

UML Class Diagram

J. Eickhoff, *Simulating Spacecraft Systems*, Springer Aerospace Technology 1,
DOI 10.1007/978-3-642-01276-1_8, © Springer-Verlag Berlin Heidelberg 2009

8.1 Objectives of Simulator Software Design

The following subsections describe the most common software engineering techniques used in spacecraft system simulators. As apparent from the explanation of the complexity of simulation-based test benches and their real-time requirements as well as from the section on underlying mathematics, the software engineering implementation of such systems becomes very challenging. In fact it is so complex, that the code for such a simulator cannot be implemented simply by simply being typed into an editor. A technology is needed for the implementation of a spacecraft simulator which supports the modeling of:

- Simulator kernel (reusable)
- Simulated spacecraft equipment models (partly new for each project)
- Simulated line connection types e.g. different data buses between OBC and other spacecraft equipment (partly new for each project)

From the requirements towards the simulator and the modeling of the real spacecraft's topology it can be derived that the simulation preferably has to be implemented applying object oriented software technologies.

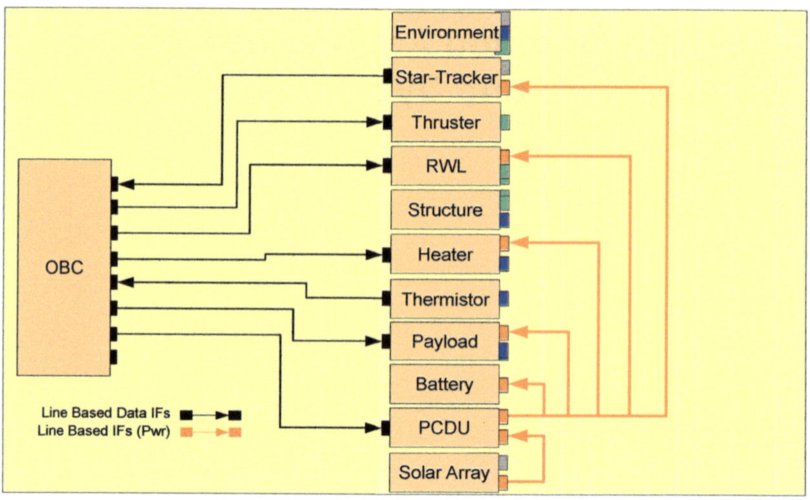

Figure 8.1: Models and their line interconnections.

Object-oriented programming languages enable to implement a software class for each type of functional element - e.g. the equipment type of a star tracker. All numerical variables are determined therein - still without values respectively, only with initialization values - and in addition all computation functions (so-called methods) are defined, including input and output functions to other classes respectively to the simulator kernel or the solver. At software initialization the number of required instances of each such such a class is determined – in the example above the number of simulated star trackers. By this technique the topology of the real system can be similarly reproduced in the source code. The maintenance of the code is also largely simplified.

The same concept can be applied for the classification and instantiation (creation of model instances in computer memory and setting of according parameter value defaults) of line types in the system model: For example there might be a class for a MIL-STD-1553B data bus and a class for analog lines. At start of the simulator an instance of the line class is created for each real analog connection in the system and respectively for each data bus connection. The equipment component models are connected - in this way reflecting the real system's topology.

Furthermore, software classes can be organized into a hierarchy using object oriented modeling. Subclasses can inherit parameter and function properties from superclasses. This implies certain functions have only to be coded once for definition of a superclass. An example for such a generic function required for all types of model subclasses e.g. can be the computation control of the equipment models in a time-step loop or iteration-step sequence. All the different equipment subclasses automatically inherit these functions through the object-oriented approach.

Furthermore, if cross-couplings are part of the real system design, they have to be considered when models and lines are instantiated - again to properly reflect the real system's topology. The following figure depicts a double cross-coupling for line connections. Each of the I/O-board instances of the on-board computer can access both the nominal and the redundant star tracker electronics via its data bus.

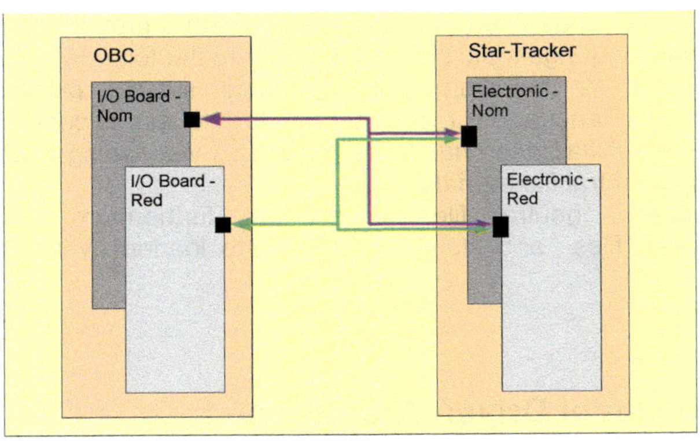

Figure 8.2: Cross-coupling of line connections.

For coding methodology and concerning the to be selected programming language itself, the following requirements can be derived:

- The software code has to be as modular as possible.
- The simulator kernel has to be reusable from project to project.
- Equipment and line models have to be reusable over multiple projects - as far

as the same equipment features are to be simulated.

- The system topology has to be modeled as close as possible to the real system's structure.
- The instantiation (generation of the instances) of equipment and line connection models has to be performed at initialization time.
- Thereafter equipment models have to be connected to each other and to the solver, especially for exact integration methods on system level.
- Specific components, which run as "Hardware in the Loop" later, may need specific interfaces to be programmed.
- Furthermore, the generation of program code is necessary for
  - memory allocation and parameter initialization functions of all components (to be performed during simulator initialization phase),
  - for memory deallocation (to be performed during simulator shutdown),
  - for calculation phase numerics and for
  - administrative functions such as data logging to file, simulator telemetry submission to control console and external model stimulation.
- Multitasking / multithreading functions might be necessary for frontend-card drivers, multiprocessing support, etc.
- In addition a frequently requested function for pure simulation testbenches (SVF type) is to provide the possibility for saving entire simulation operational states - so-called statesets - in a file and to later be able to reload them later as defined simulated spacecraft state.

The following requirements concerning the implementation and programming technology result from these explanations:

- For object-oriented software design preferably a standardized design language has to be used - not to be mixed with a programming language. For the design language topic please also refer to the following subsection.
- The use of a hardware and operating system platform independent programming language is recommended for the source code.
- The system initialization has to be performed via loading of configuration files defined in neutral file formats.
- Standardized, neutral file formats are furthermore recommended for initialization files and statesets, which are loaded by the simulator during startup.

8.2 The Model Driven Architecture

Concerning the topic of object-oriented modeling of a spacecraft system firstly a short introduction to the so called “Model Driven Architecture” (MDA) concept shall be provided. Thereafter subsequently the simulator specific implementation techniques are described in more detail.

The “Model Driven Architecture” defines a dedicated approach for model driven software development. It is based on a precise distinction between software functionality and its implementation technology. In the 90's the approach was pursued to generate software from dedicated "Computer Aided Software Engineering" (CASE) tools in an almost fully automated manner. Tools for the "System Analysis and Design Technique" (SADT)[20] and first object-oriented software design tools for Booch's and Jacobson's design methods followed this approach. Due to the partly resulting strong bindings to dedicated software design tools and the tool vendors these approaches only showed limited success.

MDA therefore follows a slightly different approach although it is not entirely new. Also in the MDA approach the source code is generated from a software design notation in which for example the classes for the equipment models and their dependencies and interactions are described. However only a practicable part of the source code is generated automatically. The generated source code then has to be completed by all non auto-generated functions and has to be compiled for the target operating system / hardware platform. It is important to bear in mind that:

- In code generation following the MDA approach, function code has to be implemented manually in all those cases where it is easier to program the algorithm directly than to implement it in the formal design notation and to auto-generate code from there. Examples for such cases are algorithms implementing physical equipment behavior in an equipment model or algorithms in a numerical solver.
- Furthermore the distinction between the software design layer, the code layer and the runtime platform in MDA is kept strict and clean.

Thus MDA separates the modeling and code implementation into a multi layer model, ranging from the more abstract level to the more specific. To be more precise the following layers exist:

- A "Computation Independent Model" (CIM) for the textual description of software functionality, e.g. for requirements specification documents describing the functionality of a spacecraft equipment model.
- A "Platform Independent Model" (PIM) layer on which the software architecture is defined by means of a standardized functional and topological notation and by means of a software design tool. By this means not only the platform independence but also the independence from a specific implementation language shall be provided.
- A "Platform Specific Model" (PSM) provides a code model for a specific target platform – for example C++. The conversation from the PIM to the PSM typically is done by means of dedicated code generators. If the code generated from design is not functionally complete, the developer has to fill in manually the corresponding parts (e.g. numerics) as described above.
- On lowest level finally the executable is placed in form of compiled code, which is linked with operating system dependent libraries.

In the process of conversion from one layer down to the next, the more abstract representation of the upper layer is converted to a more detailed and less abstract

[20] SADT also is called IDEF0. It is one design language of the ICAM family developed for the US Air Force at the beginning of the 1980s.

representation with more details. This process repeats and finally leads to an executable binary code. For each transformation process to a next lower layer, a code or model generator has to be implemented by experienced experts. For the step from PSM to executable code these tools are well known as "compiler" and "linker".

The MDA is an implementation strategy developed by the “Object Management Group” (OMG), a consortium of companies engaged in development of object-oriented tools and languages. The OMG develops open specifications to improve interoperability and portability of software systems. The OMG also develops the standard for the “Unified Modeling Language” (UML) which today is the most common software description language and also is the standard for the MDA PIM layer. UML will be described in more detail in chapter 8.4.

The following example (please also refer to figure 8.3) illustrates such a stepwise refinement from simulator software design to runtime code. The example - for code of spacecraft equipment models - is similar to the simulator development concept applied at Astrium GmbH - Satellites:

- On the CIM layer, the simulator and equipment model user and software requirements are defined in textual form in a requirements management tool.

- The simulator software design on PIM layer is based on UML.

- From UML it is possible to auto-generate ANSI C++ code directly for the PSM layer:
  - ◇ The C++ code can be manually instrumented by the missing numerics functions.
  - ◇ Or auto-generated code originating from Simulink models can be included. The conversion from Simulink to C/C++ in this case is done using the MathWorks Real-Time Workshop.
  - ◇ Alternatively auto-generated code originating from Stateflow state-machines can be included. The conversion from Stateflow also is done with the MathWorks Real-Time Workshop.

- On the lowest code and runtime layer the integrated development environment used is relatively independent of the operating system. The runtime platform again is quite variable:
  - ◇ For example PC / Linux or PC / Windows are depicted as operating platforms in figure 8.3 for the non real-time capable testbenches and
  - ◇ RT-Linux or VxWorks for the STB / EFM test beds.
- The configuration of the simulation with initialization and characterization data is done by means of files based on the standardized XML format (The topic of XML is covered in chapter 8.5).

The same approach of course can be applied to the design and coding of simulator kernel modules.

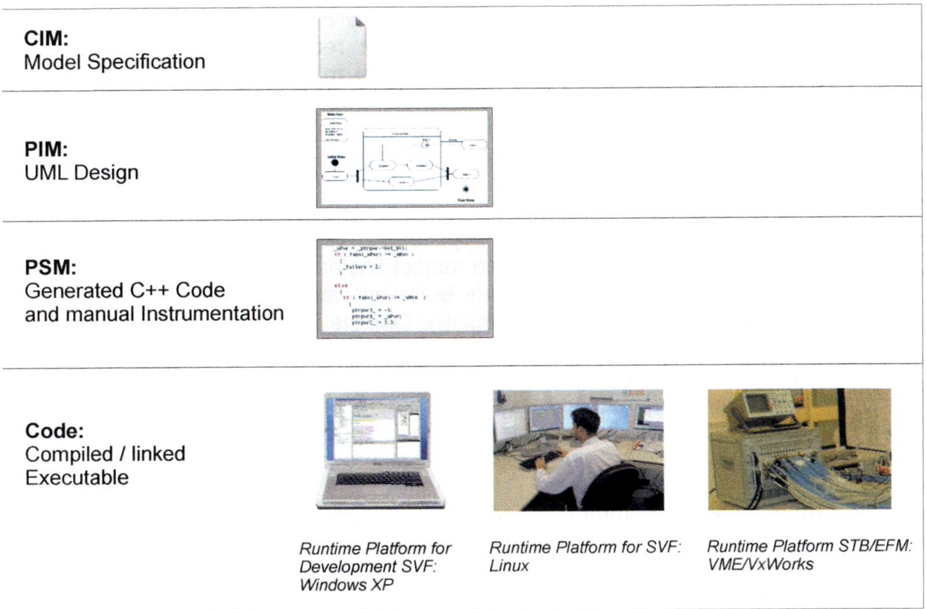

Figure 8.3: MDA concept example for simulator development.

Further reading concerning the "Model Driven Architecture" (MDA) is listed in the according subsection of this book's references annex.

8.3 Implementation Technologies - Programming Languages

As already mentioned, object oriented programming languages like C++, C# or Java are the best choice for system simulators. However it has to be considered that the simulation has to be platform independent according to the MDA paradigm which means first and foremost independent from the operating system - which practically declassifies C# as an implementation language. Furthermore, it has to be considered for hybrid testbenches that the language has to be compiled for a real-time operating system which requires availability for hardware and real-time operating system optimized compilers. As a result, in most cases C++ is selected as a technical platform. Diverse large spacecraft simulation systems are implemented in C++, for example MDVE, SimTG and others, cf. [16], [17], [19].

In latest developments Java is increasingly used as an implementation language. This language is completely independent from the operating system. Such a Java based simulation application, for academic tutorial applications, is *OpenSimKit* [23]. Since in Java a virtual machine is placed between the compiled program code and the computer operating system, Java applications implemented on one operating system can be run directly on another without recompilation. Furthermore, the language Java itself has a higher abstraction layer than C++. The programmer only has to "think in objects". He or she no longer has to use pointers to objects or functions - the use of which which often leads to bugs or memory leaks - nor has to handle references of parameters respectively of objects.

However, the virtual machine between compiled software and operating system is a significant disadvantage for real-time applications like hybrid testbenches. Such a virtual machine can generate required object instances at runtime (which occur eventually by only copying data packets between class instances or by generating simulator TM packages for the Core EGSE). Objects no longer needed - and which thus no longer are referenced - a so-called "garbage collector" deletes from memory from time to time. This garbage collection process starts according to the amount of memory to be cleaned, requires extra computing power and thus this process is running aside the simulation computation. So it can completely disturb the real-time performance of such applications even on a real-time operating system.

For this reason various commercial manufacturers and also open source teams lately developed real-time capable Java implementations which already meet sophisticated performance requirements and which are even partly certified for military applications. Please refer to [69] and [70] for commercial systems and to [71] for an open source implementation. The basic principle of all these Java real-time applications is to equip the Java virtual machine with a garbage collector which has a predictable influence on the required CPU load and where the initiation of the garbage collection process and its hibernation are exactly controllable. By this means the garbage collector can never affect processes with higher priority. The average computing performance of such Java real-time systems on average is approximately 10 to 20% lower than standard Java but the implementations comply to hard real-time requirements. The disturbances caused by garbage collection are below 100 microseconds for commercial systems using standard x86 or PowerPC processors. This implies these systems are definitely suitable today for embedded applications.

Further reading on functionalities and syntax of both C++ and Java as well as its real-time variants are listed in the according subsection of this book's references annex.

8.4 Implementation Technologies - The Unified Modeling Language (UML)

The “Unified Modeling Language” has already been cited as a powerful and appropriate tool for graphical design of software. UML is a graphical notation for structure and functional

- specification,
- visualization,

- definition and
- documentation

of object oriented software systems. UML visualizes the system design by means of graphical diagrams, however it does not apply any rules for the process of the design development, the level of modeling detail nor for software verification and testing. Nevertheless the graphical notation only is one aspect of the modeling "language". UML basically defines a set of terms and relations which are used to specify the models. The UML diagrams only give a graphical view on cut-outs of the model. Parts not "drawn" in diagrams are not fixed in the software design. The UML standard further defines a data format which can be used to transfer the models and diagrams between the UML tools of different vendors and it formalizes design optionally down to a level where automated code generation from the graphical definitions is possible.

In respect to the graphical notation the current UML V2 standard defines six different kinds of structural diagrams, the:

- Class diagram
- Composite structure diagram
- Component diagram
- Deployment diagram
- Object diagram
- Package diagram

And in addition to these UML 2.0 comprises seven behavior diagrams, namely the:

- Activity diagram
- Use case diagram
- State machine diagram
- The interaction diagrams which are the:
  - ◇ Sequence diagram
  - ◇ Communication diagram
  - ◇ Interaction overview diagram
  - ◇ Timing diagram

UML CASE tools provide the functionality to auto-generate source code from this UML PIM notation for different kinds of target languages (C++, Java, C#,...) on PSM level. Code generation is not possible from all types of diagrams because some are too abstract and imprecise. In the following paragraphs the most important UML diagram types are explained. These real world examples originate from the "Stuttgart Small Satellite Program".

Class diagrams: Class diagrams illustrate the software classes, their aggregations and inheritance architecture.

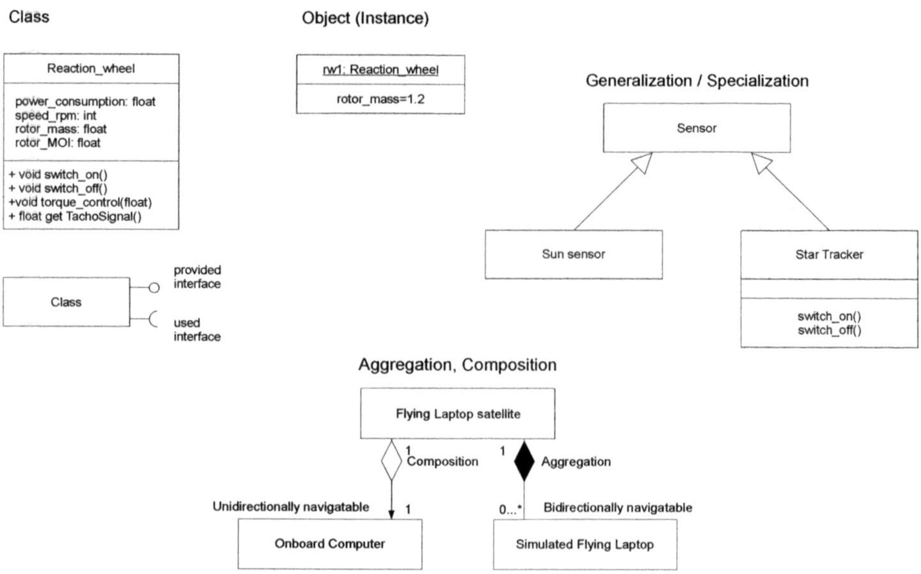

Figure 8.4: UML class diagram.

Composite structure diagrams: A Composite structure diagram is a type of static structure diagram in UML, that describes the internal structure of a class and the collaborations that this structure makes possible.

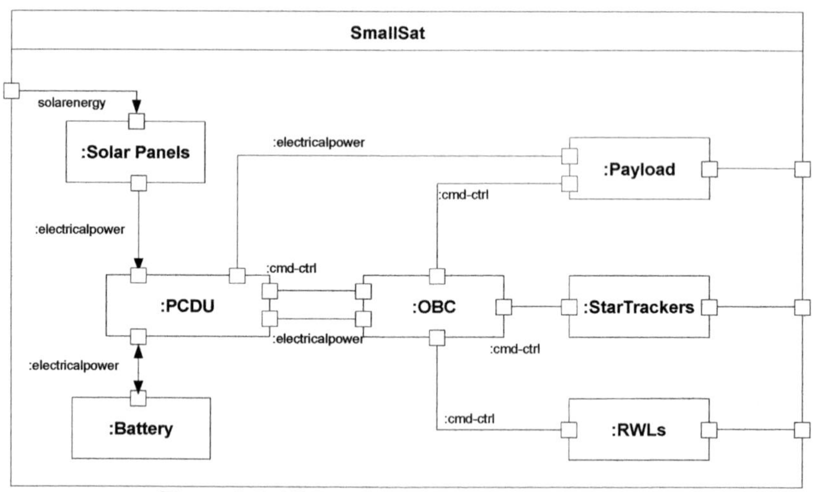

Figure 8.5: UML composite structure diagram.

Component diagrams: Component diagrams illustrate the dependencies between the major software system components on the level of modules like card drivers data bases and user interfaces.

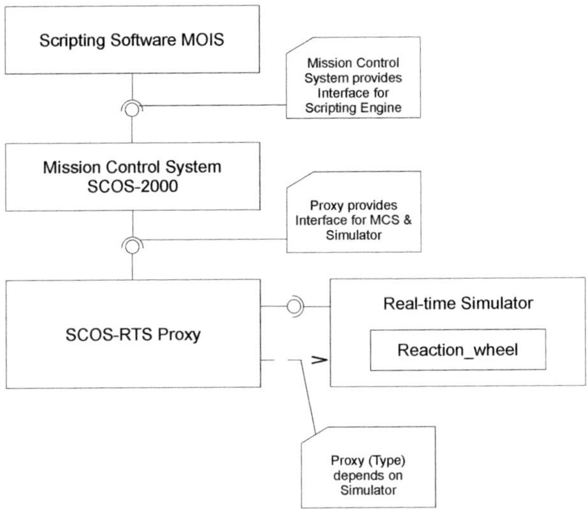

Figure 8.6: UML component diagram.

Deployment diagrams: Deployment diagrams illustrate the deployment of the infrastructure modules of the system to different computers.

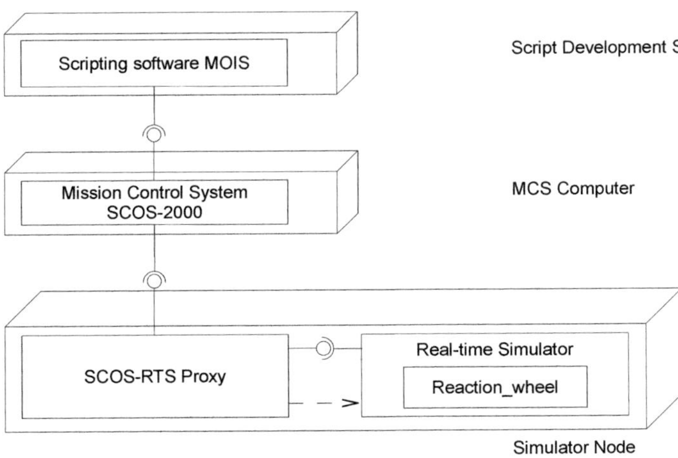

Figure 8.7: UML deployment diagram.

Object diagrams: Object diagrams illustrate the statical instance, role and template architecture[21] of the system.

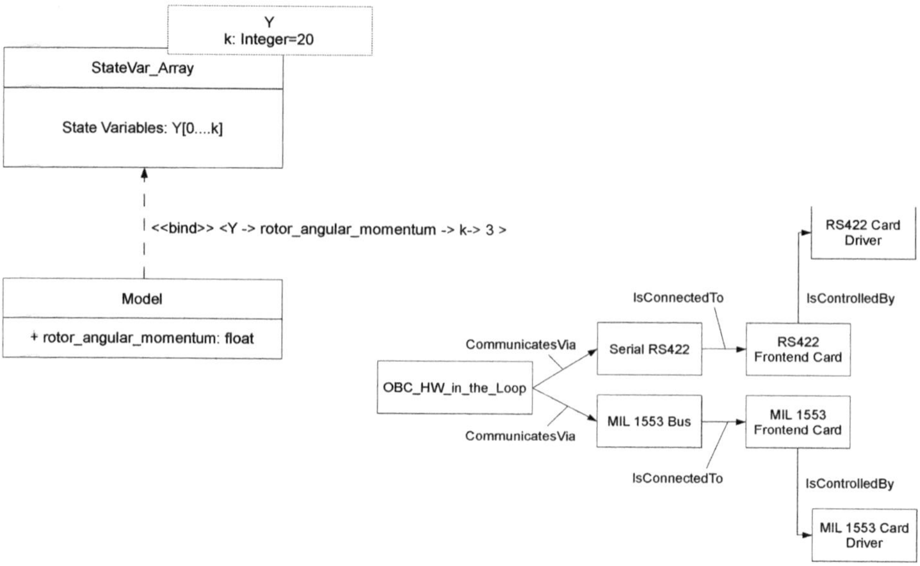

Figure 8.8: Template and associations in UML object diagrams.

Package diagrams: Package diagrams illustrate the dependencies between system components using a structuring based on software application criteria – for example structuring of the software system due to applied operating system or target hardware.

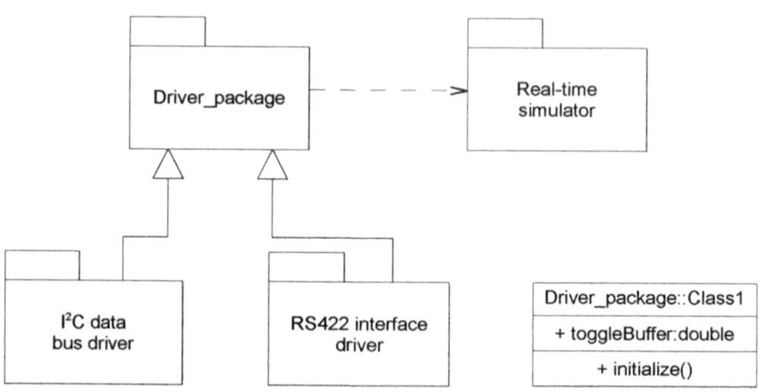

Figure 8.9: UML package diagram.

[21] Templates describe class structures and functions that allow to operate with generic types (float, double etc) without being recoded for each data type.

Activity diagrams: Activity diagrams are the first type of behavior diagrams and illustrate the process flow in the software as well the system behavior and its control structures.

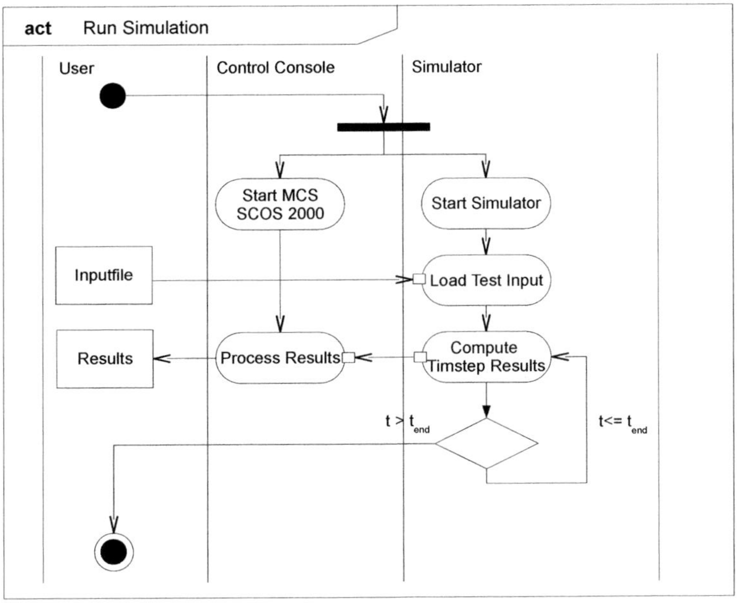

Figure 8.10: UML activity diagram.

Use case diagrams: Use case diagrams illustrate the interaction between actors and the software in application cases of the system.

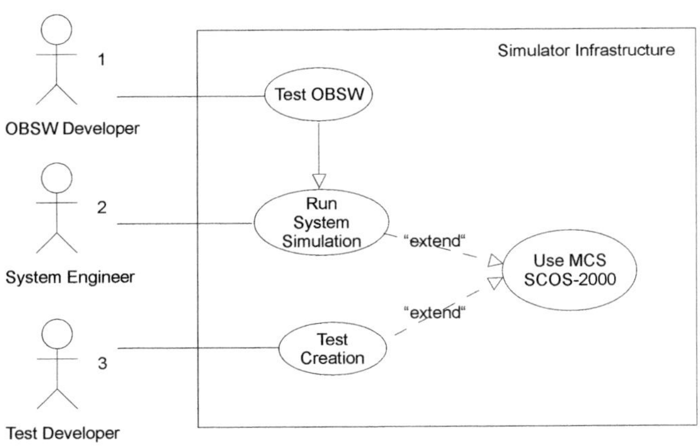

Figure 8.11: UML use case diagram.

State machine diagrams: State machine diagrams illustrate the states of system components (e.g. simulated spacecraft or its equipment) and the transitions between those states for the modeling of state machines (time discrete reactions of the component on external or internal stimuli).

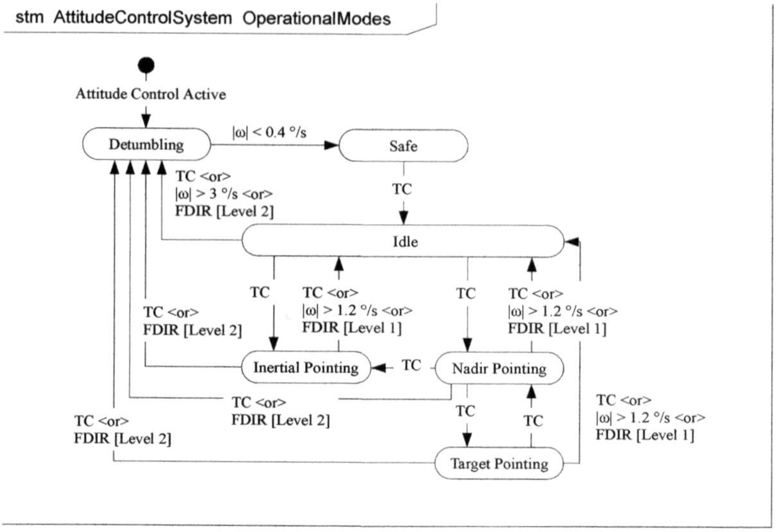

Figure 8.12: UML state machine diagram.

The example of figure 8.12 shows the a state machine modeling a fictitious satellite's operational modes as they are defined for on-board software design.

Sequence diagrams: Sequence diagrams are the first type of interaction diagrams and illustrate the interaction of system components in a chronological chain - if needed including interaction conditions.

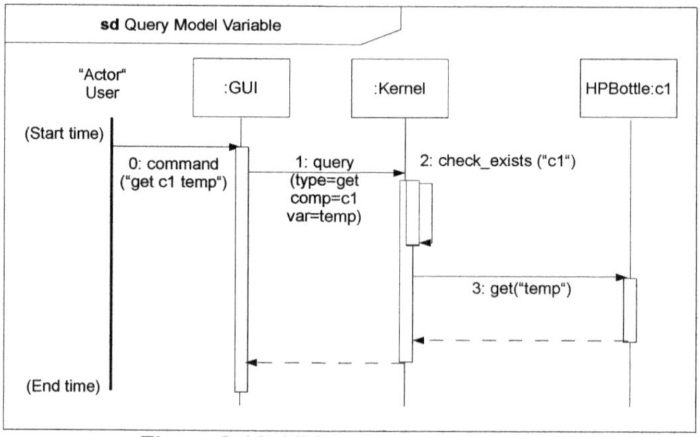

Figure 8.13: UML sequence diagram.

Communication diagrams: Communication diagrams illustrate the elementary connections between functionalities in the system and are typically used in early conceptual phases of a software development.

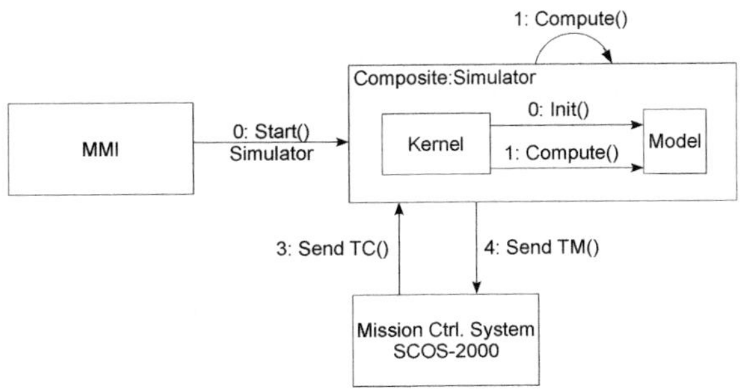

Figure 8.14: UML communication diagram.

Timing diagrams: Timing diagrams illustrate the behavior of the system in the time domain and especially depict the dynamical aspects of the system behavior. They allow a more precise, quantitative specification of the timing behavior than the other UML diagrams. Thus they are well suited for the detailed design of real-time systems.

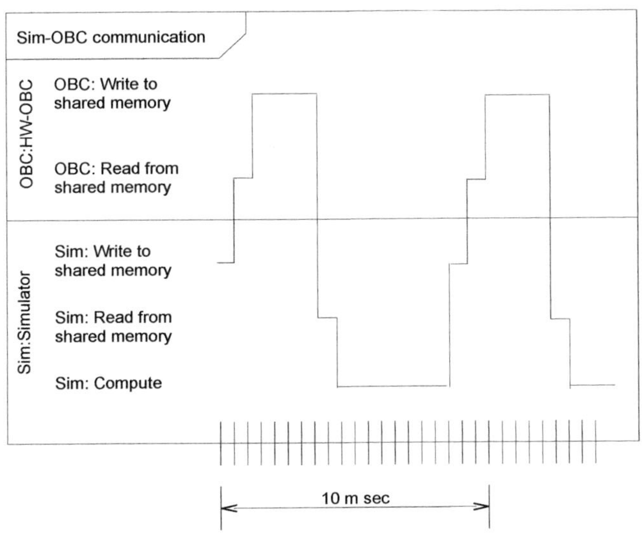

Figure 8.15: UML timing diagram.

After this quick tour on UML diagram types the specific options that are offered by UML V2 with respect to the modeling of state machines shall be outlined in a bit more detail here. This is of specific relevance since most equipment models in a spacecraft simulator comprise a time discrete part – especially those representing intelligent on-board equipment with a multitude of operational modes. This functional non-analog part is preferably represented by a state machine – please refer to the breakdown of an equipment model in figure 4.15:

- UML reflects state machines as an object-oriented variant of the Harel type.
- These are used to model the event driven behavior of corresponding units.
- Since UML 2.0 state machine diagrams are no longer limited to representing a single software class only.
- The state machine diagrams cover the whole life cycle of the class(es). This includes the construction and destruction of the corresponding software objects.
- The state machine diagrams allow a hierarchical modeling of states. This enables the modeling of systems with a complex behavior.
- Numerical algorithms can be included into the states as well as into entry and exit functions.
- State transitions can be defined to be event driven or time dependent.

State machine diagrams allow one to define:

- Checking whether or not a command execution is permitted in the current operational mode
- Checking whether or not a transition to a specific mode is permitted in the current operational mode
- Blocking of incoming commands as long as a transition is ongoing in the state machine
- Mode specific parameters of the model (characterization parameters)
- Transition durations for all possible mode transitions.
- Transition modes

In the following paragraphs now the topic of code generation from UML notation shall be treated in more detail since this is one of the major strengths of UML.

8.4.1 Code Generation from UML

The possibility to generate code from UML (if necessary for different programming languages) depends on the abilities of the applied UML tool. As long as no pure drawing tools are used, code usually can be generated at least from the following diagram types:

- All diagrams describing the static system structure (class diagrams etc.)
- State machine diagrams
- Sequence diagrams (possible since UML standard V2.0)

In order to be able to support different UML details in the particular diagrams, the code generator eventually needs to be adapted. An important feature of UML namely is its adaptability to user requirements. UML itself allows notation adaptations for user purposes, the so-called "profiling". The most intuitive and more widely means of UML profiling is the application of so-called "stereotypes" for class templates – please refer to figure 8.16. At hand of the stereotype name the code generator identifies that besides the variables and functions defined in the class diagram it has to generate certain additional functions or parameter definitions for the class. This is performed automatically during code generation by the use of template definitions in the code generator configuration and according entries in the code generator script. Function extensions by stereotypes e.g. avoid necessity for complex inheritance hierarchies.

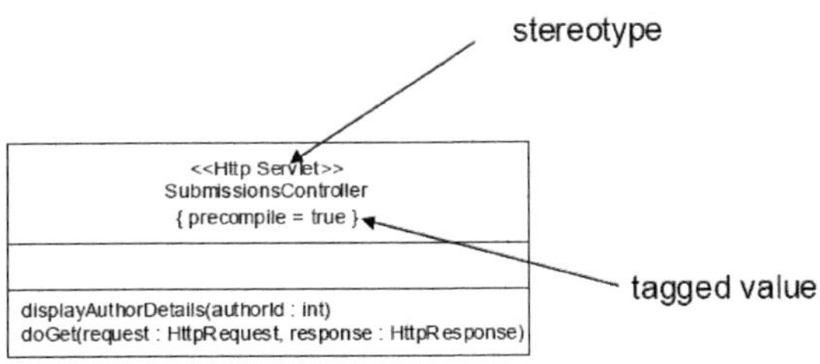

Figure 8.16: UML class definition by stereotype. Example: OMG

Another means of UML profiling is the use of so-called "constraints". Their definitions are specified in the "Object Constraint Language" (OCL) which originally was developed by IBM.

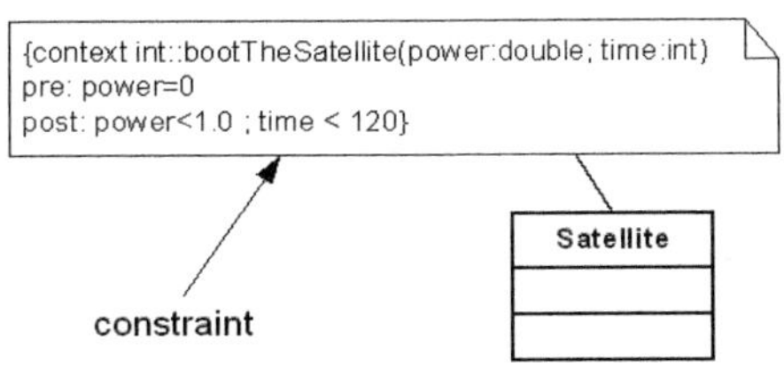

Figure 8.17: UML diagram with constraint.

Complete notation changes for special purposes are principally also allowed. However, the parsing of the diagrams and the code generator's conversion definition then have to be specified by the user, as these changes affect the so-called "metamodel" which is the "interpretation convention" behind the graphical notation.

How is this flexibility and adaptability achieved? For the answer first figure 8.18 shall be considered which depicts the 4 levels of UML modeling. The lowest level M0 represents the generated output which might be C++ code for a simulator or table structures for a data base etc. The information which already was cited in the presented UML diagram types can principally be found in the M1 level which stands for the UML based modeling of software structure and behavior. In figure 8.18 a class and an instance in accordance to the propulsion system of figure 1.12 are depicted in the M1 level (cf. [23]).

The UML tool and in particular its code generator need to "know" what a method, a class, an instance and an attribute are. The definition of this "meta information" is laid down in the metamodel level M2. Potential stereotype definitions and their functionalities are also defined here. Via this metamodel the UML diagram semantics and also the additions / changes defined by the user are defined. Finally as topmost layer there exists the M3 level, the "meta object facility" (MOF). This is the internal UML tool database layer, where the software components designed with UML and the metamodel definition (including potential user adaptations) are stored in a standardized way prescribed by the OMG's UML standard. During data exchange between two UML tools, the exchange takes place between these meta object facilities.

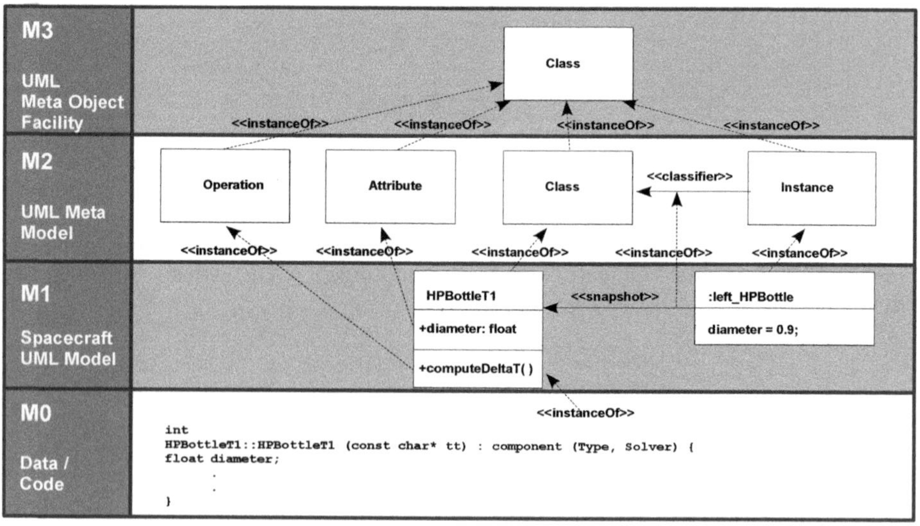

Figure 8.18: UML tool modeling levels according to OMG standard.

From these structured definitions a code generator now can consecutively parse the M1 layer elements and interpret their subparts (methods, instances, parameters etc.) by means of the meta model information - e.g. the characteristics of a propulsion system from figure 1.12 respectively [23]. The program code (e.g. in C++ or Java) can be created during this parsing and interpretation process successively by means of a modular script.

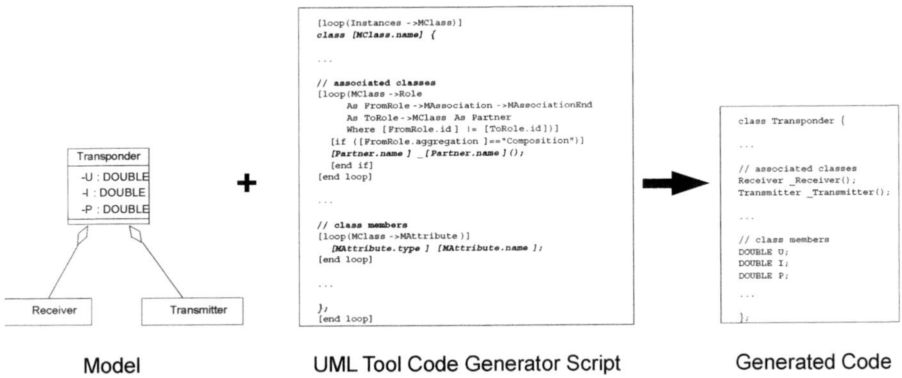

Figure 8.19: Code generation from UML diagrams - simplified illustration.

Figure 8.19 illustrates the code generation process applying another simple example. The code generator applying its control script, parses the model of a simplified spacecraft transponder. Parsing and code generation are performed by identifying the syntax elements of the UML diagram and by interpreting their semantics, identifying classes, variables, compositions etc. and finally by generating the according target language sections - here in C++. During this process also all eventually noted OCL constraints will be included into the target code. The example shows an extract from a control script in TDL script language as used by the UML tool "OpenAmeos" (cf. [78]).

Further reading on the UML language, the notation and tools is listed in the according subsection of this book's references annex.

8.4.2 Designing a Simulator Kernel using UML

The following figure 8.20 provides an overview on how complex a UML diagram can get if it covers a whole simulator kernel – only depicting the complete class hierarchy and not yet covering any functionalities. Therefore in normal software design no single class diagram for the entire kernel would be generated. Instead one top level diagram and a multitude of subdiagrams would be designed. Just for the purpose of complexity visualization figure 8.20 provides the overview of a real spacecraft simulator kernel, namely a deprecated kernel implementation of the Astrium MDVE spacecraft simulator. However with design of such a diagram the complete UML modeling of a simulator kernel is not finished. In addition to this class diagram functional diagrams like activity and sequence diagrams etc. – and if needed deployment diagrams - have to be added to complete the kernel design description.

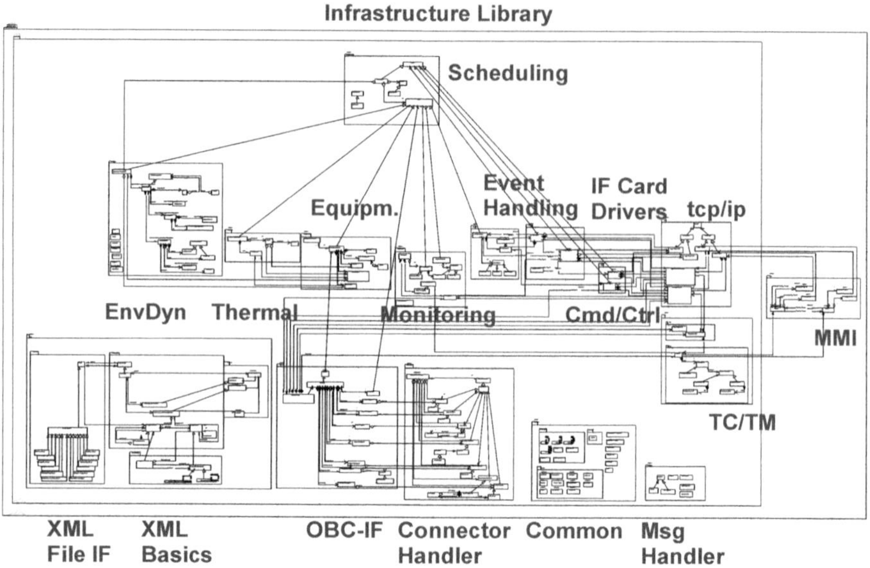

Figure 8.20: UML class diagram of a spacecraft simulator kernel. © Astrium

The kernel library in this example all in all comprises:

- The components for central functionalities like:
  - The numeric initial value problem DEQ solver
  - The space environment model including atmospheric, magnetospheric and celestial body effect computation
  - The spacecraft structure dynamics model (in this case a rigid body model).
  - A configurable thermal network model
  - Superclasses of equipment model class types and interconnection class types

- The functions for:
  - Initialization of the system (including the import of configuration files)
  - Scheduling of the simulation models
  - Interpretation of simulator commands
  - Generation of simulator result data streams
  - Handling events from the frontend-card drivers on hybrid bench setups

- The interfaces for:
  - TCP-based commanding of the simulator
  - Data transfer to a graphical user interface (MMI)
  - Reading the simulator configuration files (XML)

- ◇ Data in and output to and from the frontend-card drivers
- ◇ Data interchange with the OBC model respectively with "Hardware in the Loop" OBC in hybrid testbench variants

8.4.3 Designing Spacecraft Equipment Models with UML

The modeling of spacecraft equipment with UML shall be explained in the following paragraphs beginning with some aspects of an on-board computer model to demonstrate a more complex model structure.

The internal structure and the electric components like processor board with I/ O-controllers etc. have already been depicted in figures 5.6 and 5.8. The first one depicts processor modules, transponder interface cards and the connection to the internal MIL-STD-1553B bus (all redundant), the latter shows the connected I/O-boards (using an example of RUAG Aerospace Sweden AB and RUAG Aerospace Austria GmbH). First of all the transponder interface board (Buffered TTR called "BUTTR" in figure 5.6) is considered. The following class diagram is a detailed model of this board with subclasses for receiver, memory and interfaces-ASIC[22] (CROME[23]) as well as I/O-controllers, which functionally model the different interfaces.

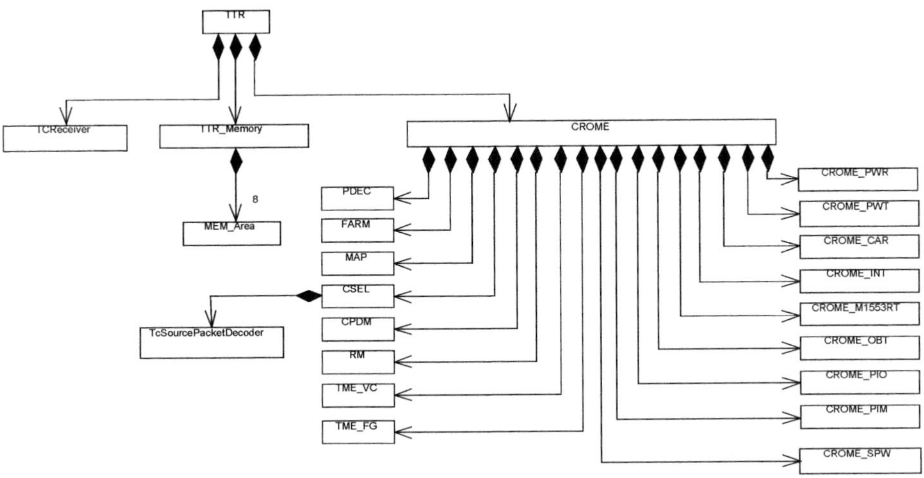

Figure 8.21: Class diagram of an OBC transponder interface board model. © Astrium

Interface registers, write and read functions as well as the timing performance and interrupt handling functions have to be defined and implemented as class attributes and member functions (methods) for each of those controller modules. A definition table tool provided to the user by the OpenAmeos UML tool with filled out entries for

[22] ASIC = Application Specific Integrated Circuit (specific electronic component developed for a single application).

[23] CROME and COCOS are brand names of a complex ASIC controller chips designed by RUAG Aerospace Sweden AB.

the member variables and functions of one single ASIC chip class is shown in the figure below. This demonstrates the complexity already arising for modeling one single on-board computer interface board. Specific simulator libraries are used for modeling the microprocessor itself of the OBC. They have already been mentioned in section 5.5 - please also refer to [27].

												FARM on 6/3/2004
	1	2	3	4	5	6	7	8	9	10	11	12
1	**FARM**			**Analysis Items**				**C++ Items**				
2	**Attribute**	**Type**	**Default Value**	**Visibility**	**Class Attr?**	**Derived ?**		**Const?**	**Visibility**	**Volatile?**		
3	lock	bool	true	private								
4	**Operation**	**Arguments**	**Return Type**	**Visibility**	**Class Op?**	**Abstract ?**	**Throws**	**Const?**	**Visibility**	**Virtual?**	**Inline?**	**Ctor Init List**
5	initProtCircuit	def_arg:int=0	void	public								
6	resetProtCircuit	def_arg:int=0	void	public								
7	resetCircuit	def_arg:int=0	void	public								
8	initCircuit	def_arg:int=0	void	public								
9	receiveTcSegment	tcSegment:unsigned char *, tcBytes:unsigned int, mapId:unsigned int	void	public								
10	decodeAdrRd	addr:unsigned int, data:int *, ws:int *	int	public								
11	decodeAdrWr	addr:unsigned int, data:int *, ws:int *, numBytes:int	int	public								
12	decodeIOPtr	addr:unsigned int, numBytes:int	char *	public								
13	cmdListRegisters	registerList:string&	void	public								
14	setLock	toStatus:bool	void	public								

Figure 8.22: Example for class parameter and function definitions. © Astrium

Besides modeling of equipment classes themselves, a functionality has to be available in a spacecraft simulator which works on instance level and which informs each model instance, which of its interface ports is connected to which harness line interconnection instance. Mapping of these harness line class instances to equipment model instances usually is performed via so-called "mapping tables" for which own class types (mapper classes) have to be defined in the simulator kernel. This mapping mechanism connecting harness line instances to equipment model instances must be initiated at simulator initialization time.

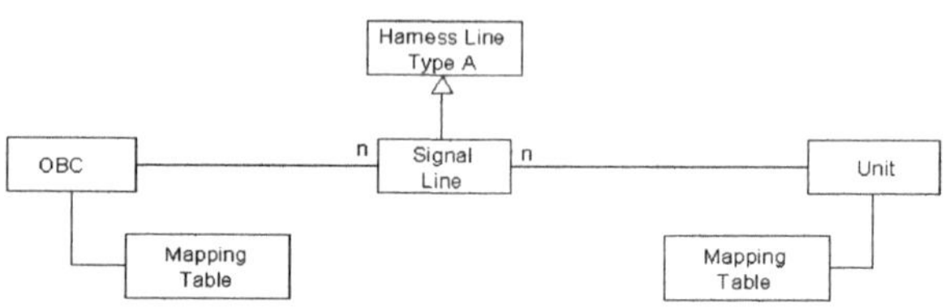

Figure 8.23: Modeling line interconnection mapper structures. © Astrium

Besides UML diagrams for equipment definition and line interconnection instantiations, other diagrams are useful and necessary for visualization and definition of the functions in equipment models as well as in the simulator kernel. One example for such behavior diagrams often used in spacecraft equipment modeling is the state machine diagram which has already been introduced in figure 8.12. Another popular tool for modeling functionalities in UML are the sequence diagrams. The example of this diagram type cited below includes:

- The processes in order to set line interconnection data (SET)
- The data report from one line interconnection (REPORT)
- The process to establish a line interconnection (CONNECT)
- The process for disconnecting a line interconnection(DISCONNECT)

Below the diagram a simplified C++ code section is depicted. Such code could be generated from such a sequence diagram using the code generator of an UML tool.

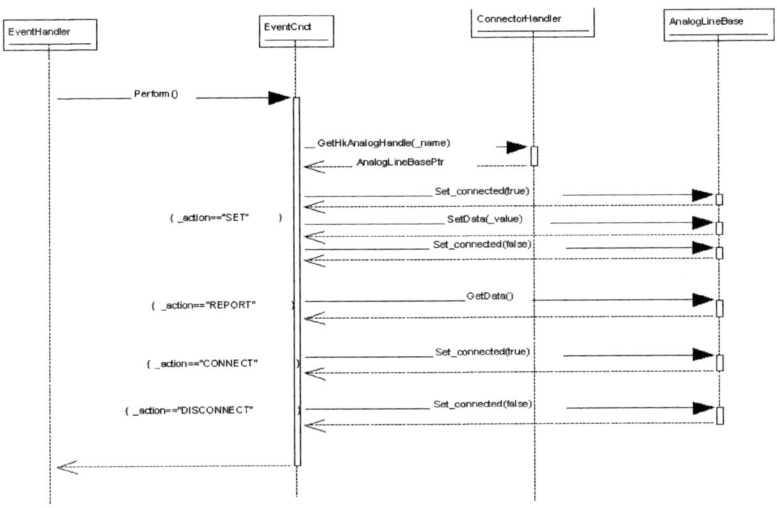

Figure 8.24: Setting and reporting of analog line interconnection parameters.

Fictitious C++ code file generated from the sequence diagram above:

```
#include ...

EventCnt::EventCnt (){
...
};

EventCnt::~EventCnt (){
...
};

int EventCnt::perform () {
  AnalogLineBasePtr =
  ConnectorHandler->GetHkAnalogLineHandler("LineName");

  if (_action=="SET") then {
. ..
  return 0;
  }
```

```
  else if(_action=="REPORT") then {
    LineValue = AnalogLineBasePtr->GetData();
    return 0;
  }

  else if(_action=="CONNECT") then {
    AnalogLineBasePtr->Set_connected(true);
    return 0;
  }

  else if(_action=="DISCONNECT") then {
    AnalogLineBasePtr->Set_connected(false);
    return 0;
  }

  else return("ERROR_unknown_cmd");
};
```

8.5 Implementation Technologies - The Extensible Markup Language (XML)

Finally the "Extensible Markup Language" (XML) must be mentioned. Via XML notation convention data can be stored in files or loaded from files supported by a formally verifiable textual notation. XML has become the de-facto standard for modern file formats in nearly all software engineering domains - not only for system simulation. In the field of system simulation files, XML format is typically used for storage / loading of configuration and initialization data of the simulator. The use of XML for such purposes shall be demonstrated in some examples.

The language syntax of XML is standardized by a W3C-Standard. The storage of data is implemented by using ASCII text files which are structured in a specific manner. These files are legible plain text files and can easily be ported between different kinds of operating systems and devices, since only being purely based on the ASCII character set. XML is very flexible, because its data structures can be created and converted according to application needs (for example from XML to HTML or vice versa). Only according syntax conversion rules have to be followed to achieve that.

To enable a software (for example a simulator) to read and write XML files, a so called "parser" has to be used. This is usually achieved by linking in a specific XML-parser library to the software - here into the simulator. The parser then enables the simulator to read an XML-file, to check it concerning consistency and completeness, to identify the content of the file and to carry out specific actions depending on the data read. This action can for example be to assign a value from the XML file to a variable of an equipment model or to generate an instance of an equipment model class.

The parser also enables the software to export data from the program to XML compatible files - e.g. simulator log files. An example for a popular open source implementation of such an XML parser library is "Xerces" for Java software or “Xerces-C++” for C++ , which also both are available for many different operating systems.

In fact to import information from an XML file with a parser, in reality two different files are needed. One is the actual XML file (with the file extension ".xml") which contains the data to be read and the other file contains information about the structure of the “.xml” file. This second file is called the “Document Type Definition”, (DTD), and this definition file itself is marked with the extension ".dtd" - please also refer to figure 8.25. Such a DTD file describes the relational class model of different associated elements. This is done by hierarchical arranged so-called “tags” that describe of a data model. The newer XML standard issues in addition define so called “XML-Schemas” (with the file extension .xsd) which are comparable to DTDs however provide diverse additional features. The examples here still stick to the older DTDs to keep them straightforward.

osk.dtd

```
<!ELEMENT OpenSimKitConfigFile      (System,
CompDefs, BranchDefs, MeshDefs, NetList,
LogOutput?)>


<!ELEMENT System    SysDesc,
SimulationCtrl)>
<!ELEMENT SysDesc  (Model,Case,Note)>
<!ELEMENT Model              (#PCDATA)>
<!ELEMENT Case               (#PCDATA)>
<!ELEMENT Note               (#PCDATA)>

<!ELEMENT SimulationCtrl  (RelAccuracy,
AbsAccuracy)>
<!ELEMENT RelAccuracy    (#PCDATA)>
<!ELEMENT AbsAccuracy   (#PCDATA)>

<!-- Definition of Component Types        -->
.....
```

Simulation Input File

```
<?xml version="1.0" standalone="no"?>
<!DOCTYPE OpenSimKitConfigFile SYSTEM
"osk.dtd">
.....
<OpenSimKitConfigFile>

<System>
  <SysDesc>
    <Model> Rocket propulsion system </Model>
    <Case>  nominal operation </Case>
    <Note>   04.12.04 </Note>
  </SysDesc>

<SimulationCtrl>
    <RelAccuracy> 0.05 </RelAccuracy>
    <AbsAccuracy> 0.0 </AbsAccuracy>
  </SimulationCtrl>
</System>

<!-- Defining the Components in the System    --
>
.....

</OpenSimKitConfigFile>
```

Figure 8.25: Characterization data for a simulation run. Example: *OpenSimKit* [23].

When importing the XML file the parser firstly checks:

- the XML language version in which the file is defined (cf. Figure 8.25):

```
?xml version="1.0"
```

- whether it is a standalone file:
  `standalone = "no"`
  or whether the syntax should be checked against a DTD respectively a schema.

- In case it has to be checked, the parser evaluates of which type the .xml file is (here "OpenSimKitConfigFile") and against which DTD / schema it has to be checked throughout parsing (here "osk.dtd"):
  `<!DOCTYPE OpenSimKitConfigFile SYSTEM "osk.dtd">`

Provided with the information from the DTD / XSD file, the parser then can read the XML file step by step or in more precise tag by tag and can verify whether the input is "well formed" w.r.t. DTD / schema and whether the input is complete. The reading order is determined by the information provided by the DTD / schema file. In the given example of figure 8.25 the parser finds the following information:

- Firstly from the DTD-entry:
  `!ELEMENT OpenSimKitConfigFile (System, CompDefs, BranchDefs, MeshDefs, NetList, LogOutput?)>`
  it can identify that the opened DTD file actually is the correct file matching the above mentioned OpenSimKitConfigFile

- Furthermore it can identify that the .xml file mandatorily contains the subsections:
  `System, CompDefs, BranchDefs, MeshDefs, NetList`
  And optionally it can contain a section:
  `LogOutput.`

- From the next entry
  `<!ELEMENT System    SysDesc, SimulationCtrl)>`
  it can identify that the subsection "System" again contains the sections:
  `SysDesc, SimulationCtrl`

- SysDesc finally contains:
  Model,Case and Note

- The entries:
```
<!ELEMENT Model          (#PCDATA)>
<!ELEMENT Case           (#PCDATA)>
<!ELEMENT Note           (#PCDATA)>
```
  are identifying Model,Case and Note to be "Parsed Character Data", which means that here the corresponding values are located as text in the .xml file.

In this manner the reading / parsing process progresses. As soon as the parser identifies an inconsistency between a specification in the DTD / schema and the XML file, for example a missing tag or a wrong hierarchy, the parser will generate a corresponding error message.

From the parsed input the parser constructs an internal structured data tree in C++ or Java from the tags it identifies. Additionally the parser generates dedicated access functions in the corresponding programming language to navigate through the data and to access the elements of the tree structure. Thus a simulator kernel can find its required input data in the data structure generated by the linked parser and can access the values and hand them over e.g. to the simulator models for spacecraft equipment. In the following paragraphs the use of XML for loading files and for configuring a system simulator and respectively equipment models will be outlined using some examples.

osk.dtd

```
.....
<!-- Definition of Component Types        -->

<!ELEMENT CompDefs(HPBottleT1 | PipeT1 |
JunctionT1 | FilterT1 | PRegT1 | SplitT1 |
TankT1)*>

<!ELEMENT HPBottleT1          (CID, Description,
Mass, Volume, SpHeatCap)>
.....
<!ELEMENT CID                 (#PCDATA)>
<!ELEMENT Description         (#PCDATA)>
<!ELEMENT Mass                (Value, Unit)>
<!ELEMENT Volume              (Value, Unit)>
<!ELEMENT SpHeatCap           (Value, Unit)>
.....
<!ELEMENT Value               (#PCDATA)>
<!ELEMENT Unit                (#PCDATA)>
.....
```

Simulation Input File

```
.....
<!-- Defining the Components in the System -->
.....
<CompDefs>
  <HPBottleT1>
    <CID> c0 </CID>
    <Description>
      Left High Pressure Bottle.
    </Description>
    <Mass>
      <Value> 28.0 </Value>
      <Unit> kg </Unit>
    </Mass>
    <Volume>
      <Value> .135 </Value>
      <Unit> m^3 </Unit>
    </Volume>
    <SpHeatCap>
      <Value> 800.0 </Value>
      <Unit> J/(kg*K) </Unit>
    </SpHeatCap>
  </HPBottleT1>
.....
</OpenSimKitConfigFile>
```

Figure 8.26: Loading information on system topology. Example: *OpenSimKit* [23].

The preceding figure 8.26 shows a simple example for the dynamical creation of equipment model instances and the initialization of these instances with characterization data. The .DTD of this example specifies that the system may be composed of the component types HPBottleT1, PipeT1, JunctionT1, FilterT1, PRegT1, SplitT1 and TankT1 only - compare the system in figure 6.7.

The DTD doesn't provide any required number of instances of each equipment type, to be part of the simulated system. This information is located inside the XML-file. In figure 8.26 the information needed for the initialization of component c0, which is of HPBottleT1 type can be easily identified. However there may be more than one of these components in the system. The entries for a system with a 2 bottle topology as in figure 1.12 would start like:

Simulation Input File

```
.....
<!-- Defining the Components in the System -->
.....
 <CompDefs>
   <HPBottleT1>
      <CID> c0 </CID>
      <Description>
         Left High Pressure Bottle.
      </Description>
      <Mass>
         .
         .
   </HPBottleT1>
   <HPBottleT1>
      <CID> c1 </CID>
      <Description>
         Right High Pressure Bottle.
      </Description>
      <Mass>
         .
         .
   </HPBottleT1>
         .
         .
</CompDefs>
         .
         .
</OpenSimKitConfigFile>
```

Figure 8.27: System topology with multiple instances of the same model type.

In the following example the dynamical configuration of a satellite "Power Control and Distribution Unit" (PCDU) model via XML is described. This not only covers the XML file structure and the import, but also the dynamical generation of C++ object instances corresponding to the content of the XML file. A satellite PCDU is composed of a digital control unit, a power bus regulation unit and a power distribution component with multiple banks of "fuses" of various types. One group of these safety devices are so-called "Foldback Current Limiters" (FCLs). The mapping between the FCLs, the cables and the consumers connected to the PCDU is not entirely fixed at implementation time of the system simulators at begin of phase C of a project. In fact they are typically changed multiple times up to spacecraft final system testing. Thus the corresponding mappings of power consumers connected to the PCDU FCLs should not be frozen in PCDU model code but should be loadable in the spacecraft simulation to make it flexible for easy reconfiguration. The mapping information preferably should be loaded from an XML-file. A PCDU model code with such loadable information on FCLs and line mappings in C++ language is provided as example below which demonstrate:

- The creation of a "SAX2" type XML parser instance
- The commands to import the XML grammar (DTD)
- The instructions to read the XML-file of the power unit

```
/**
Parsing of the DTD import of the PwrSupply XML data and generation of the
PwrSwitch elements.
*/

SAX2XMLReader* PwrSupplyConfig_XML_IF=XMLReaderFactory::createXMLReader();

// Load grammar and cache it
PwrSupplyConfig_XML_IF->loadGrammar(dtdFile,
                                    Grammar::DTDGrammarType,
                                    true);

// enable grammar reuse
PwrSupplyConfig_XML_IF->
                       setFeature(XMLUni::fgXercesUseCachedGrammarInParse,
                                  true);

// Parse xml files
PwrSupplyConfig_XML_IF->parse(PrwUnitXmlFile);
....
```

In the next step, the XML file sections of the power control unit, containing the information about to be generated FCL safety device model instances, their names, harness line connections and so forth will be analyzed. Please also refer to figure 8.28.

```
<?xml version="1.0" standalone="no"?>
<!DOCTYPE PwrSupplyConfigFile SYSTEM "PwrSupplyConfig.dtd">
<PwrSupplyConfigFile>
      ......
 <PwrSupplyConfigData>
   <PwrSupply>
     <PwrSwitchModule>
        <moduleName>LCL_1A</moduleName>
            <PwrSwitch>
              <PwrSwitchName>FCL1_1A</PwrSwitchName>
              <PwrSwitchType>FCL</PwrSwitchType>
              <PwrSwitchDescr>OBC (CTU1)</PwrSwitchDescr>
              <Map2SimHarness>
                     <SimHarnessLine>
                         <simIfId>1615</simIfId>
                         <simIfDescr>OBC (CTU1) Power Supply</simIfDescr>
                         <simIfName>pwr_ctu1_pcdu_00_FCL_1A_01</simIfName>
                         <simIfHarnessType>power</simIfHarnessType>
                     </SimHarnessLine>
               </Map2SimHarness>
                    .....
               <PwrSwitchInfoCurrent>
                   <PwrSwitchInfo>
                     <signalCode>FCL1_1A_CUR_HK</signalCode>
                     <descr>FCL1_1A Current HK TM</descr>
                     <size>12</size>
                     <wordSeqNo>6</wordSeqNo>
                     <offsetBit>0</offsetBit>
                   </PwrSwitchInfo>
               </PwrSwitchInfoCurrent>
                    .....
```

```
                <PwrSwitchInfoState>
                   <PwrSwitchInfo>
                      <signalCode>FCL1_1A_STAT</signalCode>
                      <descr>FCL1_1A Status</descr>
                      <size>1</size>
                      <wordSeqNo>63</wordSeqNo>
                      <offsetBit>0</offsetBit>
                   </PwrSwitchInfo>
                </PwrSwitchInfoState>
              </PwrSwitch>
                  .....
        </PwrSwitchModule>
       </PwrSupply>
     </PwrSupplyConfigData>
</PwrSupplyConfigFile>
```

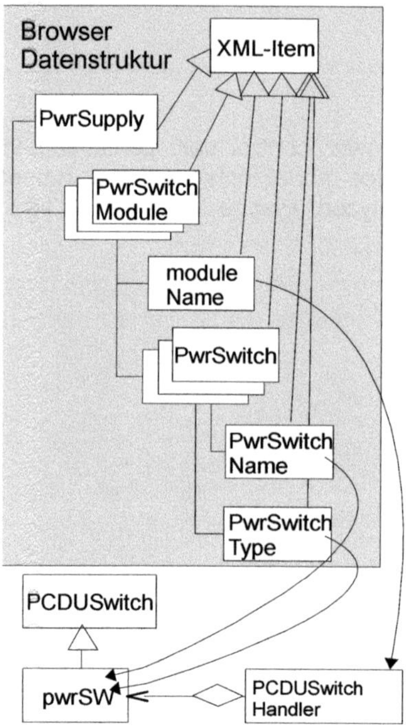

Figure 8.28: XML file structure and created classes for a PCDU.

Finally excerpts of a fictitious C++ code follow which

- create the `PwrSwitch` FCL safety module instances, defined in the above cited XML file section and
- which registers these FCL switches at the handler for the following characterization - i.e. importing of the threshold levels, the OBC commands and the connection to the harness lines.

```
....
PwrSupplyConfig_XML_IF->InitPowerSupply(PCDUSwitchHandler& psh_ //
                                        ,ConnectorHandler& ch_ //
                                        ) {

  for (i=0; i<nPwrSwitchModules; i++) {
    pwrSwitchModule=pwrSupply->GetChildElement("PwrSwitchModule", i);
    pwrSwitchModuleName =
            pwrSwitchModule->
                           GetChildElement("moduleName")->Get_textNode();

    nPwrSwitchs=pwrSwitchModule->CountChildElements("PwrSwitch");
    for (j=0; j<nPwrSwitchs; j++) {
      pwrSwitch=pwrSwitchModule->GetChildElement("PwrSwitch", j);

      // PwrSwitch
      //-----------------------------------------------------------
      // PwrSwitchName
      aChild=pwrSwitch->GetChildElement(childName="PwrSwitchName");
      if (!aChild) throw (string ("non existing child element '"
                          +childName+"' for switch "+pwrSwitchName));

      pwrSwitchName=aChild->Get_textNode();

      //  PwrSwitchType
      aChild=pwrSwitch->GetChildElement(childName="PwrSwitchType");
      if (!aChild) throw (string ("non existing child element '"
                          +childName+"' for switch "+pwrSwitchName));

      pwrSwitchType=aChild->Get_textNode();

      // PCDUSwitchHandler registration / instantiation
      pwrSw = PCDUSwitch::VirtualConstructor(pwrSwitchType,
                                             pwrSwitchName);

      // power switch instance is mapped onto module
      psh_.RegisterPowerSwitch2Module(pwrSw,pwrSwitchModuleName);
```

In addition XML files can be applied for simulation state "serialization". This functionality allows stopping of a system simulation at a desired point in time, to store its state to file and to use the file(s) later for resuming computations or to base different test case variants on a common initial starting condition. In modern object oriented programming languages like C++ and especially in Java, extensive and easy to use libraries for serialization in diverse file formats are available. The serialization requires saving the status of the simulator kernel, the status of every equipment model and of each harness line model in the state-set files. For simulator infrastructures which provide such state set saving typically XML is used for these files.

Further reading and Internet pages on XML are listed in the according subsection of this book's references annex.

8.6 Implementation Technologies - Modeling Frameworks

For the implementation of such complex software tools like system simulators today not only modern design languages like UML and efficient and powerful programming languages like C++ or Java are used, but also so-called "modeling frameworks". Such modeling frameworks are generic software environments for simplifying the development of application programs - in this case for spacecraft simulators. These modeling frameworks comprise:

- A graphical user interface
- Intelligent code editors
- Powerful build tools (like Apache Ant)
- Compiler, Linker, Debugger

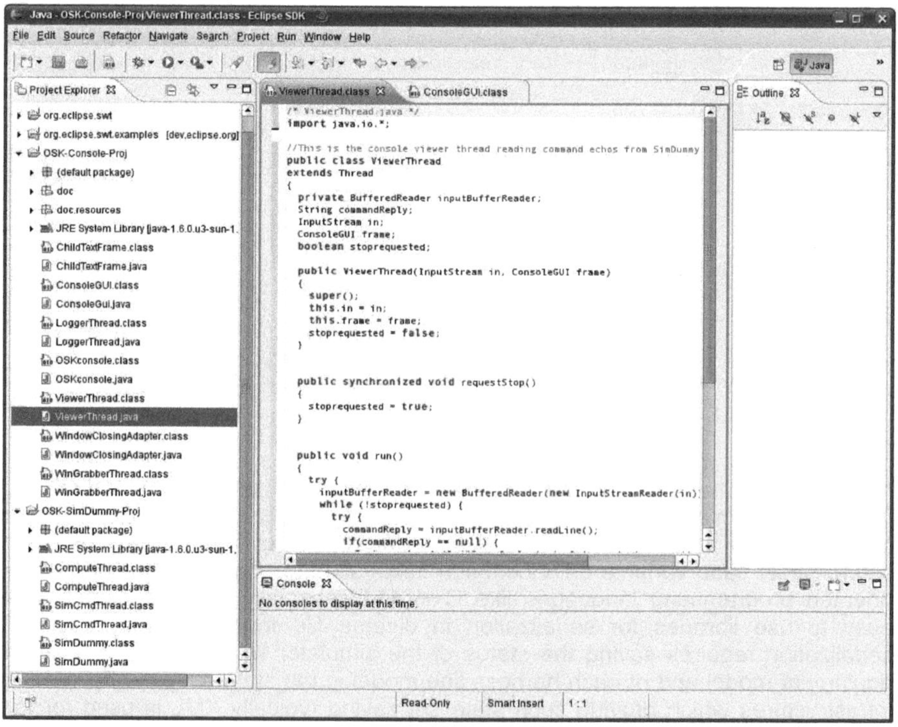

Figure 8.29: Eclipse as Integrated Development Environment (IDE).

Diverse further modules can be loaded as so-called "plugins" - e.g. for simulators the cited XML-parser or serializer modules. Logfile output interfaces, external command interfaces, the connection between debugger and simulator and the usage of modeling framework graphics as basis for simulator user interface design also can be managed via plugin mechanisms. This approach describes using the frameworks as so-called integrated development environment.

The most common modeling frameworks are "NetBeans IDE" from Sun Microsystems and "Eclipse" which originally was developed by IBM and now is further maintained by the "Eclipse Foundation" (see also figure 8.29). Both modeling frameworks are public domain and can be used for developing software based on diverse target languages - e.g. Java and C++. Only Eclipse shall be treated here, since it is the most widespread modeling framework.

Eclipse also provides a UML design infrastructure plugin for the modeling framework which makes this platform a somewhat ideal platform for simulator development.

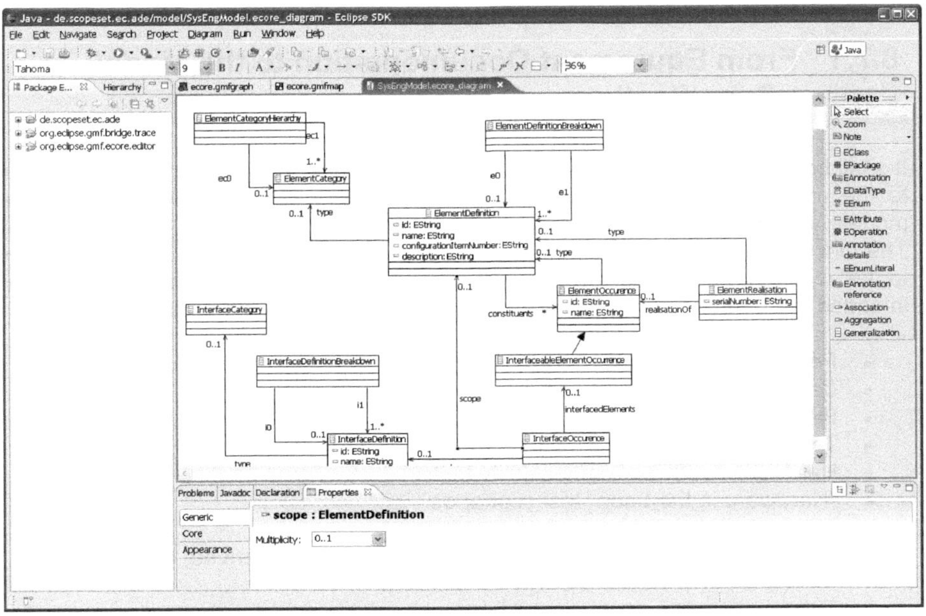

Figure 8.30: UML views in Eclipse (Eclipse UML). © ESA / ScopeSET

Furthermore such modeling frameworks support the design approach to base the simulator development using the framework itself as "Rich Client Platform" (RCP) - i.e. as software skeleton. This implies e.g. Eclipse first is used to design the code for simulator modules by means of UML. C++ or Java code can then be generated as described in previous paragraphs too. However in the RCP approach the Eclipse Framework itself with all its functions, XML-parser etc. is used as framework for the simulator - as "main" program of the simulator kernel so to say. This approach relieves the simulator developer from coding all the functions for XML-file reading, for logging and their integration with the software kernel. The developer can make use of Eclipse baseline features without recoding them and of a huge amount of plugins freely available as open-source code.

Further reading and Internet pages on Eclipse as a development platform and as Rich Client platform - Eclipse Modeling Framework, (EMF) - are listed in the according subsection of this book's references annex.

8.7 From a Model Specification to the Simulation Run

In this chapter finally the development equipment model for a spacecraft simulation will be outlined by means of an example, starting from the beginning with the model specification over the design of the software model to the integration into the simulator and to a simulation run.

8.7.1 From Equipment Documentation to the Model Specification

It shall be recapitulated that the equipment models are connected to each other and to the on-board computer model via simulated harness lines and to the system models via mathematical connections. The system equations are solved centrally. The functionalities of the, to be implemented, model code functions are driven by the to be modeled real spacecraft equipment and by the test scenarios which are intended to be run on the simulation-based testbenches. The latter for example are:

- Performance analysis scenarios
- On-board software test scenarios
- Performance verification scenarios
- Test procedure debugging scenarios
- Hardware / software compatibility tests
- "Hardware in the Loop" test scenarios

From these case scenarios the model requirements can be derived, considering:

- The functionalities / physics of the real hardware which is to be represented.
- The represented functionalities of the data protocols which are used between the equipment and the on-board computer - for example specific packet information can be empty or replaced with dummy data or counters.
- The required implementation precision of the models considering all to be modeled aspects from the real equipment's physics up to the to be modeled data transmission via protocols.

The representation of the real equipment's physics in the model actually can differ from the real equipment to a large extent. For illustration, the following example shall be given - the computation of a star tracker output - the quaternions - for the star tracker's data protocol:

The physical simulated satellite attitude and position in the simulator is calculated by the integration of the equation of motion for the structure model. In the simulation - in contrast to reality - the star tracker model can receive the attitude and position directly from the structure dynamics module where the results already are available. It only has to calculate the pointing direction in the orbit reference frame by combining the satellite's attitude and the star tracker's relative alignment to the satellite's simulated structure. This calculated viewing direction now can be made available directly via data protocol to the OBC by computing the according protocol data packets. Together with additionally simulated effects like noise, temperature dependent effects and so forth, these data packets are composed.

Thus the attitude determination is not simulated at all via the steps from a star constellation view input to the camera optics head and along through the electronics, modeling the decoding of simulated star maps down to the communication interface which would be the physical path in a real star tracker hardware. Instead the equipment model only functionally representing the real hardware and just "picks" the spacecraft attitude which is directly available in the dynamics module, adds error effects and protocol encoding.

The scope of technical functions which is to be represented in detail in a model as well as the model's interfaces implicitly are defined through the documentation provided by the supplier of the real spacecraft equipment. The essential documents here are the equipment "Interface Control Document", (ICD), and the equipment "Design Document", (DD).

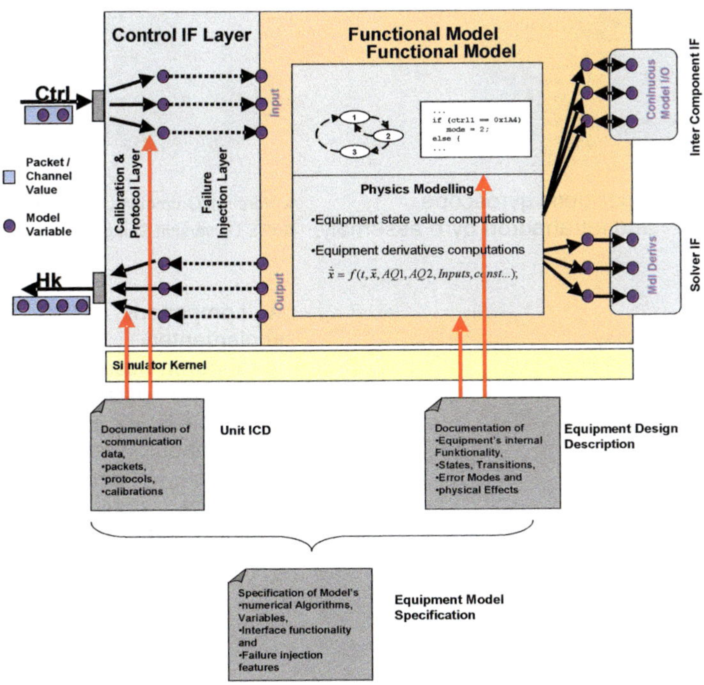

Figure 8.31: Input documentation for design and coding of an equipment model.

From the equipment ICD, the future design model interfaces towards the OBC and towards other equipment models can be derived (ports for connection to simulated harness lines). From the equipment design document the functionalities can be derived which are to be reflected as software algorithms in the model itself - eventually allowing functional simplification depending on the simulator user requirements.

8.7.2 Application Example - Fiber-optic Gyroscope

The following example shows step by step the way from the product specification down to the integrated model in a simulator. As example equipment serves a "fiber-optic gyroscope" (FOG). Such FOG sensors are used on board modern spacecraft to determine the rotational rates around the different spacecraft axes - respectively also rotation angles. The following figure shows a fiber-optic gyroscope unit and the tetrahedron mounting assembly which is required to achieve a 3 from 4 redundancy against a single gyroscope failure in orbit:

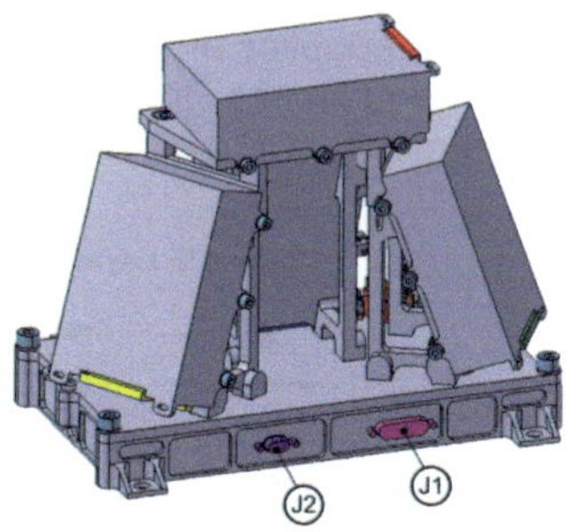

Figure 8.32: Fiber-optic gyroscope and tetrahedron gyro assembly.

© Northrop Grumman LITEF GmbH, Freiburg
© IRS Universität Stuttgart

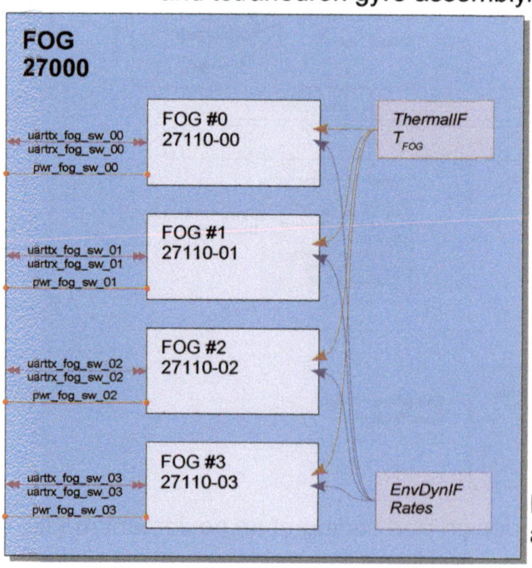

A component model has to be implemented reflecting the tetrahedron assembly including the 4 gyroscope instances and modeling the harness line connection ports on the tetrahedron frame. The presented example originates from the university small satellite project "Flying Laptop", (FLP), at Universität Stuttgart, Germany.

Figure 8.33: Required model architecture. © IRS Universität Stuttgart

The following fiber-optic gyroscope "Interface Control Document", (ICD), is an example for a product specification which serves as input for a model developer. The figure shows an extract from the table of contents:

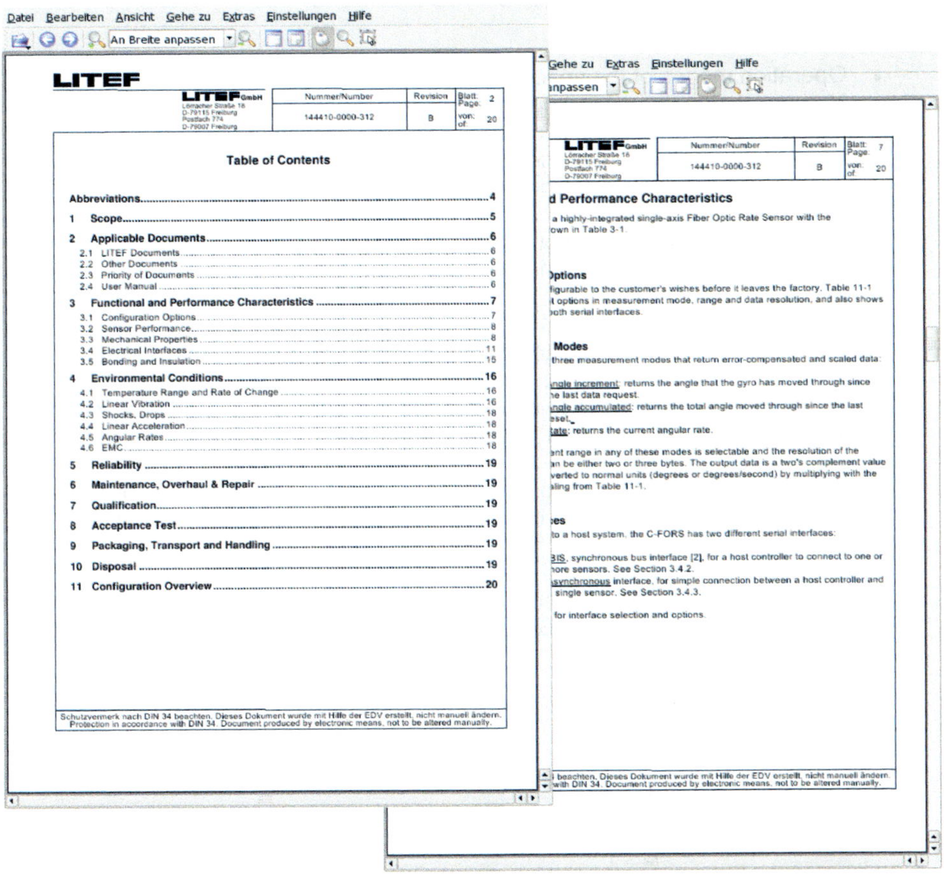

Figure 8.34: Fiber-optic gyro ICD. © Northrop Grumman LITEF GmbH, Freiburg

8.7.3 Writing an Equipment Model Specification

From this flight hardware documentation, which is provided by the equipment supplier, now a definition of a model description has to be created. An according model specification has to define precisely which functions and effects of the real equipment hardware have to be modeled by which algorithmic implementation in the satellite simulator. And those effects simplified or neglected have to be cited explicitly in the model specification document as "simplifications". This clearly identifies the application cases for which the model will be suited and the model limitations (e.g. with respect to later use in other projects). For writing the model specification the developer has to analyze:

- Operation modes of the spacecraft component
- Numerical algorithms for modeling equipment physics
- Requirements to external stimulation (failure injection etc.)
- Equipment electrical interfaces
- The equipment's physical connections – their types and numbers
- Equipment data interfaces
- The equipment input / output data
- The data protocols and formatting to be used for the component input / output data

The created Equipment Model Specification (cf. again figure 8.31) then comprises:

- All algorithms which have to be converted into UML design and later into code
- The important design and control parameters to be used by the model algorithms
- The component interfaces to data I/O-lines, control and power supply lines
- The numeric component interfaces to solvers and system models like environment and dynamics

The following example shows some basic table of contents elements of such an Equipment Model Specification – citing again the same example as above, the fiber-optic gyroscope:

1. Introduction
1.1 Scope of Document
1.2 Basic Concept

2. References

3. Fiber-optic Gyro
3.1 Scope of Document
3.2 Dynamics
3.3 Power
3.4 Thermal
3.5 Communication

3.6 Logical Operation

3.7 Failure Control

3.8 Schedule

3.9 Model Interface Layer

4. Variable List

4.1 Nomenclature Information

4.2 Special Variables

....

The steps to be performed after definition of such a model specification are (cf. also to the figure below):

- The model design (in UML),
- the generation of the framework source code (here in C++),
- the instrumentation of the framework code with manually coded algorithms,
- compilation and make process,
- testing of the equipment model (unit tests),
- the integration into the simulation environment and testing (integration tests)
- and finally to perform system tests with the entire simulation.

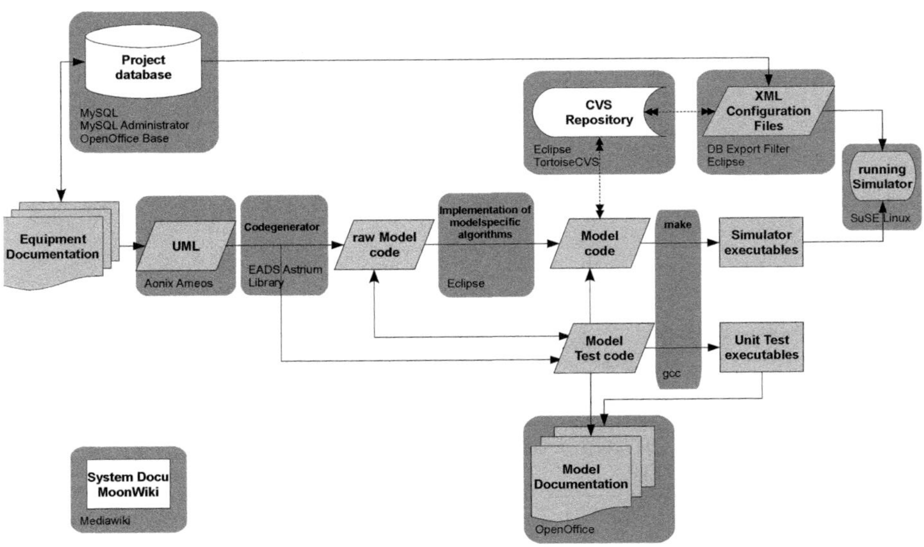

Figure 8.35: Steps of model development. © IRS Universität Stuttgart

8.7.4 Translation of the Model Specification into UML Based Design

As initial step the equipment model designer has to specify the according software classes their member variables and functions to represent the equipment. For simulators which load the number of equipment instances at initialization time, no fixed number of instances has to be specified. For simulators which fix at design time in UML the number of equipment instances in the spacecraft system, also this information has to be laid down in the UML design. In this case an instance diagram of a component defines:

- The structures of equipment classes and subclasses
- Internal member variables
- Externally visible access names of member variables
- Connection ports to connect a model instance and the corresponding simulated harness line instance

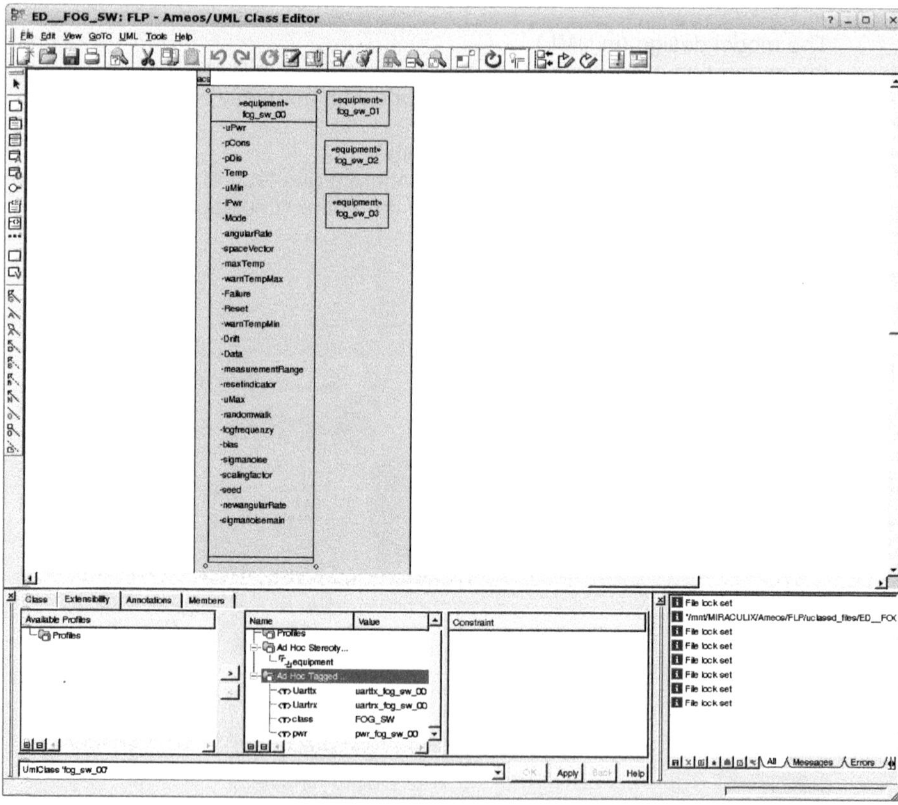

Figure 8.36: Class / instance diagram of the tetrahedron gyro mounting assembly.
© IRS Universität Stuttgart

After the design of the equipment model's basic structure in UML and eventually the corresponding instance specifications some additional model functionalities have to be specified which are:

- The read / write functions from / to simulated harness lines
- The functions for characterization parameter loading via the simulator kernel
- The functions for accessibility of dedicated component internal parameters from Core EGSE via the simulator kernel
- Functions for generation of simulator telemetry packets comprising selected variables
- Functions to register a model with the simulator kernel scheduler
- The access to simulator kernel variables such as time t
- Access functions to numeric solvers

Those system level functionalities however normally are not redesigned for each model but are achieved by specifying according class template types for the equipment model class and respectively its subclasses. The code generator driven by this template information then inserts corresponding code snippets implementing these functions into the code frame being generated from the UML diagrams information.

The connection of the equipment model with the simulated harness lines for data exchange with their counterparts in the spacecraft system simulation can simply be achieved via definition of fixed connections in the UML diagrams. This leads to a fixed spacecraft model design which has to be edited on UML level and to be recompiled / relinked each time a signal mapping changes during the project. It already was mentioned that this is a non-preferable solution since e.g. thermistor channel allocation to OBC input ports or power channel allocations to PCDU output ports are still rather floating at the beginning of a phase C spacecraft design where the first complete simulators are to be implemented.

The alternative is to achieve a model to harness line connection mechanism via dedicated model functions enabling data read / write to model ports as already cited. Those model functions connect to a dedicated mapper class during the initialization of the simulator. The mapper class again connects the equipment components. This approach is much more flexible than the one cited previously (please also refer to figure 8.23). The functions for accessing mapper classes again are either inherited from corresponding superclasses or by template expansion during code generation.

Figure 8.37 depicts a connection between a FOG and an OBC model by using a simulated harness. It also shows the according class hierarchy for the simulation of the serial lines with the corresponding predefined functions and their connection to both FOG and OBC model. It has to be kept in mind that this is a simplified description (mapper functions not shown here) and furthermore that the modeling of harness line classes and their data packets is normally not the task of the equipment model developer. Typically standardized libraries of such harness line classes are available for the model developer at his disposition when designing a spacecraft equipment model like the FOG example here.

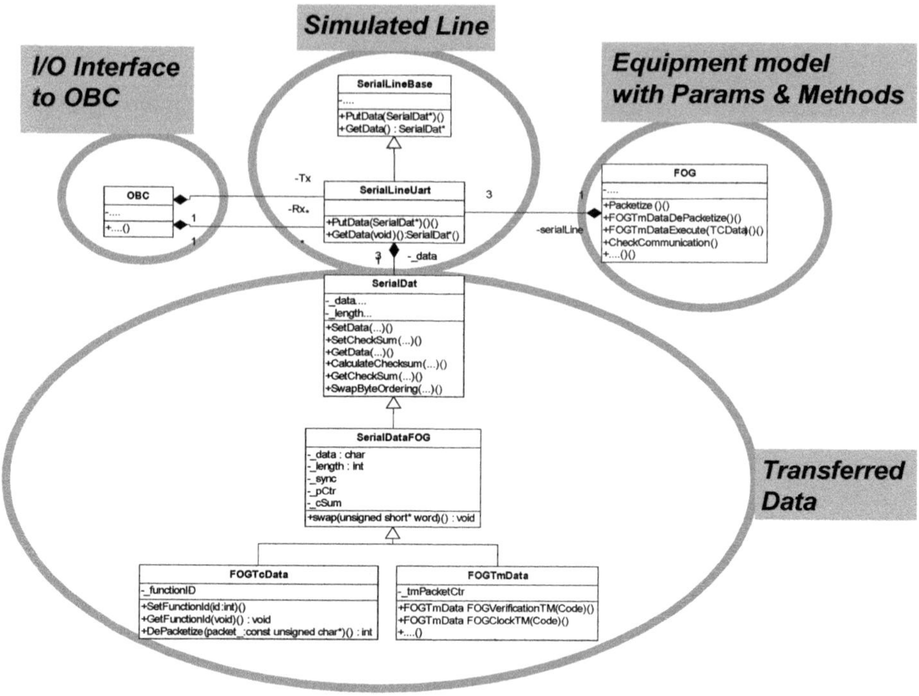

Figure 8.37: Modeled FOG equipment, OBC and transfered data structures.

8.7.5 Code Generation and Code Instrumentation

The designed UML diagrams now describe classes, instances and if necessary connections between components as well as function diagrams of finite state machines etc. - please also refer to figure 8.19. The software code for the equipment model class then can be generated from these UML diagrams applying according code generation scripts which are usually company proprietary and allow for interpretation of the simulator environments specific features, e.g. for correct template interpretations. The generated software code comprises the following code functions:

- Class definitions for the main class of the equipment type and for possible associated subclasses
- Class variables
- Variable names for external variable access
- Empty class member functions (methods) - to be instrumented manually with code modeling the equipment's physical behavior
- Connection ports
- Class handlers in order to model connection types
- Class interfaces to system modules and simulator kernel and scheduler
- Eventually component instance definitions (for each instance one class in the modeled spacecraft)
- Instance names

Thus after the code generation from UML the model code consists of the following files (C++ language presumed):

- .hpp file of the component class
- .cpp file of the component class
- .hpp files of all component subclasses
- .cpp files of all component subclasses

The figure below shows extracts of both a .hpp declaration file and a .cpp code file:

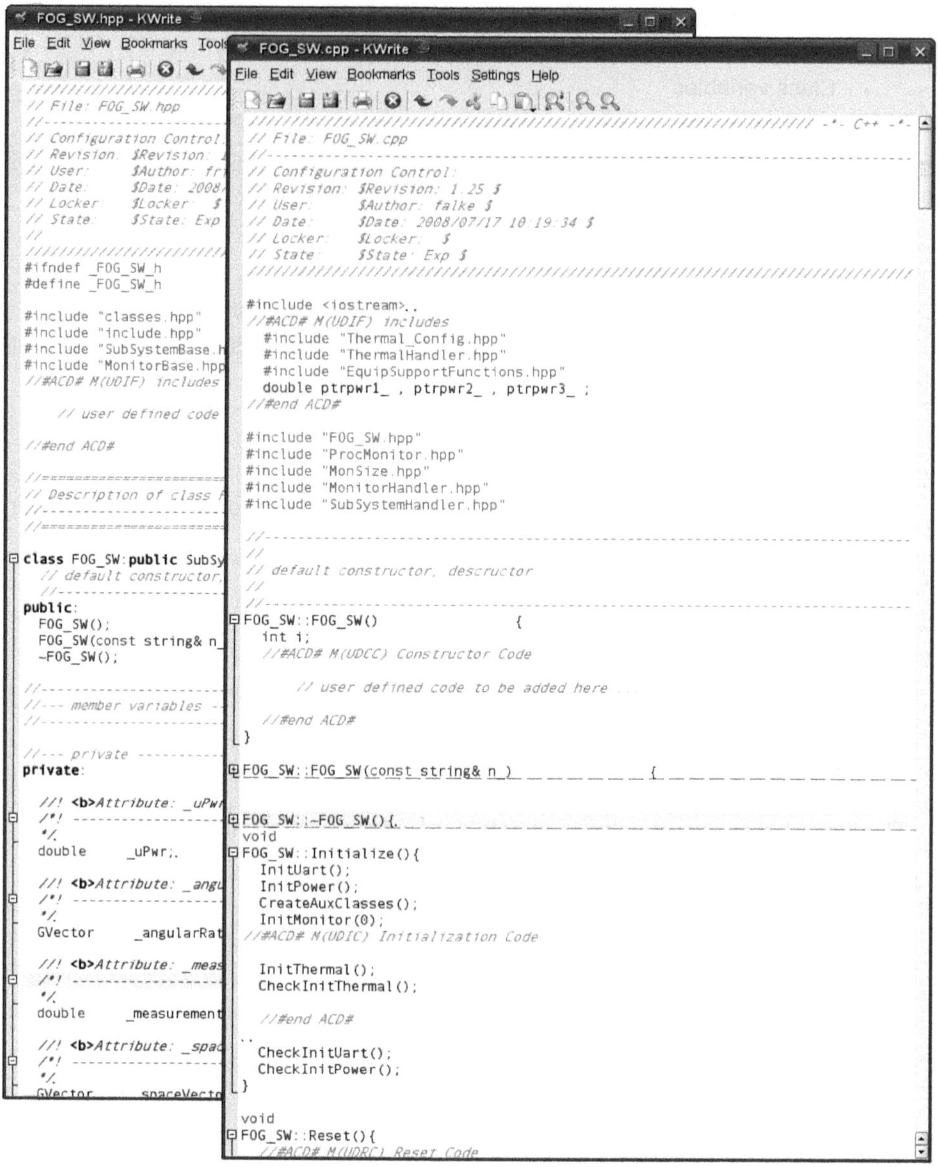

Figure 8.38: Model code declaration file (.hpp) and source code file (.cpp) – extracts.
© IRS Universität Stuttgart

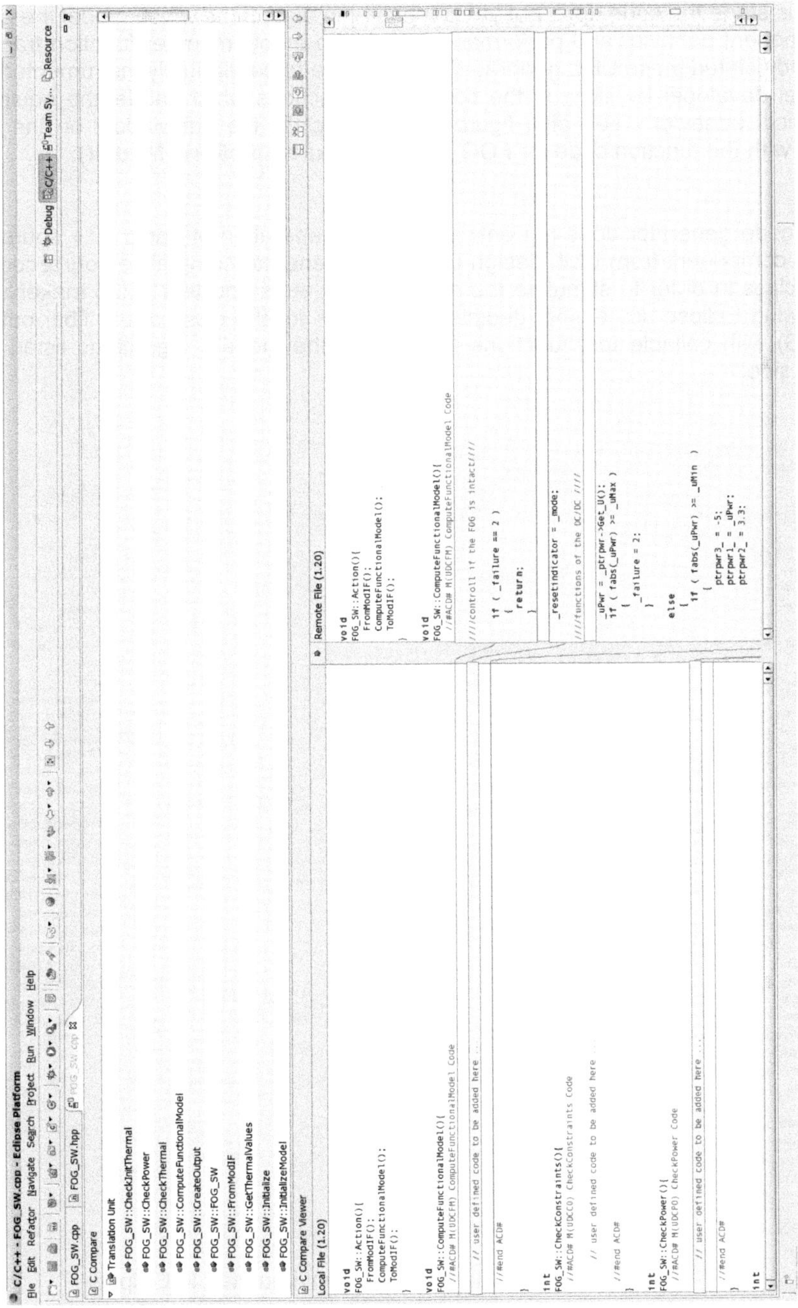

Figure 8.39: Instrumentation of generated model functions with algorithm code.

© IRS Universität Stuttgart

At this stage the code does not yet comprise the algorithms which model the physical component behavior and performance - only the empty member function frames. As already stated these UML generated frames have to be manually instrumented by the model developer by filling in the sourcecode sections which model the equipment's physical behavior - see also figure 8.39, especially the subwindow on the bottom right with the function code for **FOG_SW::ComputeFunctionalModel()**:

The code generator does not only enable the generation of basic C++ source code for a component from UML design but also to generate compatible sourcecode for a test class in order to stimulate the model in unit tests and test setup makefiles. The following Eclipse screenshot depicts a test class for the class of the fiber-optic gyro (FOG) with callable test functions for each of the model's algorithms listed on the right side:

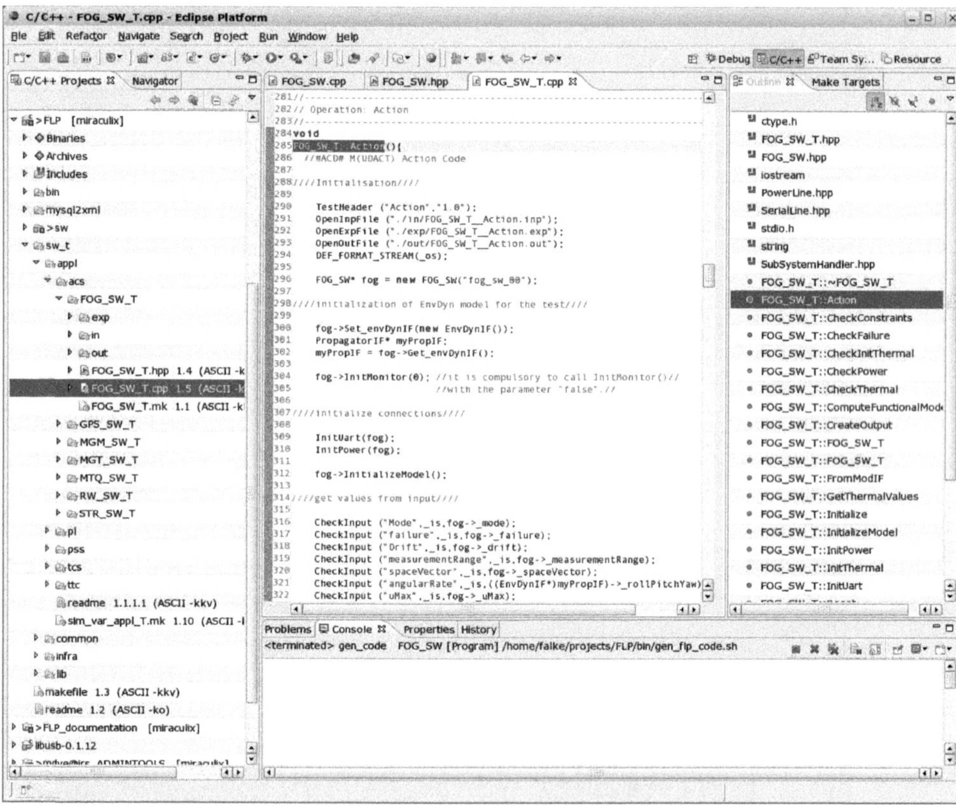

Figure 8.40: Test class and test environment in Eclipse.

© IRS Universität Stuttgart

8.7.6 Integrating the Model into the Simulator

The next steps concern the integration of the compiled and unit tested FOG model class into the overall spacecraft simulator. Virtually all simulator infrastructures provide a sort of registry file in which all equipment model classes, being part of the simulator executable, are registered.

For simulators, which fix the number of equipment instances already at modeling time and thus with the executable code, in such a registry each model instance has to be cited. The registry code is parsed at simulator startup to generate the class instances in memory[24]. For simulators, such as *OpenSimKit*, which instantiate the models at runtime, only the equipment classes are registered in the registry file.

In the example case this file is called **CreateSubSystem.cpp** and an extract is shown in figure 8.41 below.

Thereafter the equipment model code can be linked together with the simulator infrastructure. The equipment models all are grouped by type (AOCS models, power subsystem models etc.) in so-called packages. A separate makefile is generated for each package in order to compile the equipment models and to create according libraries. The fiber-optic gyro in the example here is made part of the "Attitude Control System" (ACS) library. Figure 8.42 in excerpt shows the makefile for this ACS library:

[24] In similar registries also all types of simulated harness lines are defined.

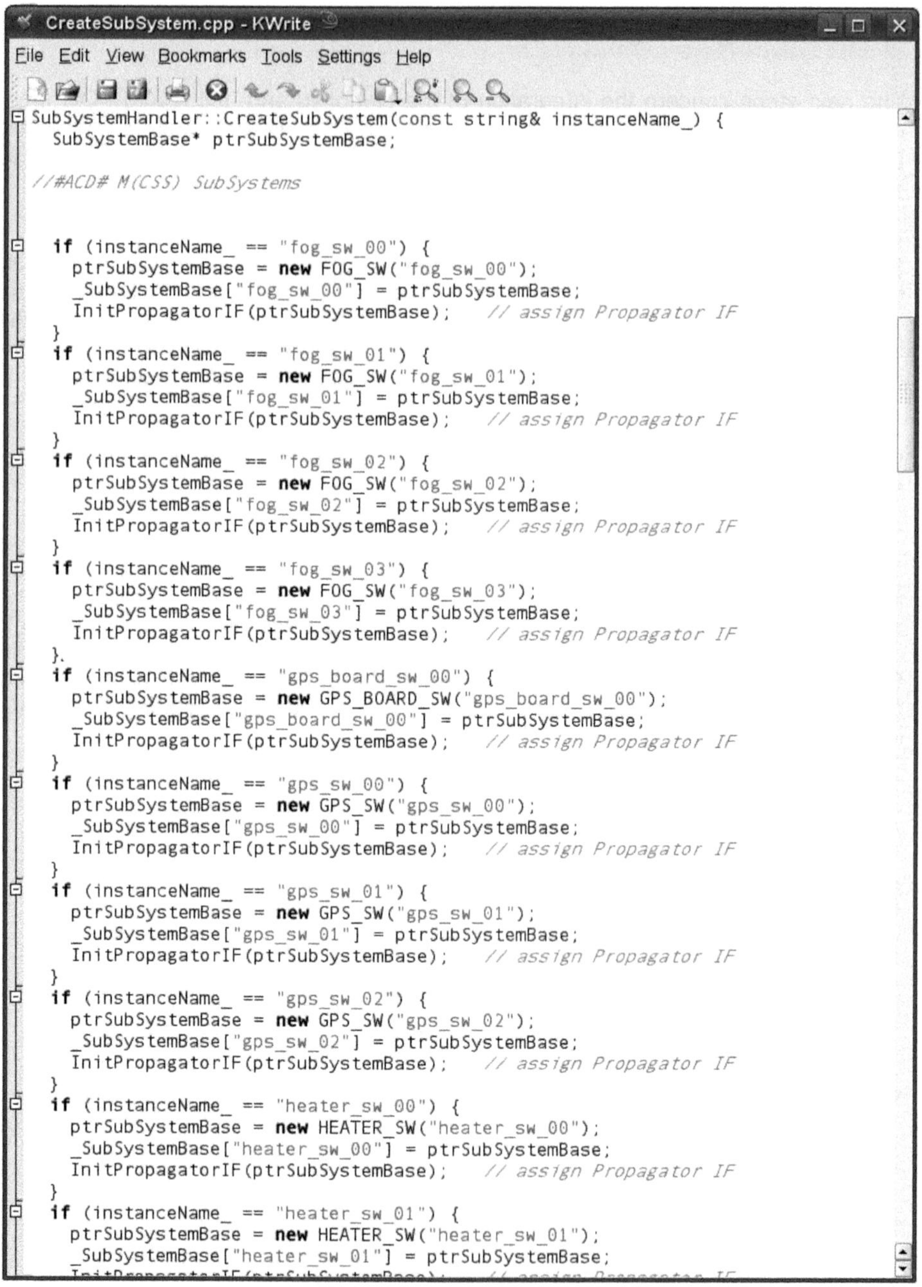


```
SubSystemHandler::CreateSubSystem(const string& instanceName_) {
  SubSystemBase* ptrSubSystemBase;

//#ACD# M(CSS) SubSystems


  if (instanceName_ == "fog_sw_00") {
    ptrSubSystemBase = new FOG_SW("fog_sw_00");
    _SubSystemBase["fog_sw_00"] = ptrSubSystemBase;
    InitPropagatorIF(ptrSubSystemBase);     // assign Propagator IF
  }
  if (instanceName_ == "fog_sw_01") {
    ptrSubSystemBase = new FOG_SW("fog_sw_01");
    _SubSystemBase["fog_sw_01"] = ptrSubSystemBase;
    InitPropagatorIF(ptrSubSystemBase);     // assign Propagator IF
  }
  if (instanceName_ == "fog_sw_02") {
    ptrSubSystemBase = new FOG_SW("fog_sw_02");
    _SubSystemBase["fog_sw_02"] = ptrSubSystemBase;
    InitPropagatorIF(ptrSubSystemBase);     // assign Propagator IF
  }
  if (instanceName_ == "fog_sw_03") {
    ptrSubSystemBase = new FOG_SW("fog_sw_03");
    _SubSystemBase["fog_sw_03"] = ptrSubSystemBase;
    InitPropagatorIF(ptrSubSystemBase);     // assign Propagator IF
  }
  if (instanceName_ == "gps_board_sw_00") {
    ptrSubSystemBase = new GPS_BOARD_SW("gps_board_sw_00");
    _SubSystemBase["gps_board_sw_00"] = ptrSubSystemBase;
    InitPropagatorIF(ptrSubSystemBase);     // assign Propagator IF
  }
  if (instanceName_ == "gps_sw_00") {
    ptrSubSystemBase = new GPS_SW("gps_sw_00");
    _SubSystemBase["gps_sw_00"] = ptrSubSystemBase;
    InitPropagatorIF(ptrSubSystemBase);     // assign Propagator IF
  }
  if (instanceName_ == "gps_sw_01") {
    ptrSubSystemBase = new GPS_SW("gps_sw_01");
    _SubSystemBase["gps_sw_01"] = ptrSubSystemBase;
    InitPropagatorIF(ptrSubSystemBase);     // assign Propagator IF
  }
  if (instanceName_ == "gps_sw_02") {
    ptrSubSystemBase = new GPS_SW("gps_sw_02");
    _SubSystemBase["gps_sw_02"] = ptrSubSystemBase;
    InitPropagatorIF(ptrSubSystemBase);     // assign Propagator IF
  }
  if (instanceName_ == "heater_sw_00") {
    ptrSubSystemBase = new HEATER_SW("heater_sw_00");
    _SubSystemBase["heater_sw_00"] = ptrSubSystemBase;
    InitPropagatorIF(ptrSubSystemBase);     // assign Propagator IF
  }
  if (instanceName_ == "heater_sw_01") {
    ptrSubSystemBase = new HEATER_SW("heater_sw_01");
    _SubSystemBase["heater_sw_01"] = ptrSubSystemBase;
```

Figure 8.41: Registration file for all equipment model instances.

© IRS Universität Stuttgart

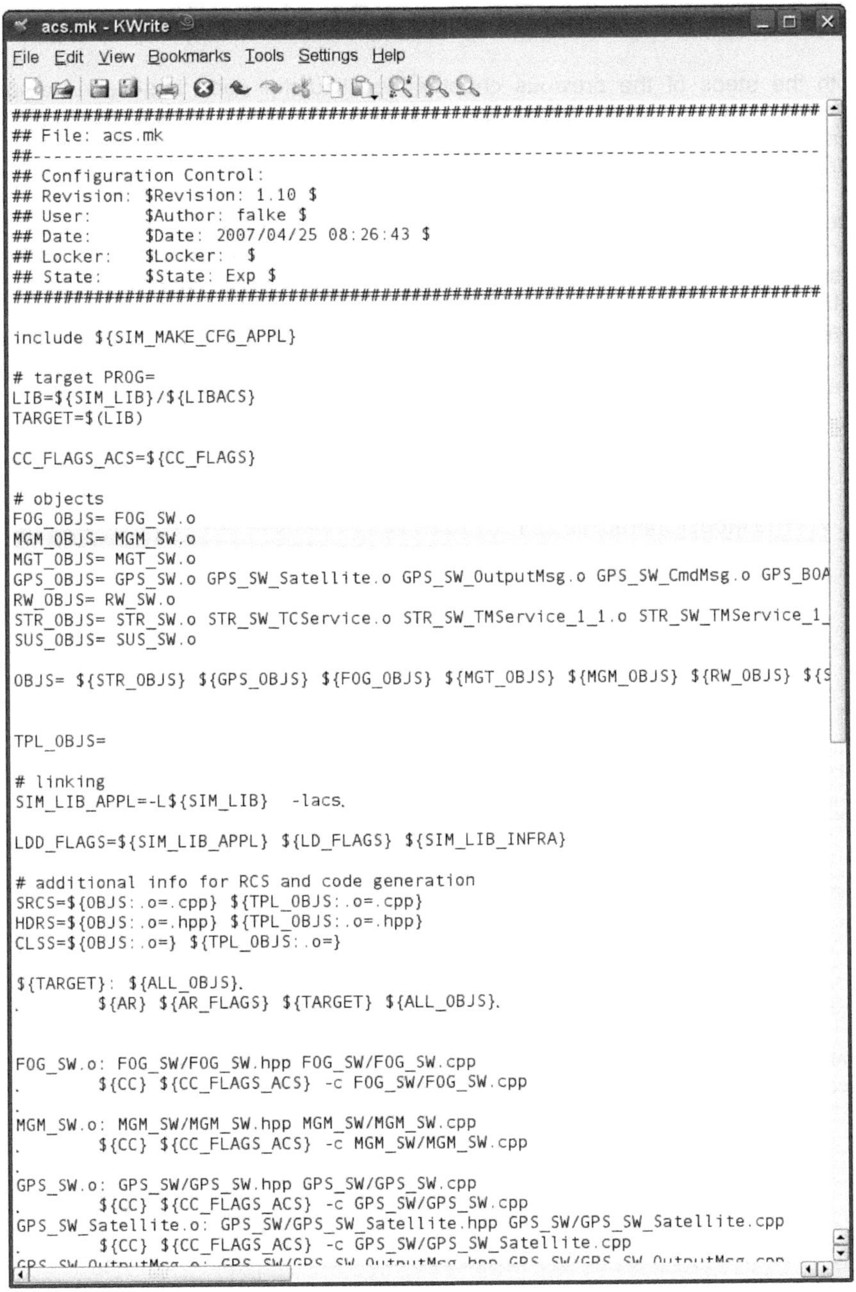

```
acs.mk - KWrite
File Edit View Bookmarks Tools Settings Help
################################################################################
## File: acs.mk
##------------------------------------------------------------------------------
## Configuration Control:
## Revision: $Revision: 1.10 $
## User:     $Author: falke $
## Date:     $Date: 2007/04/25 08:26:43 $
## Locker:   $Locker:  $
## State:    $State: Exp $
################################################################################

include ${SIM_MAKE_CFG_APPL}

# target PROG=
LIB=${SIM_LIB}/${LIBACS}
TARGET=$(LIB)

CC_FLAGS_ACS=${CC_FLAGS}

# objects
FOG_OBJS= FOG_SW.o
MGM_OBJS= MGM_SW.o
MGT_OBJS= MGT_SW.o
GPS_OBJS= GPS_SW.o GPS_SW_Satellite.o GPS_SW_OutputMsg.o GPS_SW_CmdMsg.o GPS_BOA
RW_OBJS= RW_SW.o
STR_OBJS= STR_SW.o STR_SW_TCService.o STR_SW_TMService_1_1.o STR_SW_TMService_1_
SUS_OBJS= SUS_SW.o

OBJS= ${STR_OBJS} ${GPS_OBJS} ${FOG_OBJS} ${MGT_OBJS} ${MGM_OBJS} ${RW_OBJS} ${S


TPL_OBJS=

# linking
SIM_LIB_APPL=-L${SIM_LIB}  -lacs.

LDD_FLAGS=${SIM_LIB_APPL} ${LD_FLAGS} ${SIM_LIB_INFRA}

# additional info for RCS and code generation
SRCS=${OBJS:.o=.cpp} ${TPL_OBJS:.o=.cpp}
HDRS=${OBJS:.o=.hpp} ${TPL_OBJS:.o=.hpp}
CLSS=${OBJS:.o=} ${TPL_OBJS:.o=}

${TARGET}: ${ALL_OBJS}.
.       ${AR} ${AR_FLAGS} ${TARGET} ${ALL_OBJS}.


FOG_SW.o: FOG_SW/FOG_SW.hpp FOG_SW/FOG_SW.cpp
.       ${CC} ${CC_FLAGS_ACS} -c FOG_SW/FOG_SW.cpp
.
MGM_SW.o: MGM_SW/MGM_SW.hpp MGM_SW/MGM_SW.cpp
.       ${CC} ${CC_FLAGS_ACS} -c MGM_SW/MGM_SW.cpp
.
GPS_SW.o: GPS_SW/GPS_SW.hpp GPS_SW/GPS_SW.cpp
.       ${CC} ${CC_FLAGS_ACS} -c GPS_SW/GPS_SW.cpp
GPS_SW_Satellite.o: GPS_SW/GPS_SW_Satellite.hpp GPS_SW/GPS_SW_Satellite.cpp
.       ${CC} ${CC_FLAGS_ACS} -c GPS_SW/GPS_SW_Satellite.cpp
```

Figure 8.42: Makefile for an equipment model library.

© IRS Universität Stuttgart

8.7.7 Configuration Files for a Simulation Run

With the steps of the previous chapter the simulator code has been completely created. However, before being able to start a simulation, the spacecraft simulator has to be configured with a significant amount of detailed parameter values from files - usually provided in XML format.

The information in these files is necessary in order to initialize all characterization parameters (e.g. power consumption) of all simulated model instances correctly. It is essential to characterize both simulator, OBSW and control console with compatible data records from system engineering infrastructure data bases for the entire spacecraft. The issue of simulator integration in such an entire infrastructure is further treated in section 10. The simulator configuration with respect to models in this example is implemented via 4 files which are explained in more detail below:

- The ModelDefaultFile.xml
- The ModelCharacFile.xml
- The SimHarnessFile.xml
- The SchedulingTableFile.xml

The ModelDefaultFile.xml

It is the initialization file for

- a default value setting for all parameters of all model instances in the simulated system.
- The XML structure has to be in line with the corresponding DTD.

The file is imported at simulator startup. The simulator reports according error messages when detecting inconsistent entries.

```
<!-- BEGIN instance : fog_sw_00 ======= -->
        <FOG_SW id="fog_sw_00">
            <angularRate>0.0, 0.0, 0.0</angularRate>
            <bias>0.00001281847373</bias>
            <Data>0</Data>
            <Drift>0.0</Drift>
            <Failure>0</Failure>
            <fogfrequenzy>10</fogfrequenzy>
            <iPwr>0.0, 0.0, 0.0</iPwr>
            <maxTemp>90.0</maxTemp>
            <measurementRange>16.777216</measurementRange>
            <Mode>0</Mode>
            <newangularRate>0.0</newangularRate>
            <pCons>0.0</pCons>
            <pCons_NOMINAL>0.5, 1.800, 0.2, 0.05</pCons_NOMINAL>
            <pDis>0.0, 0.0, 0.0</pDis>
            <randomwalk>0.00001783144719</randomwalk>
            <Reset>0</Reset>
            <resetindicator>0</resetindicator>
            <scalingfactor>0.000152</scalingfactor>
            <seed>1</seed>
            <sigmanoise>0.00000001</sigmanoise>
            <sigmanoisemain>0.0</sigmanoisemain>
            <spaceVector>0.000000000000000,1.000000000000000,0.0000000000000
            <Temp>20.0</Temp>
            <uMax>5.5</uMax>
            <uMin>4.75</uMin>
            <uPwr>0.0</uPwr>
            <warnTempMax>85.0</warnTempMax>
            <warnTempMin>-57.0</warnTempMin>
        </FOG_SW>
<!-- END instance : fog_sw_00 ======= -->

<!-- BEGIN instance : fog_sw_01 ======= -->
        <FOG_SW id="fog_sw_01">
            <angularRate>0.0, 0.0, 0.0</angularRate>
            <bias>-0.00001127191809</bias>
            <Data>0</Data>
            <Drift>0.0</Drift>
            <Failure>0</Failure>
            <fogfrequenzy>10</fogfrequenzy>
            <iPwr>0.0, 0.0, 0.0</iPwr>
            <maxTemp>90.0</maxTemp>
            <measurementRange>16.777216</measurementRange>
            <Mode>0</Mode>
```


Figure 8.43: The configuration data set: ModelDefaultFile.xml.

© IRS Universität Stuttgart

The ModelCharacFile.xml

It is the initialization file for the current simulation run.

- It contains only those model instance parameter values which for the actual simulation run differ from the defaults (e.g. for testing degraded battery performance).
- The XML structure is compliant to the corresponding DTD.

The file is imported after the ModelDefaultFile.xml at simulator initialization time. The simulator reports error messages when it detects incorrect content.

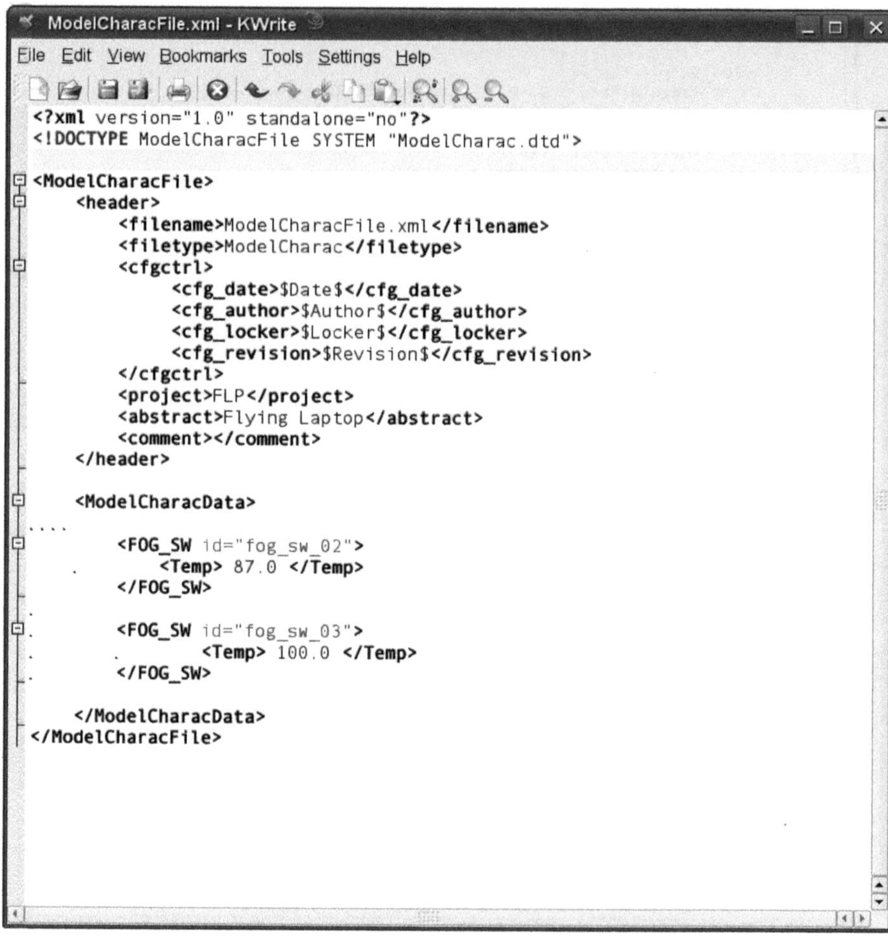


```xml
<?xml version="1.0" standalone="no"?>
<!DOCTYPE ModelCharacFile SYSTEM "ModelCharac.dtd">

<ModelCharacFile>
    <header>
        <filename>ModelCharacFile.xml</filename>
        <filetype>ModelCharac</filetype>
        <cfgctrl>
            <cfg_date>$Date$</cfg_date>
            <cfg_author>$Author$</cfg_author>
            <cfg_locker>$Locker$</cfg_locker>
            <cfg_revision>$Revision$</cfg_revision>
        </cfgctrl>
        <project>FLP</project>
        <abstract>Flying Laptop</abstract>
        <comment></comment>
    </header>

    <ModelCharacData>
....
        <FOG_SW id="fog_sw_02">
            <Temp> 87.0 </Temp>
        </FOG_SW>

        <FOG_SW id="fog_sw_03">
                <Temp> 100.0 </Temp>
        </FOG_SW>

    </ModelCharacData>
</ModelCharacFile>
```

Figure 8.44: The configuration data set: ModelCharacFile.xml.
© IRS Universität Stuttgart

The SimHarnessFile.xml

This is the file for definition of the harness line connections between equipment models - respectively between equipment models and I/O-card drivers in case of hybrid testbenches.

- It contains the definitions of all harness line connections between all models.
- The XML structure is compliant to the corresponding DTD.

- The SimHarnessFile.xml file is independent from the simulation run but is specific for a testbench setup. When the setup is changed by e.g. replacing a simulated model by hardware equipment, the routing of the according harness lines between simulated models and those connected via frontend-cards has to be adapted in the file accordingly.

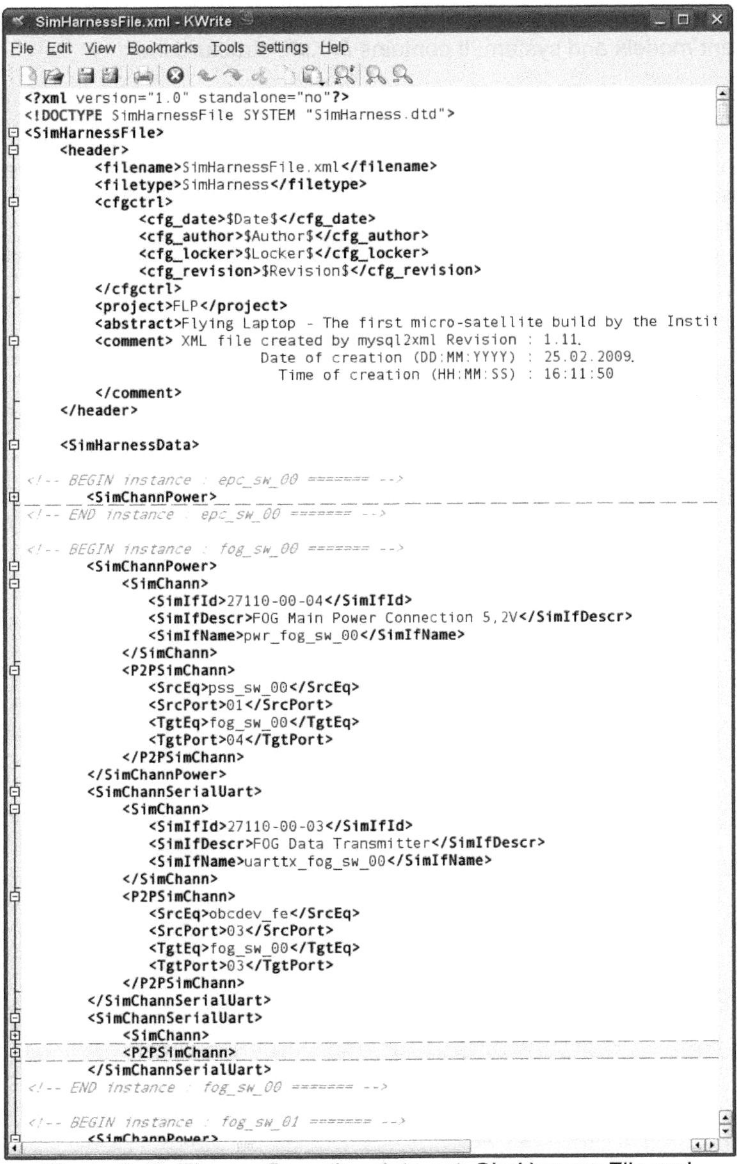


```
<?xml version="1.0" standalone="no"?>
<!DOCTYPE SimHarnessFile SYSTEM "SimHarness.dtd">
<SimHarnessFile>
    <header>
        <filename>SimHarnessFile.xml</filename>
        <filetype>SimHarness</filetype>
        <cfgctrl>
            <cfg_date>$Date$</cfg_date>
            <cfg_author>$Author$</cfg_author>
            <cfg_locker>$Locker$</cfg_locker>
            <cfg_revision>$Revision$</cfg_revision>
        </cfgctrl>
        <project>FLP</project>
        <abstract>Flying Laptop - The first micro-satellite build by the Instit
        <comment> XML file created by mysql2xml Revision : 1.11.
                          Date of creation (DD:MM:YYYY) : 25.02.2009.
                            Time of creation (HH:MM:SS) : 16:11:50
        </comment>
    </header>

    <SimHarnessData>

<!-- BEGIN instance : epc_sw_00 ======= -->
        <SimChannPower>
<!-- END instance : epc_sw_00 ======= -->

<!-- BEGIN instance : fog_sw_00 ======= -->
        <SimChannPower>
            <SimChann>
                <SimIfId>27110-00-04</SimIfId>
                <SimIfDescr>FOG Main Power Connection 5,2V</SimIfDescr>
                <SimIfName>pwr_fog_sw_00</SimIfName>
            </SimChann>
            <P2PSimChann>
                <SrcEq>pss_sw_00</SrcEq>
                <SrcPort>01</SrcPort>
                <TgtEq>fog_sw_00</TgtEq>
                <TgtPort>04</TgtPort>
            </P2PSimChann>
        </SimChannPower>
        <SimChannSerialUart>
            <SimChann>
                <SimIfId>27110-00-03</SimIfId>
                <SimIfDescr>FOG Data Transmitter</SimIfDescr>
                <SimIfName>uarttx_fog_sw_00</SimIfName>
            </SimChann>
            <P2PSimChann>
                <SrcEq>obcdev_fe</SrcEq>
                <SrcPort>03</SrcPort>
                <TgtEq>fog_sw_00</TgtEq>
                <TgtPort>03</TgtPort>
            </P2PSimChann>
        </SimChannSerialUart>
        <SimChannSerialUart>
            <SimChann>
            <P2PSimChann>
        </SimChannSerialUart>
<!-- END instance : fog_sw_00 ======= -->

<!-- BEGIN instance : fog_sw_01 ======= -->
        <SimChannPower>
```

Figure 8.45: The configuration data set: SimHarnessFile.xml.

© IRS Universität Stuttgart

The file is imported at simulator startup. The simulator reports error messages when detecting incorrect line connections. It In such case it will refuse startup.

The SchedulingTableFile.xml

This is the file for configuration settings of the simulator scheduling details for equipment models and system. It contains in XML structure the

- cycle times, and
- relative time intervals of all model computations,

as explained in section 7.3. The file is imported at simulator startup. The simulator will refuse startup in case inconsistent content is detected.

```
<?xml version="1.0" standalone="no"?>
<!DOCTYPE SchedulingTableFile SYSTEM "SchedulingTable.dtd">
<SchedulingTableFile>
    <header>
        <filename>SchedulingTableFile.xml</filename>
        <filetype>SchedulingTable</filetype>
        <cfgctrl>
            <cfg_date>$Date$</cfg_date>
            <cfg_author>$Author$</cfg_author>
            <cfg_locker>$Locker$</cfg_locker>
            <cfg_revision>$Revision$</cfg_revision>
        </cfgctrl>
        <project>FLP</project>
        <abstract>Flying Laptop</abstract>
        <comment></comment>
    </header>

  <SchedulingTableData>

    <Scheduler>
        <Period>0.010</Period>
    </Scheduler>

    <TCO type="cfm">
        <Name>EnvDynPropagator</Name>
        <Offset>0.000</Offset>
        <dt>0.010</dt>
        <Status>0</Status>
        <MPStatus>0</MPStatus>
    </TCO>

    <TCO type="equipment">
        <Name>fog_sw_00</Name>
        <Offset>0.020</Offset>
        <dt>0.050</dt>
        <Status>0</Status>
        <MPStatus>0</MPStatus>
    </TCO>

    <TCO type="equipment">
        <Name>fog_sw_01</Name>
        <Offset>0.020</Offset>
        <dt>0.050</dt>
        <Status>0</Status>
        <MPStatus>0</MPStatus>
    </TCO>

    <TCO type="equipment">
        <Name>fog_sw_02</Name>
        <Offset>0.020</Offset>
```

Figure 8.46: The configuration data set: SchedulingTableFile.xml.

© IRS Universität Stuttgart

8.7.8 Simulation Run

The completely configured simulator can now be started. The following figure shows a live demonstration of the simulator start and simple interactions with the FOG model. The simulator command window can be identified in the upper left corner. All received commands are logged there together with the command input the models receive and together with their submitted output (partly as hexadecimal data packets).

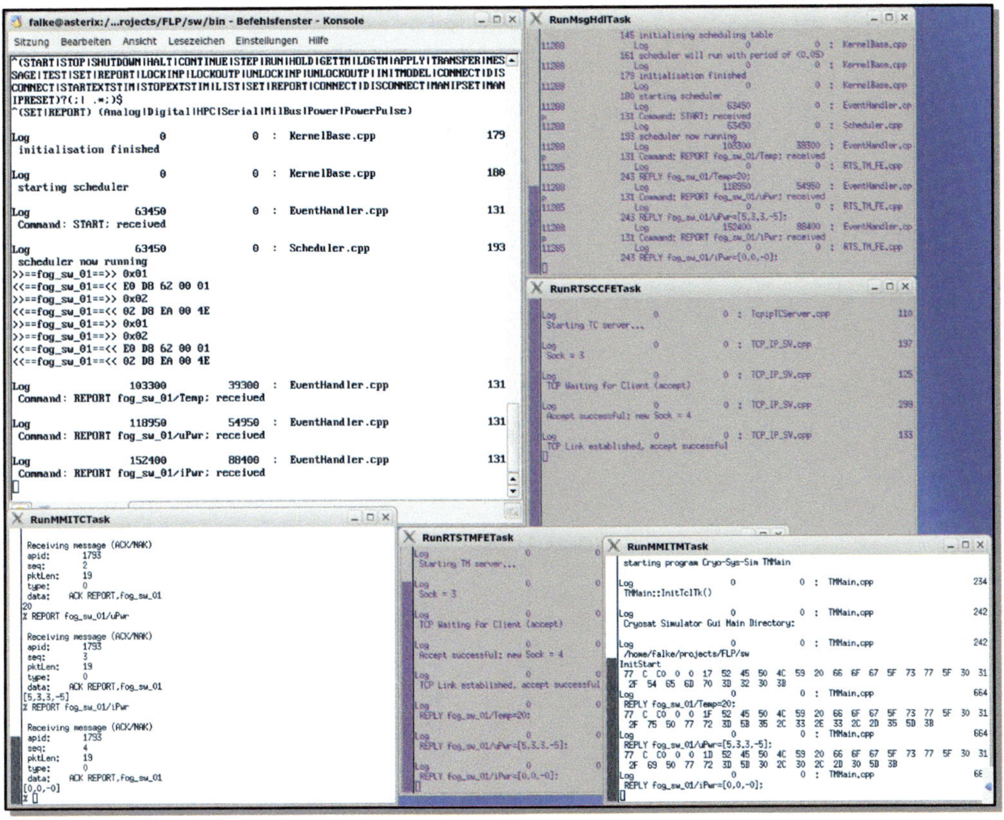

Figure 8.47: Log windows from different threads of the running simulator.
© IRS Universität Stuttgart

More straightforward for a beginner to read is the log window in the upper right corner. It is the simulator-message-handler-task window. Its output provides information about the clear text commands which have been received by the simulator from the control console. The result values (e.g. FOG rotational rate vector) responded by the simulator are also displayed. However, cyclic simulator telemetry is not included here. The other windows contain log output reporting protocol connection statistics between simulator and control console.

9 Simulator Development Compliant to Software Standards

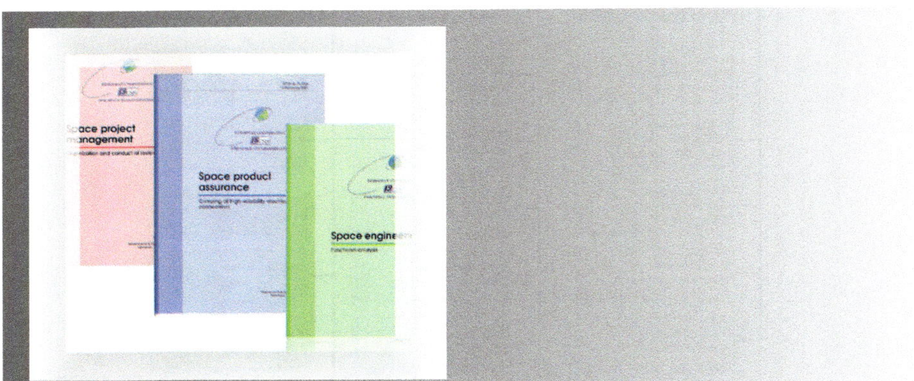

J. Eickhoff, *Simulating Spacecraft Systems*, Springer Aerospace Technology 1,
DOI 10.1007/978-3-642-01276-1_9, © Springer-Verlag Berlin Heidelberg 2009

9.1 Software Engineering Standards – Overview

By means of the simulation based testbenches, amongst other components, the spacecraft on-board software is verified - first as pre-verification on the SVF, then with the hardware / software compatibility tests on an STB and finally in an EFM configuration. Since the on-board software is one of the most critical elements of a spacecraft, immediately on the spacecraft customer side, questions concerning software quality and verification state of the simulators and testbenches arise. Therefore the compliance to an according set of software standards of the spacecraft customer always will be an essential requirement for testbench development.

Such software standards prescribe diverse development guidelines for software in a space project - here to be interpreted accordingly for the simulators, respectively all software elements of testbenches. Prescribed usually are the development approach, the development phases, the review milestones, documentation to be delivered for each milestone such as development plan, requirements, architecture and design documentation, test plans, reports, and all their content structure. Several software standard families exist:

ECSS Standards

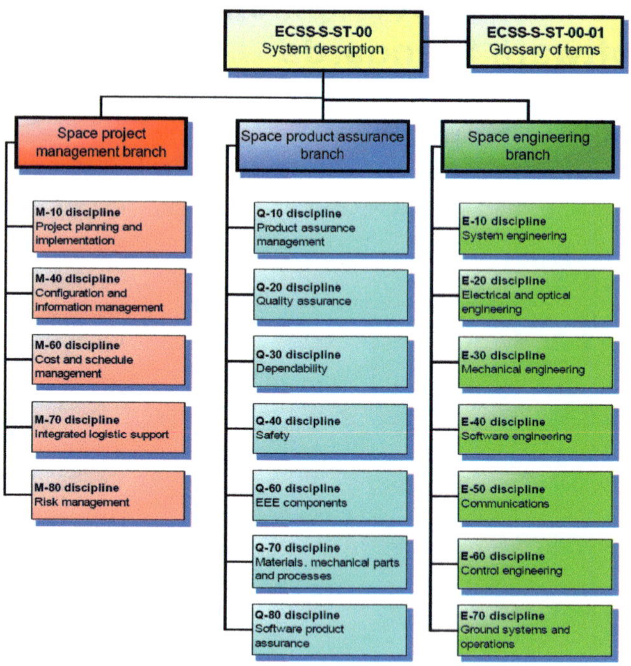

Figure 9.1: Family of ECSS standards. © ECSS

For European space projects there exists an entire suite of standards for spacecraft development in general, the so-called ECSS standards. These standards are elaborated and published by the European Cooperation for Space Standardization, (ECSS). This commission includes members from the European Space Agency, (ESA), diverse national agencies and from industrial partners. Relevant for software development and thus for simulators and testbenches are especially the standards:

- ECSS-E-ST-40 Software engineering, and
- ECSS-Q-ST-80 Software product assurance.

Please also refer to figure 9.1. The ECSS standards are a family of cross-referencing documents which is very exhaustive and precise but also rather unhandy to read since the reader has to follow cross-references from one document to the other. The standards currently are under revision. The completely revised document set shall be available during 2009.

Aeronautical Software Standards (Aerospace) - DO178B

DO178B defines the guidelines for aeronautics software. It was developed by the Radio Technical Commission for Aeronautics, Inc. (RTCA) and was accepted as certification standard for aeronautics software by the US Federal Aeronautics Association FAA (see Advisory Circular AC20-115B). Albeit de facto it meanwhile is a world wide applied standard for aeronautic software and its development.

- DO178B primarily treats the software development itself. Within the development process diverse accompanying quality and test documents are to be worked out. So DO178B to a certain extent is the counterpart to both ECSS-E-ST-40 + ECSS-Q-ST-80.
- DO178B for space business always is applicable for systems which in parallel are interfering with aeronautics, such as
  - ◇ "quasi-airplanes", like Space Shuttles, or commercial spaceships like "Spaceship One", and
  - ◇ aeronautics support systems like GPS or Galileo (especially their payload software and their ground segment software).

Standards for general Software – ANSI/IEEE

In the space business generic software standards only are applicable for support tools where a tool problem or failure would not induce a disturbance of the spacecraft itself. Such equipment e.g. can be certain ground support equipment, handling equipment etc. The IEEE standards for software development are:

- ANSI/IEEE-729 Glossary of Software Engineering Technology
- ANSI/IEEE-1058 Software Project Management Plan
- ANSI/IEEE-830 Software Requirements Specification

- ANSI/IEEE-828 Software Configuration Management Plan
- ANSI/IEEE-1012 Software Verification and Validation Plan
- ANSI/IEEE-1016 Software Design Description
- ANSI/IEEE-730 Software Quality Assurance Plan
- ANSI/IEEE-1028 Software Reviews and Audits
- ANSI/IEEE-829 Software Test Documentation

Software Standards for dedicated Space Projects

For certain large-scale space projects also eventually dedicated software standards can be defined. In most cases they are derivates or combinations of various specific standards or combinations of diverse national standards. Examples are

- the Columbus Software Development Standard, (CSDS), and
- the Galileo Software Standard (GSWS).

The Galileo Software Standard (GSWS) for the European satellite based navigation system e.g. comprises all the following domains of

- software engineering,
- software quality assurance, and
- software configuration management

in "a single book" which makes them much simpler to read and to understand than the ECSS counterpart although from the point of requirements they impose on ground software, testbenches and simulators they are rather comparable. GSWS is a closed and complete pure software standard, however it to a large extent neglects the topics of hardware / software integration. For simulators however this is essential w.r.t. configuring a simulation and the Simulator-Frontend with respect to performance, I/O-line mappings and characterization settings for each I/O-line of a simulator in the hybrid testbench. Here the standard is accordingly to be "interpreted" by the testbench development team, especially for how and what is to be tested.

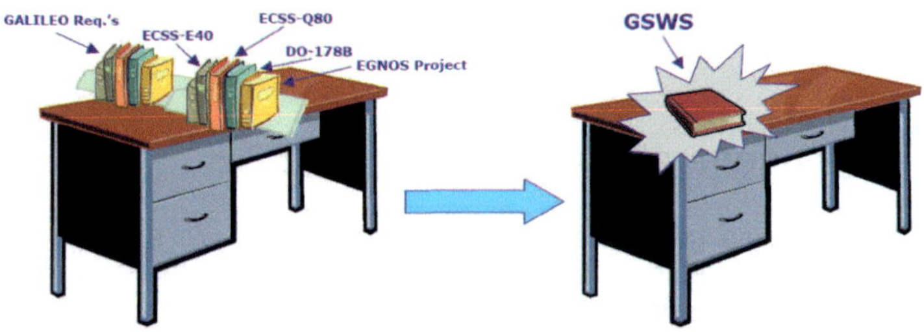

Figure 9.2: Galileo Software Standard as a closed single book standard. © ESNIS

The Galileo Software Standard comprises a common requirements set for all software development, integration and test phases in the frame of the Galileo navigation system program. Furthermore operations and maintenance topics are treated as well as the full scope of software product assurance topics. GSWS is a common standard for the:

- Space Segment, (SS), which encompasses all elements on board the Galileo navigation satellites.
- Ground Control Segment, (GCS), comprising all components inside the ground stations for control and housekeeping of the 30 satellites.
- Ground Mission Segment, (GMS), comprising all components inside the operator stations by which the Galileo payloads of the satellites are operated. This includes signal generation, security codes handling, cyclic code updates, leap time corrections of the atomic clocks aboard etc.
- Test User Segment, (TUS), comprising all elements for test of Galileo receivers and car navigation systems under realistic conditions before full in-orbit availability of the spacecraft.

Further reading and Internet pages concerning software development standards is provided in the according subsection of this book's references annex.

9.2 Software Classification According to Criticality

The requirements towards software development, testing and documentation as well as formal acceptance, which are prescribed by a software standard, usually depend on the software criticality for the space mission. Usually on-board software for safety critical systems is ranked with the highest criticality level, such as control software for ECLS Systems, manned spaceship control software or navigation software used for Airplane guiding such as Galileo navigation payload software. Software for ground equipment, such as a Power-Frontend has lower criticality ranking and for example less extensive testing is required.

Also for the simulators used for system design and verification the criticality level has to be agreed with the spacecraft customer. Since on-board software tests are not exclusively tested on pure simulation testbenches (SVF) but also on hybrid benches (STB) and finally both on-board software tests and hardware testing are also performed on FlatSat EFM configurations with all hardware in the loop, usually the simulators in the testbenches can be negotiated to be ranked to the lowest criticality level. Later when performing tests on pure "Hardware in the Loop" setups, potential bugs in formerly used simulators would show up. The table below lists the ranking levels cited from the Galileo Software Standards - called “Development Assurance Levels”, (DAL), in the GSWS.

Table 9.1: "Development Assurance Levels" in GSWS.

	SW-DAL Definition
Level A	Software whose anomalous behavior would cause or contribute to a failure resulting in a catastrophic event. (Loss of mission)
Level B	Software whose anomalous behavior would cause or contribute to a failure resulting in a critical event. (Endangering mission)
Level C	Software whose anomalous behavior would cause or contribute to a failure resulting in a major event.
Level D	Software whose anomalous behavior would cause or contribute to a failure resulting in a minor event.
Level E	Software whose anomalous behavior would cause or contribute to a failure resulting in a negligible event.

A similar classification is available in the DO178B called "Certification levels" and in the ECSS-E40, called "Class A" to "Class E" (where in the newer ECSS-E-ST-40 the Class E has been removed).

9.3 Software Standard Application Example

The most important characteristics of a simulator software and testbench development in accordance with a software standard shall now be explained taking the relatively compact Galileo software standard as example. For the developer of such simulation based testbenches always the problem exists, that such software standards typically are written by the agencies having in mind on-board software and accordingly relevant topics. Hardware / software integration problems, cabling, grounding and electrics topics which might affect also simulator software, card drivers for Simulator-Frontends etc. are mostly not in focus and have to be managed by the testbench developer himself. Wherever the entire system of simulator + testbench hardware etc. is affected, in this chapter the terminology applies the keyword "system" (e.g. System acceptance review), wherever purely the simulator software is concerned, the keyword "software" is used (e.g. software unit test plan). The development phases for simulator development and according intermediate reviews are depicted in the figure below. They are similar in GSWS and ECSS-E-ST-40:

SRR	= System Requirements Review	IRR	= Integration Readiness Review
PDR	= Preliminary Design Review	SW-QR	= System Qualification Review
DDR	= Detailed Design Review	SW-AR	= System Acceptance Review

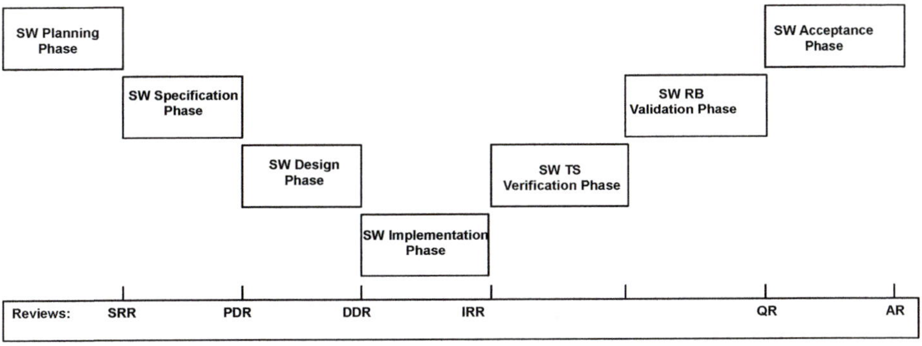

Figure 9.3: Software development process and review milestones. © ESA

For the stepwise approach, the required review milestones, the required documentation as well as document structures and content, the product assurance etc. each software standard has its own "Engineering Requirements". Some software standards replace the IRR by a "Test Readiness Review", (TRR).

Software Engineering Requirements of a Software Standard:
The engineering process requirements are specified in the software standard. Each requirement has a unique identifier number. For each requirement it is marked for which criticality level (in GSWS the DALs) it is mandatory, for which reviews it is mandatory and which documents it affects. At start of project the developer must provide a compliance matrix stating in how far one is fully, partly or non compliant to these engineering requirements (cf. figure 9.5). All deviations must be justified. At project end one must provide a compatibility matrix stating how one was compliant and which documents, review minutes, product assurance reports etc. prove the compliance. An example for such an engineering requirement is given below.

[GSWS-RV-0430] - [A:m B:m C:m D:m E:m]
Reviews shall be planned and described in the Review Plan (RVP, see A.41) that is delivered before every review. § [all] - (RVP)

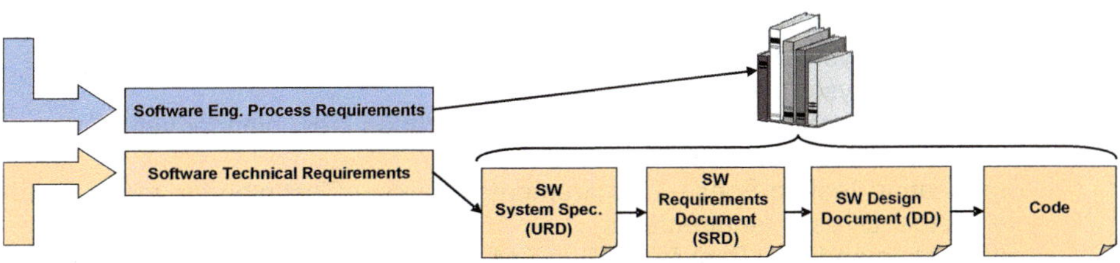

Figure 9.4: Software and process requirements driving development process.

Technical Requirements towards Simulator or Testbench:
For the simulator software, and also for hardware in hybrid testbenches, technical requirements are to be specified. The system architecture and detailed design and

code have to be developed so that the later "product" simulator / testbench fulfills these technical requirements which has to be verified during the integration and test phase (please also refer to figure 9.8 which will be explained in more detail later). An example for a technical requirement is given below, including the reference to higher level spacecraft requirements where it was derived from.

 # REQ 999: *Parameter Override by Command*
The simulator operator shall be able to override any numerically computed state variables during simulation by one-shot overrides or by continuous override. This applies for state variables in models, input parameters from OBC model to equipment models and output data from equipment models to OBC model.
GAL Originated from SATSIMREQ-42, AVREQ331, DEVSVF-1900
#END#

Engineering process requirements and technical requirements together form the development baseline for the to be developed software elements of simulator and testbenches. From the engineering requirements and the technical requirements the set of all the documents results, which are to be written either once for the overall testbench suite in the spacecraft project (e.g. development plan) or specifically per testbench (e.g. integration test reports).

Technical simulator / testbench system documentation:

The purpose and basic content of the required technical documents becomes evident already from their titles:

- Testbench User Requirements Document
- Core EGSE User Requirements Document
- Power and TC / TM-Frontend Equipment User Requirements Documents
- Special Checkout Equipment User Requirements Documents
- EGSE Interface Control Document (ICD)
- Simulator Software Unit Test Plan
- Unit Tests Reports for all Simulator Elements (e.g. equipment models)
- Testbench System Integration Test Plan
- Testbench System Integration Test Report
- Verification Testing Spec – Technical Specification (Test plan for verification of all software requirements)
- Verification Testing Spec – Requirements Baseline (Test plan for verification of all testbench user requirements)
- Integration and Test Plan for STB / EFM
- Integration and Test Report for STB / EFM
  System Configuration Files for all releases of all testbenches
- System Operations Manuals for all testbenches
- Documentation and Certificate of Conformance (CoC) concerning compliance to the general CE standards
- Acceptance Data Package

Development process documents:

At least for newcomers the content of some of the required engineering process documents does not become so easily apparent. Therefore here the most important software engineering process documents are cited with their main content elements bulleted:

- The Software Development Plan of the Testbench Infrastructures defines:
  - ◇ The development responsibilities, roles and identified key persons
  - ◇ Applied technical infrastructures (from used simulator kernels down to software configuration handling tools and document handling tools)
  - ◇ The development schedule (often aligned to the spacecraft OBSW development) and the simulator / testbench reviews and their scope
  - ◇ The risk management w.r.t. available resources and w.r.t. application of new development approaches or invented technologies
  - ◇ The technical safety aspects (mostly not relevant for simulator based testbenches, rather for installations like engine test stands etc.)
  - ◇ The interfaces to associated project partners (if relevant), e.g. in case of trans-company development approaches
  - ◇ The management of subcontractors
  - ◇ The management of software problem logging and correction tracing
  - ◇ The cycles of progress reports to the spacecraft customer (in most cases the same as for the spacecraft development process itself)
  - ◇ The applied software engineering approach and engineering environment (e.g. MDA / UML)
  - ◇ Compliance to the software standard for design, coding, testing as well as eventual deviations with justifications (must be formally agreed by customer in SRR)
  - ◇ Definition of software documentation
  - ◇ Information on number and variant of installations
  - ◇ Maintenance strategy (relevant only for operations support simulators)

- The Procured SW Justification File
  - ◇ defines all software components which are reused from predecessor projects and will not be requalified (e.g. simulator kernel software, Simulator-Frontend card driver software).
  - ◇ It defines the technical status of such reused elements,
  - ◇ reports their technical qualification performed within the predecessor projects,
  - ◇ justifies the selection of these reused software components,
  - ◇ includes a qualification plan for reused but adapted software components,
  - ◇ includes a procurement plan for procured software elements, and
  - ◇ eventually comprises a maintenance plan for reused hardware / software components from predecessor projects.

- The Software Verification and Validation Plan
  - ◇ defines the simulator development project's organization structure,

- the foreseen software verification approach with the
  - verification of administrative procedures,
  - the responsibilities of personnel and resources in the project (including tool resources),
  - the schedule of verification activities,
  - a description of the software verification activities themselves
  - and the general verification / reverificatioon approach.
- Besides this it describes the simulator / testbench validation approach
  - concerning simulator / testbench operational suitability including test approach, test concept from lower to upper level and regression testing approach for software upgrades.
  - It describes the approach for qualification of the simulator equipment models, and
  - the sequence of validation activities.

* The Software Product Assurance Plan comprises all facts concerning an independent quality assurance. This includes:
  - Firstly a description of the software development approach (V-model, waterfall approach, spiral model or other),
  - together with the corresponding product assurance, (PA), tasks for each development phase (requirements definition phase design phase, implementation phase, test phase, acceptance).
  - Furthermore for each review milestone according PA tasks are defined, such as documents to be reviewed.
  - Applicable practices for "measurement" of software quality in the form of metrics to be applied are defined (e.g. average number of open SW problem reports, average cycle time between identified SW problem and problem fix),
  - and also metrics for quality ranking of software documentation.
  - Additionally described are the countermeasures in case of violation of requirements or prescribed development processes,
  - the formalities for handling of additional requirements arising later in the project, and
  - formalities for requirements to be waived during development.
  - The document furthermore confirms the criticality ranking of the software (GSWS DAL, ECSS Class etc.)
  - and describes methods for monitoring of software element procurement.
  - Finally the PA plan comprises statements w.r.t. warranty (in most cases not applicable for simulators and testbenches) and for PA tasks during maintenance.

* Software Review Plan:
  - For each of the review milestones being defined in the software development plan, a review plan has to be prepared and has to be distributed to the review members 4-6 weeks prior to the milestone itself.
  - In the first place it provides an overview on the as designed / built status of the simulators / testbenches.
  - Further it defines which documents are to be reviewed for each milestone

and are subject to acceptance (e.g. design document, test plans / reports).
- ◇ The review plan defines the agenda of the meeting, the chairman, the reviewers, other participants (e.g. PA), the location and furthermore,
- ◇ duedates for comments (e.g. on to be reviewed documents) or for review item discrepancies (RIDs) and the duedate for RID answers back to the customer.
- ◇ Furthermore the review success criteria, as well as,
- ◇ the status of open / closed action items from the previous milestone are included.

- Compliance-Matrices against Requirements of the Software Standards:
  - ◇ By these matrices it is demonstrated that the requirements towards the software engineering process imposed by the software standard are properly fulfilled.
  - ◇ For a content example please refer to figure 9.5.

- A Software Configuration File:
  - ◇ It is generated for each released simulator software revision, also for intermediate versions and release candidates.
  - ◇ It contains the version information for each software element (kernel, equipment models etc.) of the actual release, together with:
    - ► a changelog listing all deltas to the previous version,
    - ► a list of functional limitations of the current release, and
    - ► a list of all open SW problem reports, non-conformances and those closed by the delivered release.
  - ◇ Furthermore usually an installation guide is included for the users which have to upgrade their SVFs or STB / EFM in the project.

- Software / System Verification Report:
  - ◇ The SVR (please also refer to figure 9.8) finally sums up in large overview tables which software and user requirements are fulfilled by
    - ► equipment in hardware (e.g. procured stimuli or frontend-equipment) and how this equipment complies to the requirements, or,
    - ► which required simulator software functions are verified on unit level by which tests,
    - ► or are verified on integration level by which tests,
    - ► which software requirements are verified on system level by which tests, and
    - ► which user requirements are verified on system level by which tests.

    So by the SVR it is demonstrated that all the user requirements towards the simulator / testbenches are properly fulfilled.

- Software Maintenance Plan:
  - ◇ It covers all aspects of further software maintenance after final delivery. For simulators / testbenches this only is of relevance if they are used further after spacecraft launch - e.g. for simulators in spacecraft ground segments.

Requirements	Txt	DAL	Rev	Doc	Justifications / Comments
114: ## [GSWS-SWENG-1140] - [A:m B:m C:m D:m E:m]Forward and backward traceability matrixes between TS (SRD, SWICD) and SDD (ADD section) shall be inserted in the SDD§ [SW-PDR] - (SDD)	C	C	C	C	
115: ## [GSWS-SWENG-1150] - [A:m B:m C:m D:m E:m]The Contractor shall identify the requirements in the TS and the RB that cannot be validated on its own environment, and shall agree with the next-higher-level Contractor to be validated by the next-higher-level Contractor.§ [SW-PDR] - (TS)	C	C	C	C	
116: ## [GSWS-SWENG-1160] - [A:m B:m C:m D:m E:m]The planning of software development and testing facilities design, implementation and testing shall be established and documented.§ [SW-PDR, SW-DDR, SW-TRR] - (SDP, Sw-environment documentation)	C	C	C	C	
117: ## [GSWS-SWENG-1170] - [A:m B:m C:m D:m E:m]The contractor shall define the Software Coding Standards (SCS), as part of the SDP.§ [SW-PDR] - (SDP)	C	C	C	C	
118: ## [GSWS-SWENG-1180] - [A:m B:m C:m D:m E:m]SCS shall select a SW programming language according to TABLE 6 A.SOFTWARELANGUAGE DALA DALB DALC DALD DALE Ada Allowed Allowed Allowed Allowed Allowed Assembler Allowed Allowed Allowed Allowed Allowed C Allowed Allowed Allowed Allowed Allowed C++ Allowed Allowed Java Allowed TABLE 6 A: Allowed Coding Languages vs. SW-DALs§ [SW-PDR] - (SDP)	C	C	C	C	
119: ## [GSWS-SWENG-1190] - [A:m B:m C:m D:m E:--]The SCS shall justify the programming language used to develop the SW based on maintainability, safety, interoperability, adaptability, portability and Galileo requirements on product selection and utilisation.§ [SW-PDR] - (SDP)	NA	NA	NA	NA	NA - MDVE is DAL E
120: ## [GSWS-SWENG-1200] - [A:m B:m C:m D:m E:m]SCS shall select Ada and/or C for flight software applications.§ [SW-PDR] - (SDP)	C	C	C	C	C++ used in MDVE - no flight SW
121: ## [GSWS-SWENG-1210] - [A:m B:m C:m D:m E:m]SCS shall justify the selection in case Assembler, C++ or Java are selected.§ [SW-PDR] - (SDP)	C	C	C	C	
122: ## [GSWS-SWENG-1220] - [A:m B:m C:m D:m E:--]SCS shall contain the definition of the programming language coding rules.§ [SW-PDR] - (SDP)	NA	NA	NA	NA	NA - MDVE is DAL E
123: ## [GSWS-SWENG-1230] - [A:-- B:-- C:-- D:-- E:m]SCS shall contain the definition of the language subset if any is chosen for DAL E.§ [SW-PDR] - (SDP)	C	C	C	C	
124: ## [GSWS-SWENG-1240] - [A:m B:m C:m D:m E:--]SCS shall define a safe subset of the language for DALs A, B, C and D.§ [SW-PDR] - (SDP)	NA	NA	NA	NA	NA - MDVE is DAL E
125: ## [GSWS-SWENG-1250] - [A:m B:m C:m D:m E:--]SCS shall contain the definition of defensive programming rules.§ [SW-PDR] - (SDP)	NA	NA	NA	NA	NA - MDVE is DAL E
126: ## [GSWS-SWENG-1260] - [A:m B:m C:m D:m E:--]SCS shall contain the definition of coding policy for error and fault handling (fault propagation, logging, announcement and recovery), including rules to handle the propagation of errors from procured Operational Software.§ [SW-PDR] - (SDP)	NA	NA	NA	NA	NA - MDVE is DAL E

Figure 9.5: Compliance matrix against software engineering process requirements.

© Astrium

The engineering process requirements of the applied software standard furthermore define, for which of the review milestones which of these many cited documents are to be provided as intermediate revision or as final baseline. Such a documentation overview is shown in table 9.2.

Table 9.2: Allocation of to be generated software documents to review milestones.
(Adapted from GSWS)

File	*Document*	*SRR*	*SW PDR*	*SW DDR*	*SW IRR*	*SW QR*	*SW AR*
Requirements Baseline	SW System Specification (URD) + SW IF Requirements Document	B					
Technical Specification	SW Requirements Document + SW IF Control Document		B				
	SW Operations Manual		R	R	R	R	B
Design Justification File	SW Verification & Validation Plan		B				
	SW Unit Test Plan			R	B		
	SW Integration Test Plan			R	B		
	Validation Testing Specification			R	B		
	SW Acceptance Test Plan					B	
	Acceptance Data Package + Installation Report + SW Verification Report					R	B
	Procured SW Justification Document	R	R	R	R	R	B
Design Definition File	SW Design Document		B	B	B		
	SW Configuration Document	R	R	R	R	R	R
Management File	SW Development Plan + SW Configuration Management Plan	R	B				
	Review Plan	R	R	R	R	R	R
Product Assurance File	SW Product Assurance Plan	R	B				
	SW Product Assurance Report	R	R	R	R	R	R

For design office type tool infrastructures (SDO, CDF etc.) and for the commercial tools, which are used during spacecraft design phase like Matlab / Simulink / Stateflow (cf. table 3.2 - Steps 1 and 2) normally no simulator tool software development is performed which has to follow customer software development standards. Therefore here also no requirements documents have to be generated and no software tool verification is to be performed.

When starting a simulator / testbench verification infrastructure development for a spacecraft project the first step is to identify user requirements concerning the simulator / testbenches and to formalize them in a user requirements document - sometimes also called "Software System Specification" or "User Requirements Document". The user requirements in most cases are structured according to:

- General requirements
- Requirements imposed to specific testbench types (e.g. to SVF or hybrid benches)
- Requirements imposed to the simulator equipment models

The next step is to identify the simulator / testbench elements which will be reused from previous projects respectively from company software pools. These are card drivers, simulator kernel, standard models for the diverse simulated equipment interconnections (data bus models, power line model) and also standard models e.g.

for space environment. These are to be cited accordingly in the a.m. Procured Software Justification File.

The user requirements then have to be further broken down into three classes:

- Requirements to externally procured software / hardware elements
- Requirements to pure hardware equipment for the hybrid benches
- Software requirements for the pure software parts of the testbench(es)

The first requirements type concerns e.g. equipment like Power-Frontend(s), TM/TC-Frontend, Core EGSE Equipment, SCOEs and stimuli equipment. For all these procured items dedicated requirement documents are to be written to assure they all fit properly together later when integrated into the overall testbench. For such externally procured equipment, the supplier is responsible to prove full compliance to all the imposed requirements and he has to provide a verification matrix how the requirement compliances have been inspected, tested or verified otherwise.

For all the pure hardware equipment which is procured directly by the spacecraft manufacturer (such as the test harness of a hybrid bench, test rigs etc.) the technical requirements are to be laid down in a dedicated test equipment specification, which also is a requirements document).

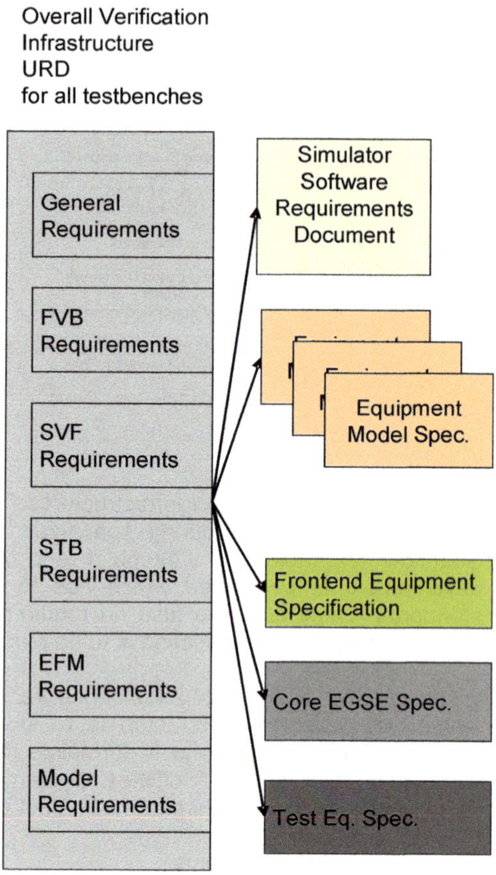

Figure 9.6: Allocation of user requirements to hardware / software requirements.

A large scope of the testbench functions is however implemented in software, the key element thereof being the spacecraft simulator infrastructure and the equipment models. Therefore from the user requirements, corresponding software requirements are derived. Specifically for these again the allocation of general requirements to the simulator infrastructure and specific requirements to the equipment models of the spacecraft shall be considered. Please refer to figure 9.7.

From the user requirements towards the software infrastructure, generic software requirements arise, tackling topics such as data interfaces, log file formats, performance requirements and compatibility to certain control console interfaces. These are collected in the simulator software requirements document. Further user requirements exist concerning the spacecraft equipment models (e.g. concerning physical effects to be considered) and from these for each equipment model type so-called "model specifications" are derived. These model specifications can be considered as annexes to the simulator software requirements document and together they represent the overall testbench software requirements document.

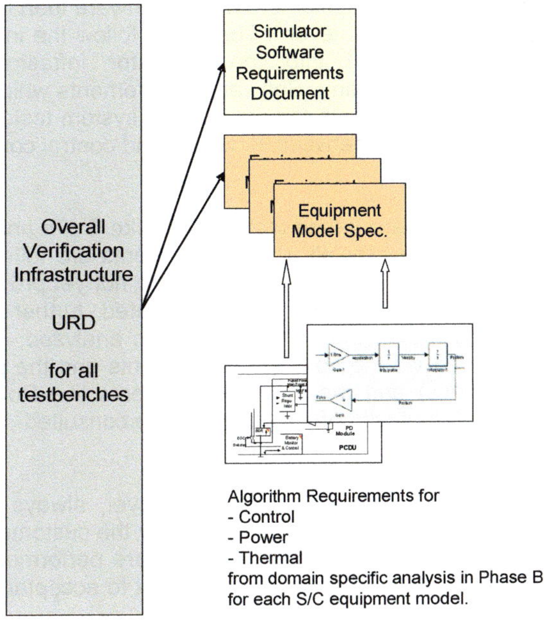

Figure 9.7: User, software and algorithm requirements.

To handle all the model specifications as separate documents has proven good practice because:

- Each document then easily can be reviewed by the corresponding real equipment's expert in the project - the so-called cognizant - and by the system engineering team.
- Secondly when managed as separate documents they can be updated individually as soon as real hardware of one equipment type changes.
- Thirdly they can individually be reused in the next project if the next spacecraft is equipped with the same hardware equipment.

The documentation typically used for definition of a model specification are the real equipment's interface control document (ICD) and the design document (DD) and eventually some technical notes (TNs) and the user manual (UMAN). Eventually in spacecraft development phase B dedicated control algorithms (for thermal, AOCS or power control for the equipment) are developed by means of design tools (cf. step 2 in table 3.2). In such cases the pre-verified algorithms will directly be taken over into the model specifications as so-called algorithm requirements. In case where tools like Simulink are applied, directly C-code can be generated from the designed / verified algorithms and this code can be taken over 1:1 into an according sourcecode file which is cited in the model specification.

The simulator and equipment model software then is designed and coded and then is tested - first on unit level[25]. Thereafter follow the integration tests, e.g. of equipment models embedded into the simulator infrastructure, their configuration and interconnection testing. Software requirements which could neither be verified in unit nor in integration tests require specific system tests for their verification - e.g. test of proper interaction between simulator and control console / Core EGSE.

For pure software based testbenches like SVFs and operations simulators, then it is analyzed in how far all user requirements are fulfilled (see figure 9.8). In case any such user requirements exist which are not yet proven on unit, integration or already performed system test level, dedicated further system level tests are to be performed. For hybrid testbenches it is analyzed, in how far the user requirements are covered via the software components and the pure hardware elements (verified test harness) and the procured elements like Power-Frontend etc. For these the verification matrices of the suppliers are consulted.

For all tests, from unit to system level, always test plans and procedures are prepared which have to be accepted by the customer during according test readiness review milestones and then the tests are performed and test reports are generated with the results which again are subject to acceptance in a qualification / acceptance review. The overall aggregation of

- which user requirement is broken down into which software requirement,
- and test equipment requirement,
- and procured equipment requirement,

and is verified via which

- unit tests,
- integration tests,
- system tests
- respectively supplier tests,

together with the according verification methods and applied procedures is included in large tables in the system verification report. For this aggregation process again please refer to figure 9.8.

[25] For simulator development in most cases a "unit" can be considered as equal to en entire equipment model.

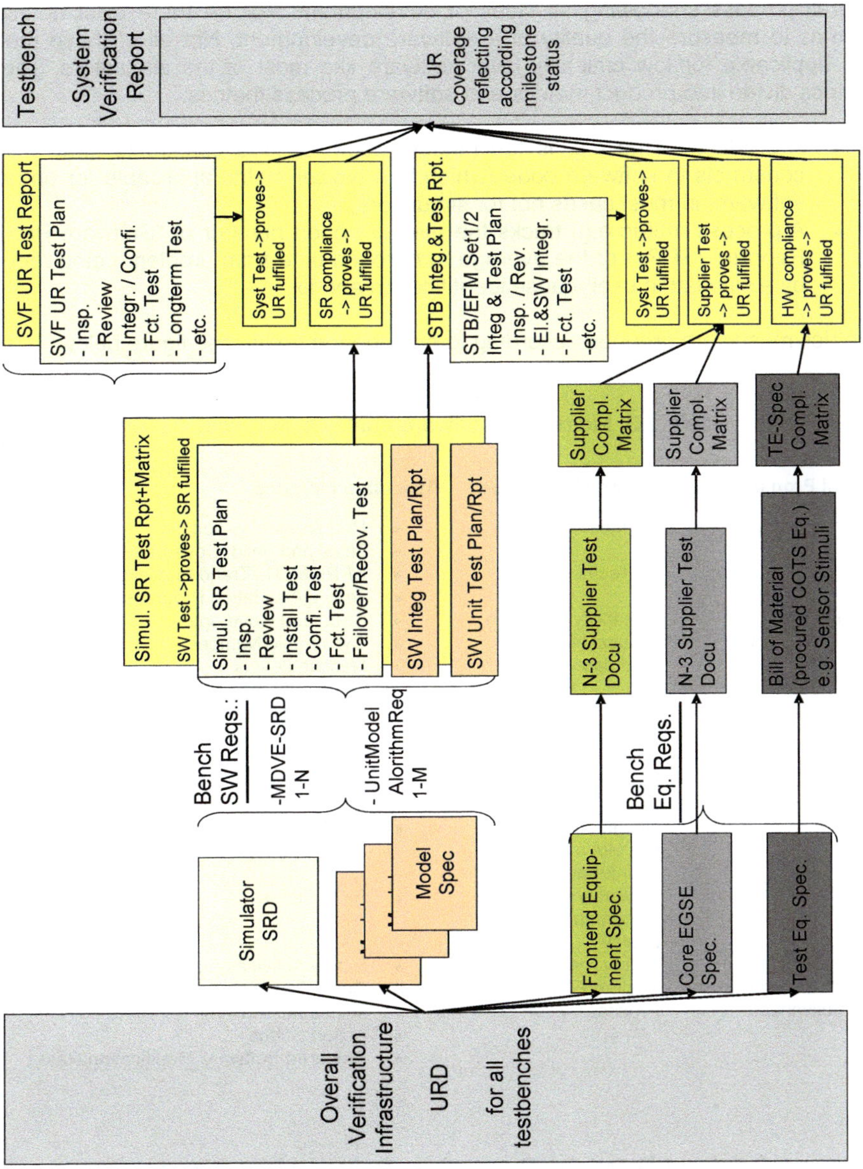

Figure 9.8: Testbench requirements and their verification.

The product assurance departments of contractor and spacecraft customer accompany the development and verify the proper generation of documentation, implementation and version control of the software elements and performance of tests - specially all tests with flight hardware in the loop. Product assurance (PA) is an overall process shadowing all steps of development. For PA there exist dedicated metrics to measure the quality of a software development. Not all of these metrics are applicable for low criticality level software like most of the simulators. The PA metrics divide into product metrics and software process metrics.

- A product metric e.g. is the nesting depth of loop constructs or IF / Then constructs in software code (which is a typical metric applicable for on-board software, in most cases not for simulators).
- A process metric e.g. tracks the average open number of "Software Problem Reports" (SPRs) or the average turnaround time from problem being reported by a simulation user, cause identification and fix.

Metrics mostly are grouped according to engineering goals, see figure below:

Table 9.3: Metrics for software quality assurance. © ESNIS

Goal Properties	Related Properties	Associated metrics
Functionality	Completeness	• Requirements Allocation • Tests and Valid. Coverage Completeness.
	Correctness	• SPRs/NCRs Trend Analysis • Testing/Validation Progress
Reliability	Reliability Evidence	• Structural Coverage
Maintainability	Analyzability	• Cyclomatic Number
	Modularity	• Nesting Levels • Modularity Size Profile • Number of Exits • Number of Entries
	Adaptability	• Average SPR/NCR Turn Around Time
Documentation Quality	Requirements Quality	• Requirements Stability
	Documentation Development and Maintenance	• Code Comment Frequency
	Operation-related Documentation quality	• RIDs Status
Suitability for Safety	Safety Evidence	• Safety Activities Adequacy
System Engineering Effectiveness	System Engineering Process evidence	• Code Size Stability • Milestone Tracking • Action Status • Procured Software Modification Rate

9.4 Critical Path in Spacecraft Development

The Critical Path

In a classic spacecraft project where still an engineering model, (EM), is built and which is not following a simulation based approach, usually the OBSW represents the main element on the critical path in the schedule. The OBSW development cannot be started before spacecraft design (including equipment design) is consolidated. OBSW testing cannot be started before OBC prototype availability and the software has already to be available for engineering model integration testing.

When applying the model based development approach which replaces the spacecraft engineering model by simulations, the simulator becomes the schedule critical element. This is because the SVF has to be available right in time for first OBSW tests, and this is even earlier than in the former EM based approach, since in such a project approach the SVF is intended to allow OBSW testing already even before OBC prototype availability. The second schedule critical element is the in-time availability of the first hybrid testbench (STB) with installed simulator and verified test harness to connect the OBC prototype as soon as being available. The time slots between available stable spacecraft equipment definitions for development of SVF and STB and their required availability are extremely short.

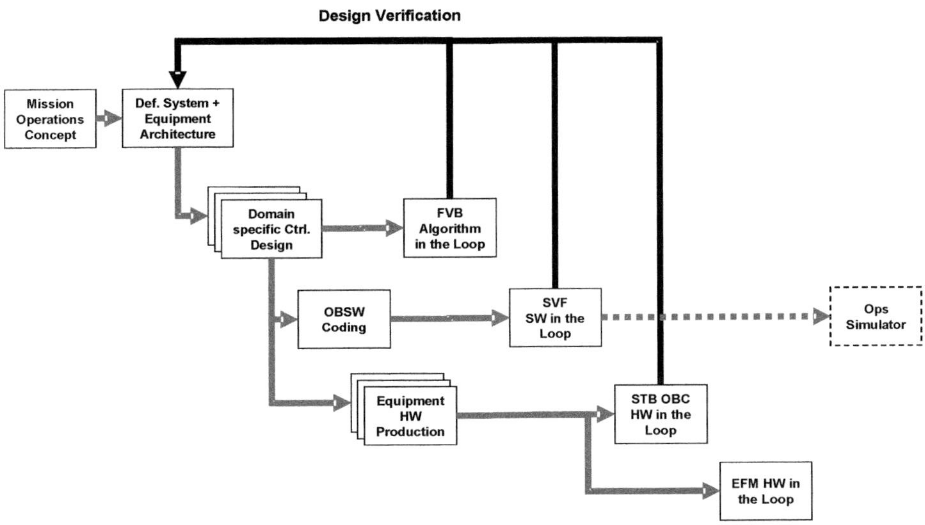

Figure 9.9: Testbench configurations in the spacecraft development flow.

Mitigation Strategies

To mitigate the risk, and to solve the problems resulting from potential late design consolidation of dedicated spacecraft components and resulting late freeze of dependent testbench components, a stepwise development approach has proven to be an adequate means. The approach parallelizes SVF development, OBSW development, STB development and Spacecraft Assembly, Integration and Tests (AIT). The stepwise testbench development usually is aligned with the OBSW versions developed and verified in the spacecraft project. This OBSW e.g. for satellite projects is developed in 3 to 4 stages with enhancing functionality as could be e.g.:

- Core OBSW with operating system, data handling functionality and TC/TM functions
- OBSW with added AOCS functionality
- OBSW with further added functionality for satellite platform control
- OBSW with further added functionality for satellite payload control

The following figure explains such an approach. The timely staggered and overlapping availability of the testbenches allows testing a new OBSW on the newest SVF, in parallel to test the previous OBSW on STB and furthermore assembly, integration and test (AIT) activities on the EFM. Furthermore operations procedures for the spacecraft for control in orbit can be tested both with the OBSW versions installed on SVF and STB.

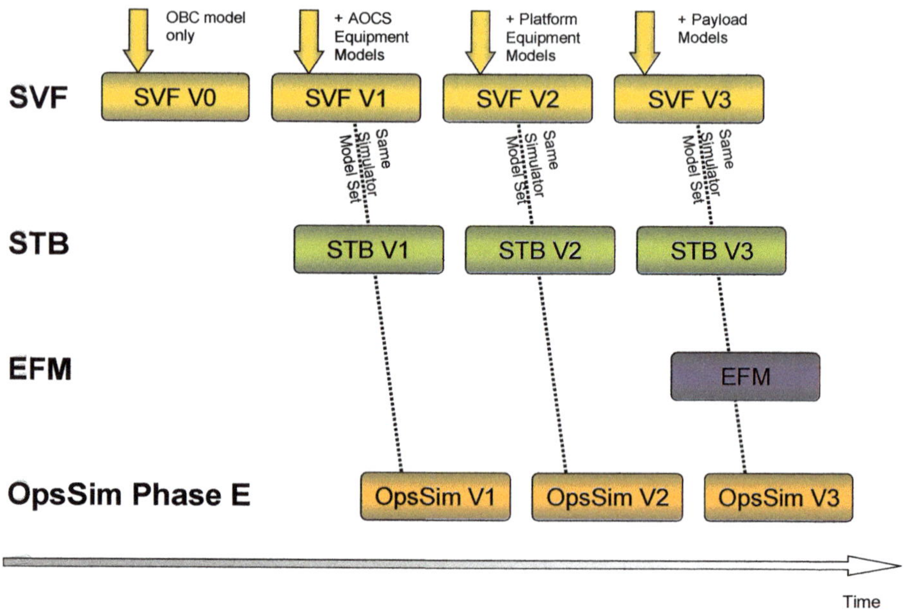

Figure 9.10: Development approach with multiple stages for each bench. © Astrium

By this approach the testbenches can be built up step by step. Always one version is under development while the previous is already in use. The schedule is optimized at its best and the testbenches get back away from the project's critical path. However the project's software standards imply preparation of a large amount of documentation and furthermore each development step has to be closed by a review milestone. So the staggered approach with its various versions for each testbench would be blocked already by the multitude of necessary reviews or if omitting them, the development would no longer be compliant to the development standards and would be refused by the spacecraft customer. But there exists a rather straightforward way out of this dilemma. In most cases at the kickoff meeting for the testbench development with the spacecraft customer it can be agreed that review milestones are allowed to be combined. An example is depicted in the following figure which shows the initial system requirements review (SRR) and preliminary design review (PDR) meetings being held conventionally with the required documents being delivered. Detailed design review (DDR) and test readiness review (TRR) of SVF V1 already here combined. Of course for such a combined milestone also the documentation set has to comprise the content for both. But since the subsequent milestones for a large number of documents only require enhanced revisions, the document upgrades then only are to be done once. The next example step then is the acceptance review (AR) for SVF V1 to be combined with the TRR of V2 and so forth up to the final version. Thus the number of reviews is significantly reduced and so is the paperwork update effort and document signature looping effort.

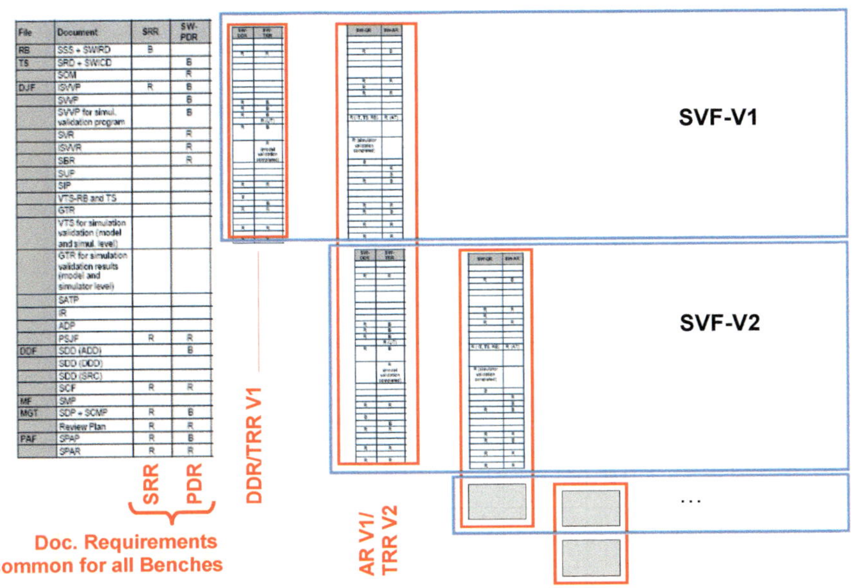

Figure 9.11: Documentation approach for combined review milestones. © Astrium

9.5 Testbench Configuration Control vs. OBSW and TM / TC

The previous chapters already clarified that in-time availability of the testbenches and their correct functionality is of essential importance. Correct functionality however does not only mean to implement the simulators and testbenches with good software quality and free of bugs. The additional challenge is to always keep the as built status of the spacecraft and the simulators and their configuration data sets in line. And moreover everything must also be aligned with the on-board software releases and the TC / TM data sets for command of the simulated spacecraft.

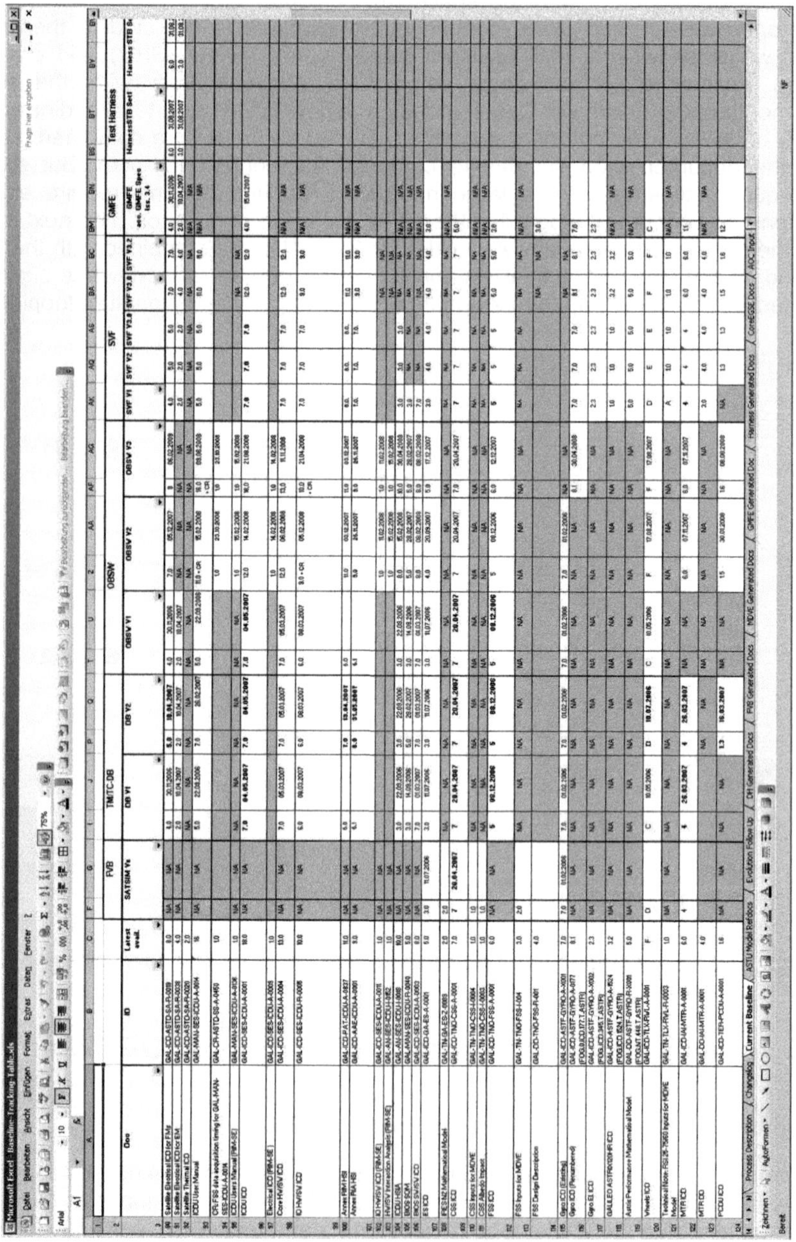

Figure 9.12: Baseline tracking matrix. Adapted from Astrium example

Therefore a permanent monitoring of the spacecraft equipment design updates, which occur in a real project, is necessary. Furthermore a version tracking of all key spacecraft and equipment design documents and their issues versus Simulink algorithm design releases, TC/TM-database releases, OBSW releases and SVF / testbench releases is of essential importance. This can easily be achieved in a simple spreadsheet matrix as it is depicted above (see figure 9.12). The matrix must be actualized on a daily basis in the project. Every OBSW developer and simulator developer can trace which versions of documents the other side used. If an OBSW and a TC/TM-database and a simulator revision all share the same documents and same document issue state, they are compliant and the OBSW can be run on the according simulator release and be commanded from control console by TCs / TMs loaded from the compliant database.

The characterization data of real spacecraft equipment at start of project are only available as data from equipment supplier documentation (specification data or as designed data). Later when the real equipment hardware is delivered, their 'as built' characterization data are available for the spacecraft manufacturer from real equipment test campaigns at supplier side. Thus the simulators and testbenches in a project must be flexibly configurable w.r.t. these characterization data and the data themselves must be configuration controlled in databases. More details on such a central data infrastructure covering the entire project is treated in chapter 10.2.

9.6 Testbench Development Responsibilities

For the technical engineering of central spacecraft components of functionalities in a project organization there are key responsible system engineers assigned, sometimes also called "architects". They report to the project manager and are guided by the senior system engineer. There exist e.g. architects for the on-board software, for spacecraft electrics, for the payload, for the spacecraft operations concept etc. The development of the simulators and testbenches as design and verification infrastructure are a similar essential task in a spacecraft project when using this model based design approach. Thus in such projects on the same level as the other architects a so-called Functional Verification Infrastructure Architect (FVI-Architect) has to be assigned. His responsibilities include the shadowing and management of

- the internal development of the simulators and the spacecraft equipment models,
- the externally procured infrastructure components such as Power-Frontends, control consoles such as Core EGSEs,
- definition and procurement of further hardware test equipment like test harness,
- version control of spacecraft characterization data for the simulators,
- configuration control of simulator and testbench hardware and software,
- overall integration and setup of the simulators and testbenches, and
- as item of key importance, the coordination of all testbench development

activities from design office (SDO) down to EFM / FlatSat to be aligned with the:

- verification plan and schedule of the OBSW, and with,
- integration activities of spacecraft hardware in AIT.

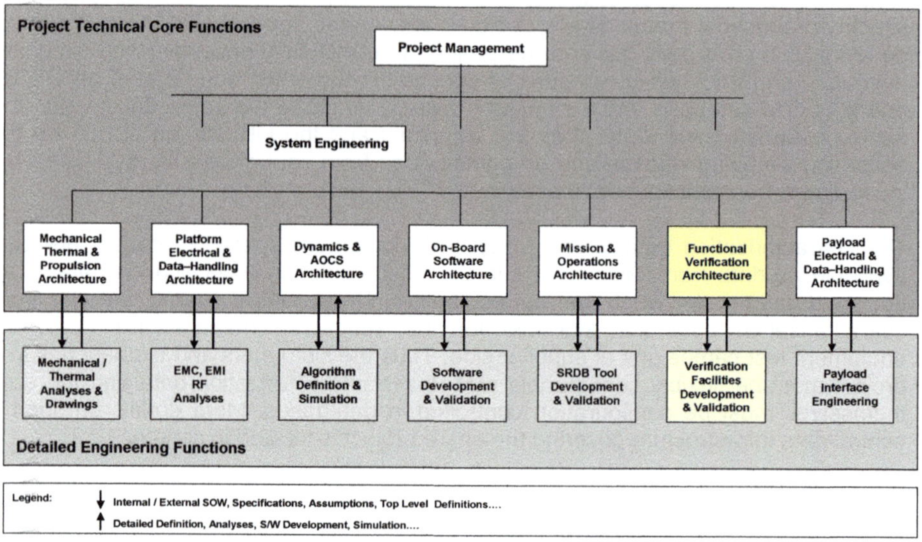

Figure 9.13: FVI architect in spacecraft project organigram. © Astrium

9.7 Lessons Learned from Projects

The development of a system simulator and its application e.g. for OBSW verification forces the spacecraft project team already in early phases to consistently define

- the system functions and their distribution towards spacecraft equipment,
- the operational behavior of the spacecraft equipment, its models and the parameterization,
- the data buses, electric, thermal and mechanic interfaces between spacecraft components,
- telecommands and telemetry on spacecraft level as well as for command and control of each equipment, and
- the basic system verification concepts, test approaches for both spacecraft and the testbench infrastructures.

The application of this model based development approach significantly changes the engineering processes compared to older classic engineering model based spacecraft projects. This requires higher precision in handling and versioning of engineering data and information, and it requires a higher team integration and optimized information exchanges over all engineering disciplines. The system simulation allows the dynamic modeling of the spacecraft and its operational behavior for spacecraft software and design concept verification. The testbench development approach is driven by the model and test philosophy chosen for the spacecraft, e.g. what is to be verified by simulations and what is to be verified in hardware tests. This spacecraft verification concept drives the required testbench types, their number and their features. An adequate concept of the simulator and testbench infrastructure allows the extensive pre-verification of OBSW on an SVF which avoids blocking of the limited OBC hardware models. It allows furthermore the pre-verification of AIT test procedures and flight procedures on SVFs also without blocking real flight hardware.

The validation of proper system modeling in the simulators and testbenches themselves is achieved by system and interface tests in the spacecraft integration campaign in EFM configurations. Finally the simulator infrastructure can be used in the ground control centers for operator training and as an flight software maintenance facility.

All in all the approach allows an early system modeling. However the simulation based approach also generates new development infrastructure components on the project's critical path if not managed properly. Especially the SVFs have to be available early enough for first OBSW tests. The SVFs must be accurate enough w.r.t. the equipment modeling and coverage of all relevant effects and features. And the STB infrastructures must be available in time for tests of OBSW and OBC and simulated spacecraft in the loop.

10 Simulation Tools in a System Engineering Infrastructure

Herschel © ESA

J. Eickhoff, *Simulating Spacecraft Systems*, Springer Aerospace Technology 1,
DOI 10.1007/978-3-642-01276-1_10, © Springer-Verlag Berlin Heidelberg 2009

System simulators, especially those integrated in SVFs and hybrid testbenches, in the real engineering process of a spacecraft in fact are no standalone tools. They are integrated into an entire system engineering infrastructure from which they receive their characterization data. Verified system characteristic parameters are stored back into the engineering environment. Such abstract simulation result characteristics e.g. can cover:

- Pointing precision or station keeping precision of a spacecraft
- Time for detumbling after launcher separation
- Payload dynamic characteristics
- Mode and attitude dependent power budgets

In this simplest case only characterization parameter values are exchanged between system engineering environment and simulator. For simulator tools which allow to dynamically load the spacecraft topology definition at simulator initialization, like *OpenSimKit* [23], the following types of exchanged information between engineering infrastructure and simulation come on top:

- Number and type of model instances in the system to be simulated
- Number and type of model interconnection line instances in the system
- Network list for model / line interconnections

This however requires to be able to provide this type of system topology input to the simulator from the system engineering environment, and not only pure characterization parameter values. This on the one hand imposes the requirement for a suitable tool to store and handle all this type of information in a real spacecraft project, e.g. a system engineering database. On the other hand it firstly requires the ability to correctly model this functional topology of the real spacecraft completely, formally and correctly in a standardized notation.

Furthermore it would be highly efficient to directly receive all spacecraft equipment based information in the same formal notation already from the equipment suppliers to be able to merge it with pure top level system information in the engineering environment or database. This implies:

- Modeling the number and type of electrical and data interfaces of a component.
- Storage of characterization data like power consumption etc. which can be operational mode dependent, and
- which implies necessity for digital definition of equipment's operational modes, commands, as well as all events inducing any mode transitions.

This leads to the necessity for complete modeling of the entire system assembly tree, the system topology (which component connected to which other by which line), the modeling of component's modes and all characterization data in a standardized notation. The modeling of the overall spacecraft is done at spacecraft prime manufacturer level. The modeling on equipment side should be done on supplier level. An adequate standardized notation for this problem is available with the "System Modeling Language" (SysML).

10.1 The System Modeling Language (SysML)

Inspired by the concepts, semantics and notation of UML, since 2001, a markup language called "System Modeling Language", (SysML), for modeling real world systems has been developed by an industrial consortium - the so-called SysML Partners. Many other industrial key players have joined the initiative in the mean time and have contributed to a consolidated language standard. The notation is rather similar to the "Unified Modeling Language", (UML), used for software engineering. SysML also provides a graphical notation which even shares some diagram types with UML. Since entire real systems, their topology and functionality can be described via SysML formally, this opens the gate to a formal spacecraft model inside the mentioned engineering infrastructure. From this, both

- the characterization information about the spacecraft can be extracted e.g. for initialization of simulators, and
- even a first UML description of spacecraft and its equipment can be derived for modeling and implementation of equipment models for the simulator.

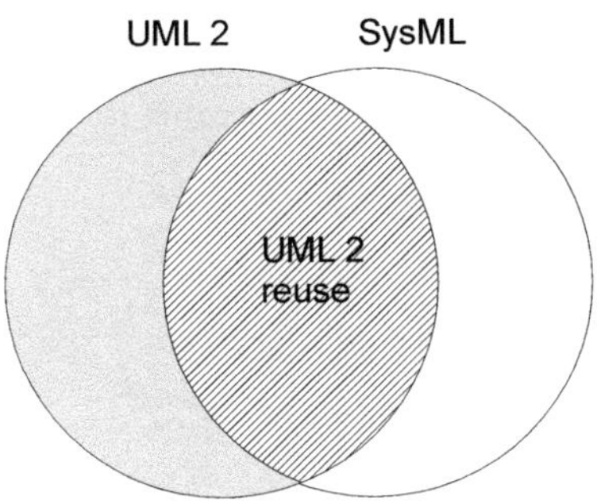

Figure 10.1: SysML to UML relation.

SysML 1.1 comprises the following diagram types:

- Requirements diagrams
- The structure and topology definition diagram types:
  - ◇ Block definition diagrams
  - ◇ Internal block diagrams
  - ◇ Parametric diagrams
  - ◇ Package diagrams
- The behavior and performance definition diagram types:
  - ◇ Activity diagrams
  - ◇ Sequence diagrams
  - ◇ State machine diagrams
  - ◇ Use case diagrams

Due to the phletora of notational details and variants of elements which may be included in the diverse diagram types, the reader interested in applying SysML is pointed to the according literature [94] and [96]. On the following pages the diverse diagram types of SysML - according to SysML standard issue 1.1 - are explained in a simplified overview without theoretic background on language architecture, metamodel, language formalism etc.

Requirement diagrams: This type of diagrams allows the formal representation of requirements towards a system (spacecraft or subunit) and the definition of dependencies between them - e.g. the dependency between certain spacecraft user requirements and derived equipment or on-board software requirements.

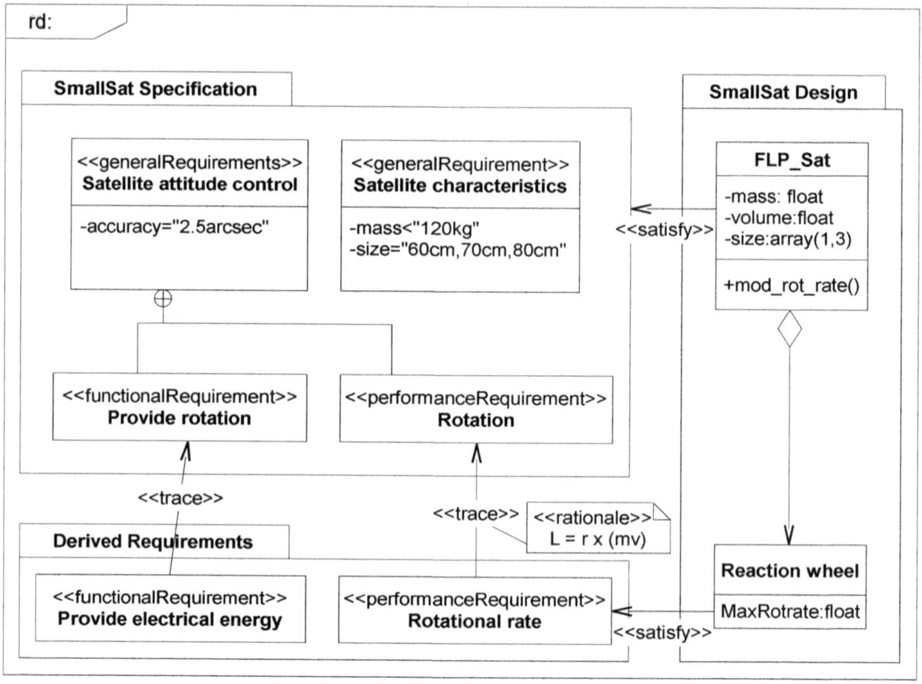

Figure 10.2: SysML requirement diagram.

Block definition diagrams: These diagrams are derived from the UML class diagrams and their notation elements correspond to their counterpart elements in UML, such as:

- Generalizations
- Associations
- Aggregations
- Compositions

Similar as in UML, SysML allows variables, values and methods and even OCL constraints to be allocated to represented system elements (please also refer to the UML diagrams 8.4 and 8.5).

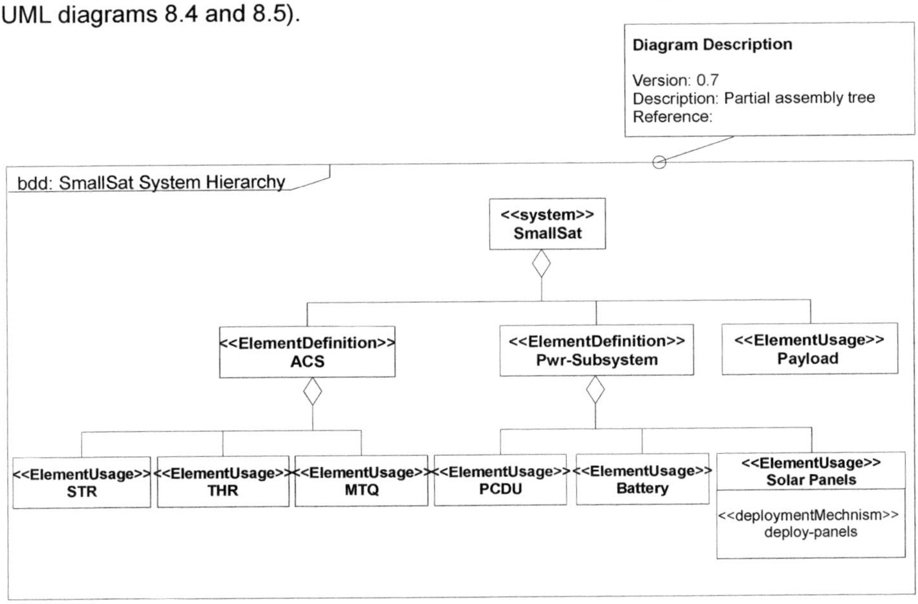

Figure 10.3: SysML block definition diagram.

Internal block definition diagrams: The internal block definition diagrams are derived from the UML composite structure diagrams. They show static breakdowns of the system. However in one type they optionally enhanced by user roles to define user / system information exchanges. In another variant they show the interfaces between equipment for exchange of influences, information or other interdependencies. For each type an example is provided hereafter.

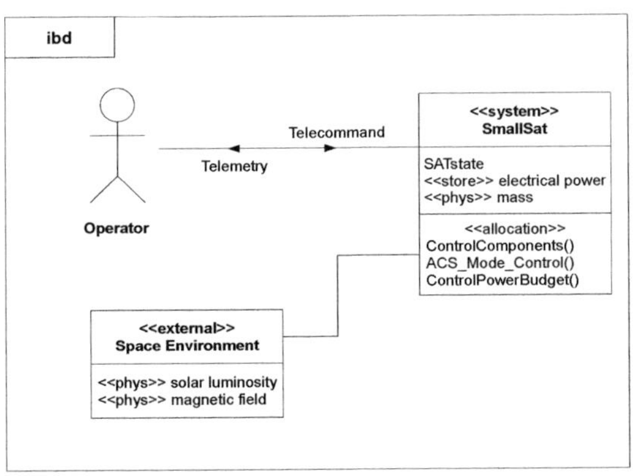

Figure 10.4: SysML internal block diagram with the role of operator.

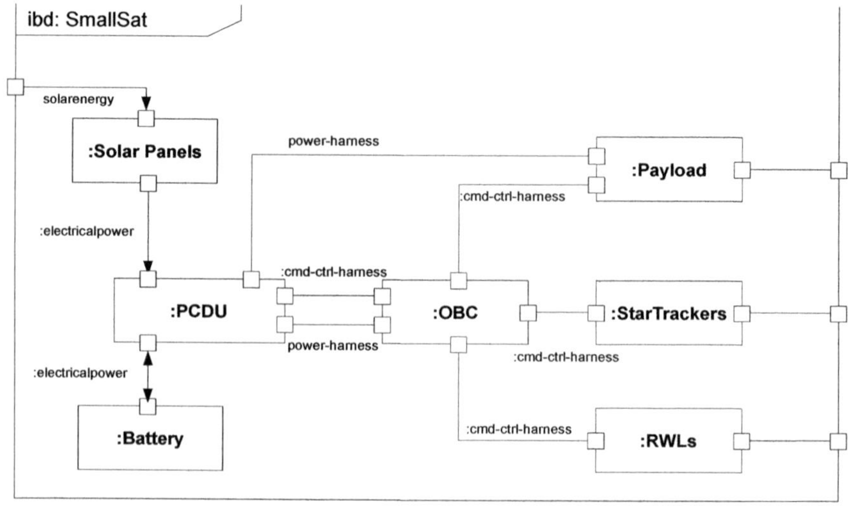

Figure 10.5: Simplified internal block diagram of a satellite topology.

Parametric diagrams: These diagrams explain the dependencies between system components, their parameters and eventual computation or transformation functionalities. For these parameters not only the names (length, power consumption, speed etc.) are specifiable, but also the parameter units (meters, watts, and so on).

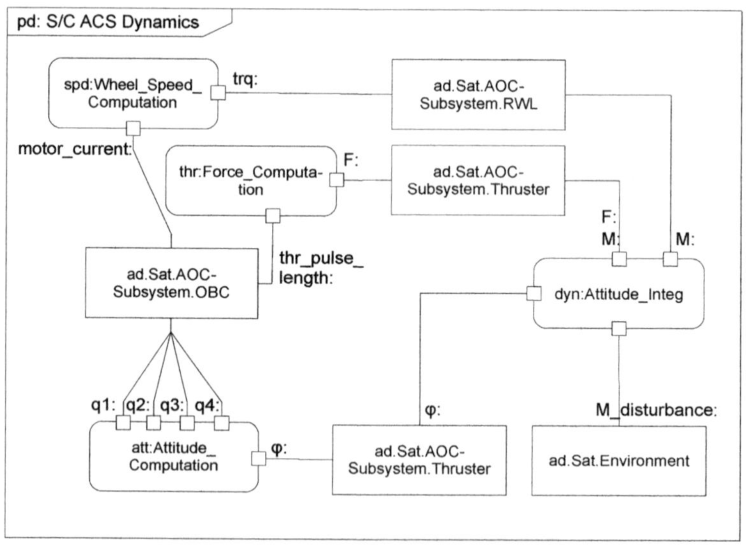

Figure 10.6: SysML parametric diagram.

Package diagrams: This diagram type shows the dependencies between major system components such as e.g. the structuring due to application criteria. The diagram notation and symbolic meanings are equal to the according UML diagram variants.

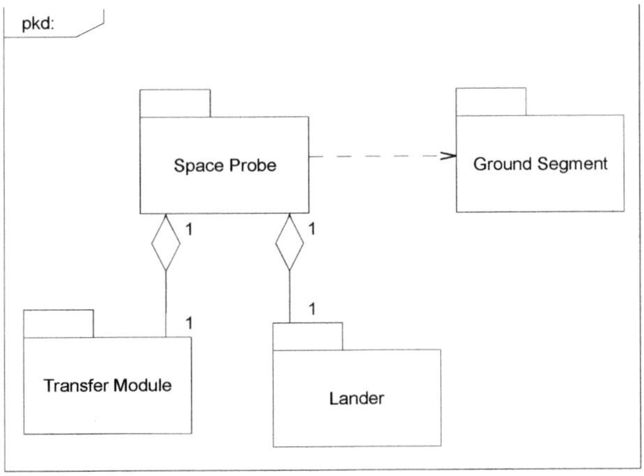

Figure 10.7: SysML package diagram.

Activity diagrams: These diagrams show the process flow in a system as well as system behavior and control structures. These diagrams in SysML are enhanced compared to their UML counterparts. For the diverse details please refer to the SysML standard [96].

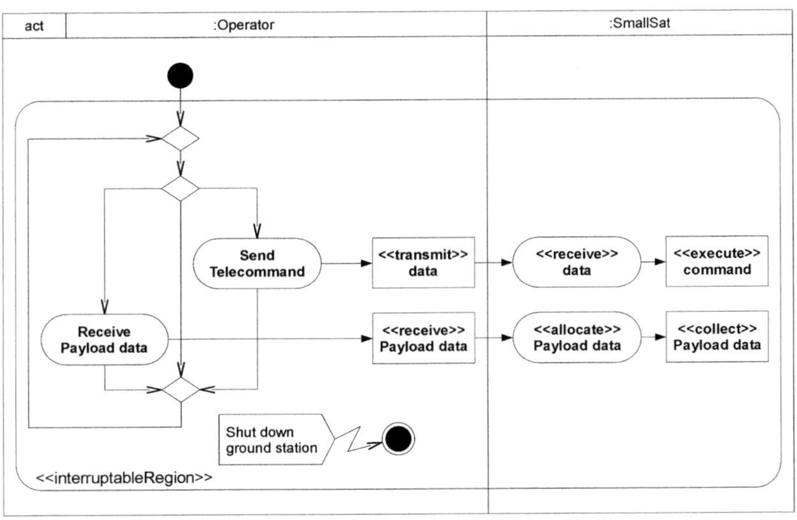

Figure 10.8: SysML activity diagram (including "Swim Lanes").

Sequence diagrams: These diagrams depict the dynamic interactions of system components in real operation sequences. The notation is identical to the corresponding diagram type in UML. As in the UML variants, here also sequences can include time information, condition terms and sequences can be specified on system element class level or on occurrence level (called instance level in UML).

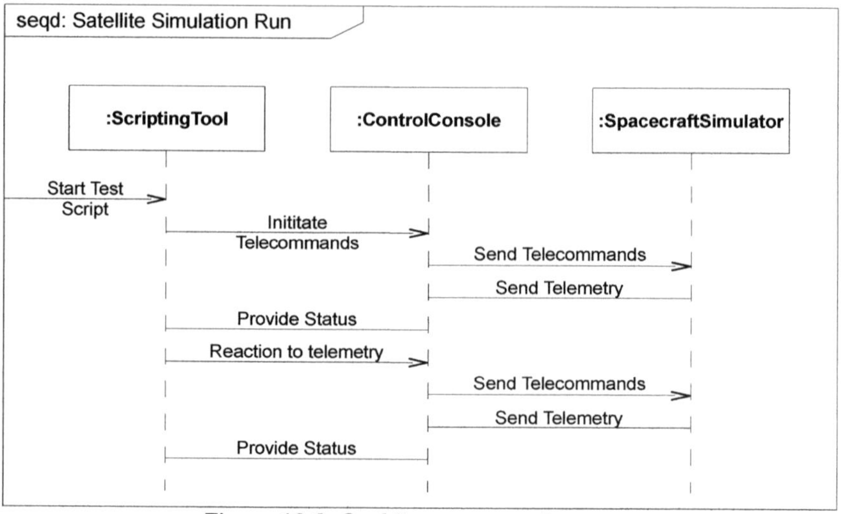

Figure 10.9: SysML sequence diagram.

State machine diagrams: These diagrams show the real spacecraft's system and equipment states and possible transitions. Furthermore the notation in SysML allows to specify the transition triggers (e.g. commands, events, failures), the transition times and submitted output (e.g. telemetry) to other units. The notation again is the same as for state machine diagrams in UML.

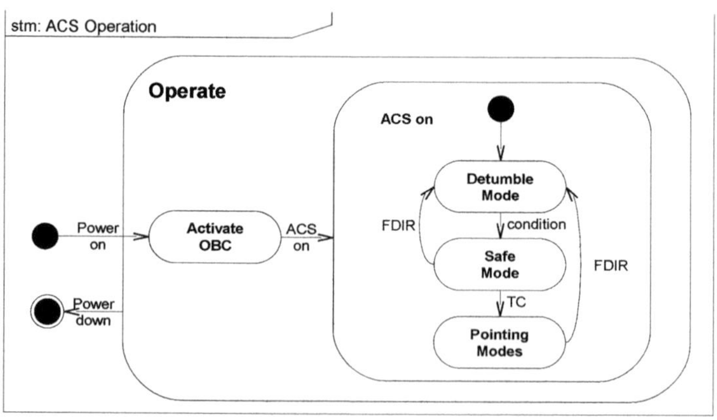

Figure 10.10: SysML state machine diagram.

Use case diagrams also exist in SysML. Their notation is comparable to the ones in UML w.r.t. notation. The types of diagrams which can be used to generate software code from or UML diagrams from[26] typically are:

- Block diagrams
- Sequence diagrams
- State machine diagrams

Code generation from SysML is performed by similar code generators and methods as applied for UML tools, please refer to chapter 8.4. The term "code" however here covers a wider scope since it is not limited to programming sourcecode but also can comprise data models (e.g. UML as already mentioned), data base structures, data export / import scripts and XML or other format data files themselves.

Further reading and Internet pages concerning system engineering standards and languages are listed in the according subsection of this book's references annex. In particular here the reader is pointed to the SysML definition standard itself [96].

10.2 System Engineering Infrastructures

These modeled spacecraft system data, equipment descriptions and characterizations, harness interconnection specifications and telecommand / telemetry definitions in latest space projects are managed via sophisticated system engineering infrastructures. This includes version control of all information - not only of spacecraft and equipment paper documentation. Vice versa the information is managed digitally and the paper documents are generated as reports from the engineering infrastructure databases. The latter contain the entire digital modeling of the spacecraft design.

The idea of such a digital data modeling has been well known and common practice for years for handling mechanical CAD model data. For the digital modeling of telecommand / telemetry data compliant to CCSDS and ESA Packet Utilization Standard (PUS) also since years digital standard models have been available. What has however recently just been achieved is build up a digital representation of the spacecraft which both covers topology, characterization and functional modeling - see the achievements in the ESA "Virtual Spacecraft Design" studies since 2006 (e.g. cf [100]). In industry since the ESA projects "Bepi Colombo" (Mercury encounter mission), "Earthcare" and "Sentinel 2" e.g. Astrium GmbH - Satellites applies a system engineering infrastructure somewhat similar to the architecture depicted in figure 10.11. Also this infrastructure allows digital modeling of all spacecraft aspects (electric, mechanic, operational). It permits to export from there characterization data sets and configuration information for mass and power budget tools, for the harness design tool, for electrical test equipment in AIT, simulator testbenches and for Core EGSE ([109]).

[26] Assuming availability of an appropriate code generator.

System Engineering Infrastructure

Since for specific engineering and modeling tasks, already dedicated tools and data containers exist (such as CAD tools for mechanical design), a system engineering infrastructure is not a monolithic single database application. Instead it is an integration of specialist system models, enhanced by containers for information not held in classic tools - with the key paradigm to avoid any doubling of reference information. The implementation of such a system engineering infrastructure typically comprises multiple tools in a layered structure. On the top level of such an infrastructure typically a so-called "System Engineering Database", (SEDB), is placed - see also figure 10.11.

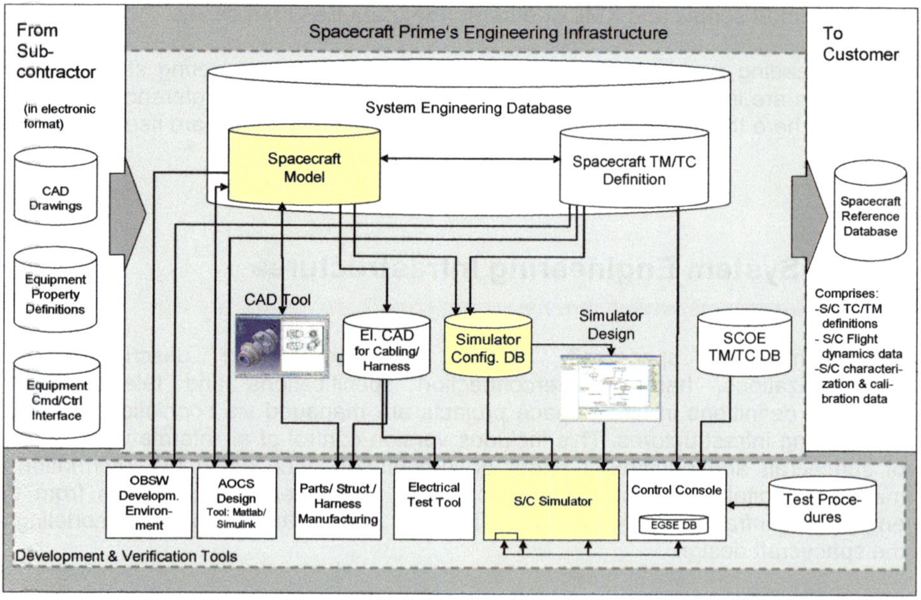

Figure 10.11: System engineering infrastructure and internal data containers.

Inside such an SEDB first of all the abstract spacecraft representation is stored. More precisely the spacecraft design is modeled in a SysML based representation and the model elements are parameterized. In detail the following information is included:

- Product trees of spacecraft and virtual spacecraft installations (i.e. test-benches). In multi satellite missions this includes trees for all instances of the spacecraft series - since they might slightly differ.
- Operational modes of spacecraft and all equipment.
- Physical characteristics of spacecraft, which might be mode dependent, e.g. moments of inertia are dependent on whether solar arrays are deployed or not.

- Physical characteristics of each equipment - also equipment mode dependent.
- System configurations (undeployed / deployed, begin / end of life etc.)
- Functional definitions of line interconnection between equipment on board.
- Equipment telecommand and telemetry data and protocol packet definitions.
- Operational PUS service handler configurations for on-board software configuration.

Besides this spacecraft model, the telecommand and telemetry definition data of the spacecraft for ground / space communication (respectively between simulated spacecraft and control console) should be included in the SEDB.

On the intermediate level there exist databases / tools which appear to be more common and which serve to further detail the top level information in the system engineering database. These are:

- The mechanical tool for "Computer Aided Design" (CAD)
- The electrical CAD tool for harness design
- The simulator database for handling all models and configuration data of the simulators in the testbenches
- The database for handling SCOE telecommand / telemetry definitions
- The management tool for handling test procedures

In the lowest layer of the engineering infrastructure the "data consumers", the client tools are placed such as:

- The OBSW development environment
- The functional phase B algorithm design simulators (Simulink, PSpice etc.)
- The verification testbenches from SVF to EFM
- Diverse measurement and automated test equipment applied in AIT

This listing for the lowest level does not claim to be exhaustive since via diverse data exporters also diverse applied simple budget engineering tools (mass, power, link budgets) on spreadsheet technology etc. can be supplied with data from the system engineering infrastructure.

Vertical Refinement and Horizontal Consistence

The logic behind such a digital modeling is to always keep the most detailed information in the lowest layer tools / databases. As soon as some type of information is required by multiple "clients", it must be positioned in the next higher layer. For example the information on the component breakdown of the spacecraft (the assembly tree) is of relevance both for the simulator (the equipment model instances have to represent the assembly tree item occurrences) and for the electrical harness design. Thus it is placed inside the SEDB above electrical CAD tool and simulator database. Similarly the information which equipment is connected to which other by which type of power or command / control interface, also is of relevance for multiple clients namely for the harness database, OBSW development and the simulator database. Electrical harness design is a descriptive example for the complex interaction of the databases and data containers:

- On the top level in the SEDB the assembly tree of the spacecraft is defined. And for each component / equipment the electrical interfaces are defined, i.e. their number and their types (e.g. for a payload 1x MIL-STD-1553B bus, 2x analog / thermistor, 1x RS422 serial, 1x power 50V).
- In addition on the top level of the system engineering infrastructure in the SEDB the equipment interconnections are defined, i.e. which line (ID) from which component port of the example payload is routed to which port of the OBC in the spacecraft respectively to which port of the power control and distribution unit (PCDU).
- This information is propagated from the SEDB to both the electrical CAD tool, to the OBSW CASE tool and to the simulator database.
- From the simulator database the simulator configuration files for equipment models and the functional interface models are generated. Since the simulator implements the line interconnections only functionally, this level of detail is sufficient for simulator configuration. Details on electrical connectors and pin allocations are not necessary here.
- The same top level information about equipment interconnections is imported into the electrical CAD tool for harness design. Here the harness modeling however must be detailed further. For each cited MIL-STD-1553B bus line, analog thermistor or power line it must be specified in detail on which connector of the example payload, respectively the OBC and the PCDU they are placed and on which pins signal, complement, ground, shield and redundant lines are placed.
- The information later is required to correctly manufacture flight harness and test harness for the testbenches from these design data. Furthermore this information is required later in AIT for configuration of automated electrical test tools to identify on which connector / which pins a certain signal can be acquired during electrical test.

Furthermore it is of importance that the workflow over the spacecraft development process and the phases B – E[27] is a constant refinement of the design information for all domains. This topic will be taken up again later in chapter [12]. The infrastructure has to support this accordingly and has to assure consistency of the engineering data to each other on every layer depicted in figure 10.11. For example the telecommands sent from the Core EGSE to the spacecraft in either of the testbench states must be compliant to the on-board software version, the on-board data bus protocols between OBC and spacecraft equipment and the electrical design of the harness between OBC and platform / payload equipment.

[27] In phase A of a spacecraft project usually such system engineering infrastructures are not yet applied.

Data Model of a System Engineering Database

The essential data flow for the domain of system simulation in such a system engineering infrastructure already is marked in yellow color in figure 10.11. It extends from the spacecraft model in the system engineering database via the simulator database down to the configured simulator in the spacecraft testbench. As stated the important difference to a pure database is, that the system engineering infrastructure elements as e.g. SEDB or simulator database not only have to contain simple data like variables and values. Instead they have to model the data interdependencies, both w.r.t. topological aspects as well as concerning functional aspects. This no longer can be achieved by simple table-based database concepts.

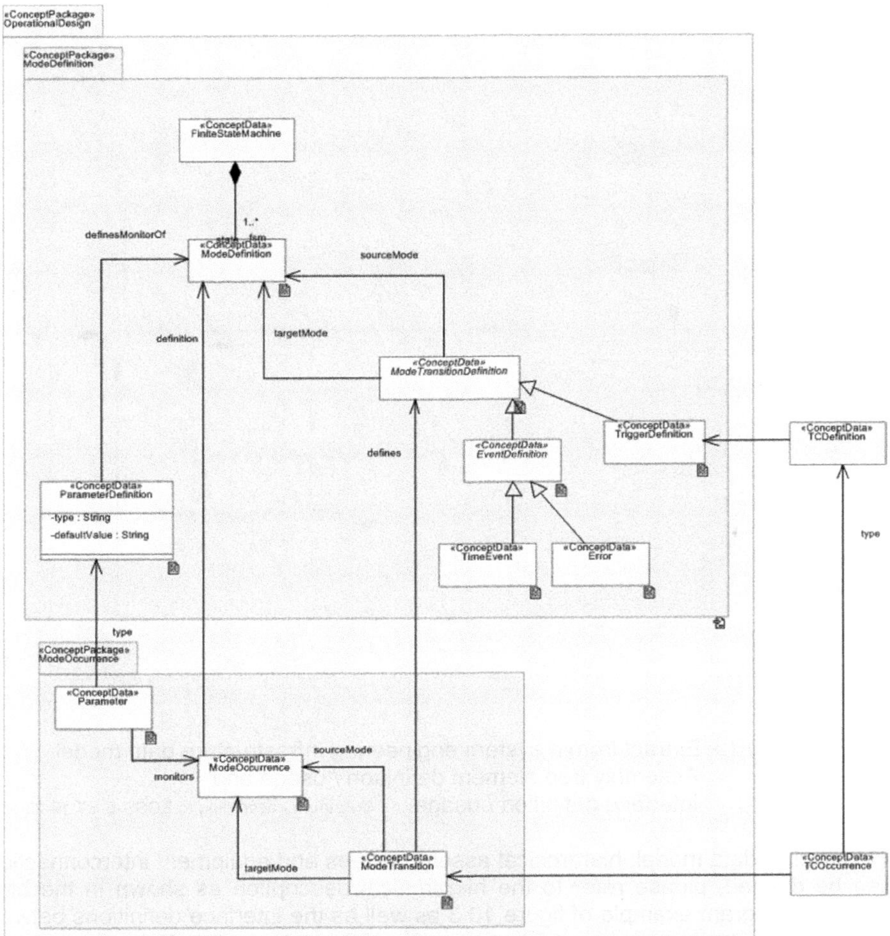

Figure 10.12: Extract from a system engineering infrastructure data model. State machine based equipment model. © Astrium - according to ECSS-E-ST-10-23 draft.

Instead SysML based modeling architectures are necessary, as e.g. developed under the ESA projects "Virtual Spacecraft Design" [100]. The following figures show extracts of such model implementations to make these abstract statements easier to grasp. The two figures below show example extracts from a system engineering infrastructure data model following the ECSS-E-ST-10-23 draft standard. Only spacecraft modeling aspects are included here. Topics like data versioning, user access rights etc. are descoped to not overcomplicate the examples. Figure 10.12 shows the SysML metamodel for an equipment definition and a mode transition definition with all trigger types and mode dependent equipment parameters. The data model corresponds to a metamodel on UML level - please refer to the M2 layer depicted in figure 8.18. The next figure shows the modeling of equipment definitions and their instantiation (usages) on the right side as well as the interlinked representations for interface definitions and instances on the left side.

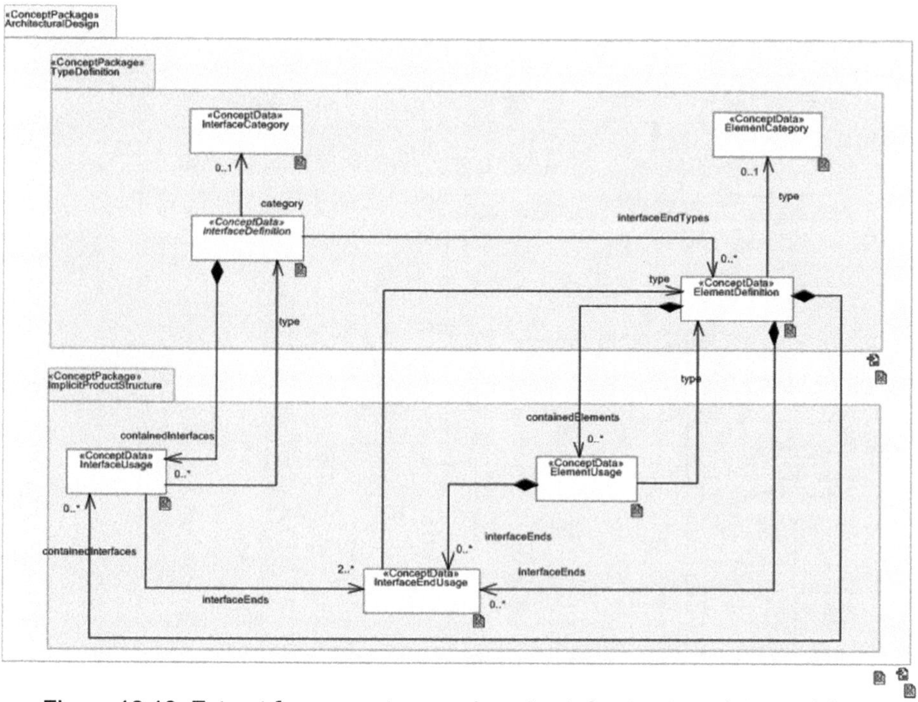

Figure 10.13: Extract from a system engineering infrastructure data model. Assembly tree element definition / usage and interface definition / usage. © Astrium - according to ECSS-E-ST-10-23 draft.

With such a data model, hierarchical assembly trees and equipment interconnections can be defined, please refer to the hierarchical description as shown in the block definition diagram example of figure 10.3 as well as the interface definitions between equipment instances as specified in an example in figure 10.5. This detailed level information about the spacecraft on instance level then corresponds to the M1 layer shown in figure 8.18.

10.3 Standards for Data Exchange Between Engineering Tools

To exchange spacecraft architectural and functional information between the tool of an engineering infrastructure or to exchange such engineering data with subcontractors applying their own tools, the use of according file or modeling standards is mandatory. In case simulator data are exchanged between tools, or in case even entire simulator equipment models are exchanged between subcontractor supplying the real equipment and the spacecraft prime integrating the testbenches, this situation gets even more constrained through suitability of applied model exchange standards.

At least for pure data exchange between tools this set of problems is not new and a topic for all project oriented developments from ship building industry via chemical plant building towards aerospace engineering. The first classical challenges of this kind arose concerning data interchange between computer aided design (CAD) tools of different manufacturers as well as between CAD tools and digitally programmed machine tools in the 1980's. A similar problem occurs during data exchange between CAD tools and structure mechanics solvers such as finite element (FEM) tools.

For a network of system engineering tools as shown in figure 10.11 the electronic data exchange is of essential importance. Concerning this topic a short overview on existing and actual data exchange standards shall be given in brief.

Data Exchange via XMI

The simplest approach can be followed if 2 tools inside the engineering environment follow the same basic design approach. An example for this case e.g. are the spacecraft SysML model within the system engineering database and the model representation in the simulator database. Both have to represent the entire spacecraft, the SEDB for all disciplines, the simulator database for the limited scope of functional simulation, however enhanced with numeric information like solver interfaces etc. Both can base their data model on SysML / UML. From the Meta Object Facility (MOF, cf. figure 8.18) of the SEDB's spacecraft model the functional and topological representation of equipment models, interconnection models and parameterization data can directly be exported and can be reimported into the simulator database's MOF. Prerequisite are the export / import filters for an according data exchange file format. The latter in SysML and UML is the "XML Interchange Format" (XMI), a specific XML variant - see [99]. From the SysML representation of the spacecraft topology and functions in the simulator database then two types of output can be generated:

- The XML configuration data files for the simulator. They directly can be extracted from the SysML model by an appropriate XML export filter.
- The top level UML classes for the simulator UML design tool can be generated from the SysML representation in the simulator database. Therefore the

information about equipment classes, harness line classes, state machines and the entire instantiation can be exported to the simulator UML design tool.

Since both SysML as UML are based on a standardized MOF representation and XMI notation, such an export to the simulator design tool is a practicable approach. In the cited cases no information has to be manually reentered into any new tool which reduces information loss, errors and data inconsistencies. Furthermore it simplifies data and model configuration control to a large extent.

Data Exchange via STEP

More complex is the situation in the generic case of data / model interchange between engineering tools since not all tools support the still rather new SysML / UML / XMI technologies already. However since as stated the data exchange problem is not new, already for a lot of data interchange scenarios older standards exist. Since 1980 the International Organization for Standardization (ISO) is working on a set of data exchange standards, the "**ST**andards for **E**xchange of **P**roduct Model Data" - STEP (ISO 10303). Here first it has to be clarified that STEP is not a standardization for system modeling like SysML and UML, but represents a family of "format notations" plus application guidelines etc. for data exchange between engineering tools. Of course a certain level of "modeling" also is required for sake of data interchange, but no dynamic functional implementation as known with e.g. the simulators.

The STEP standard family is a quite complex assembly of application principles, domain dependent descriptions, application guidelines etc. It shall be simplified here a bit for tutorial reasons. First of all there exist so-called "parts" which cover all types of system modeling definitions, e.g. (non exhaustive list) :

- CAD network models,
- CAD surface models,
- FEM models,
- kinematics models,
- thermal network models,
- electric network models,
- electronic board layout models,
- numeric control data models for plant control,
- numeric control models for digitally programmed machine tools,
- visualization models,
- and finally the important part for system engineering.

The other component of the STEP standards are the so-called "Application Protocols". STEP Application Protocols determine what STEP-conformant tools can do. The protocol relevant for the domain of system engineering is AP233. The diagram below shows in a simplified manner all topics formalized by AP233. These days still not all standard chapters for all topics are available, in particular not yet for the lower layers in figure 10.14. In AP233 are / will be specified:

- Data definitions
- Physical description of components
- Definitions concerning system and component properties
- Description of system architecture and breakdown into components
- Requirements
- Assignment of requirements to components
- Allocations of properties to components
- Function trees and flows - functional breakdowns
- State machine descriptions
- Causal chains - especially for system failure description

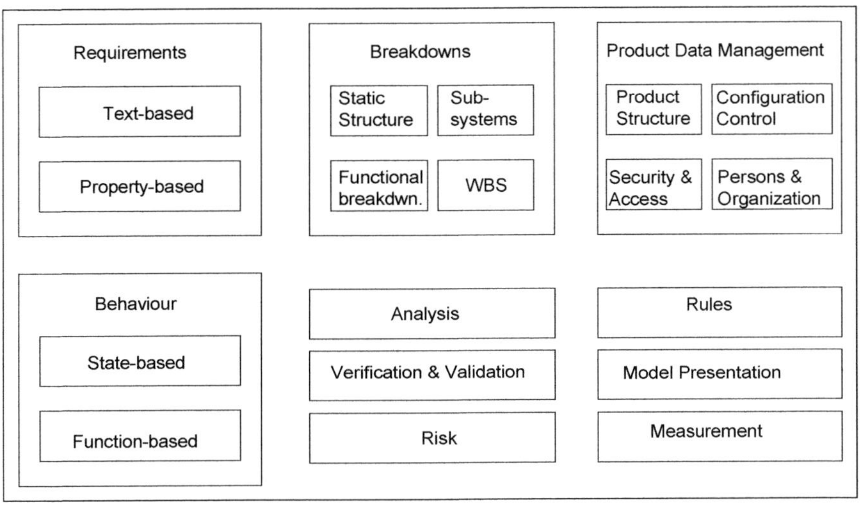

Figure 10.14: Elements of STEP/ISO 10303 AP233 for exchange of system engineering data.

With these notations the most important information on system requirements, topologies and simplified behavioral information can be modeled inside a tool. For the file based data exchange itself STEP provides an own notation language standard: EXPRESS (ISO 10303 Part 11). For the file based data exchange between two STEP compliant tools EXPRESS corresponds to the file format notation as XMI does between two SysML tools. To explain how a static system breakdown structure (see figure 10.14) can be noted in EXPRESS, a cut out of the system from figure 1.12 shall serve as example:

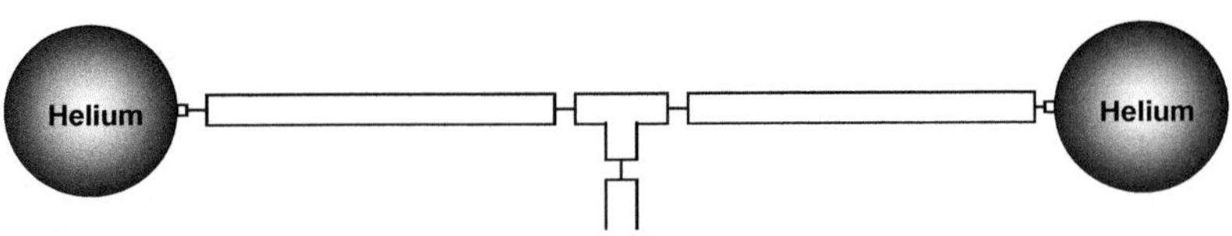

Figure 10.15: Cut out of the rocket propulsion system.

The corresponding EXPRESS notation then looks as follows:

```
SCHEMA rocket_propulsion_system;

ENTITY component;
      ABSTRACT SUPERTYPE;
      comp_mass: REAL;
      comp_heat_capacity: REAL;
END_ENTITY;

ENTITY HPBottle;
      SUBTYPE OF (component);
      He_pressure: REAL;
      He_mass: REAL;
      He_temp: REAL;
      He_massflow: REAL;
END_ENTITY;

ENTITY Pipe;
      SUBTYPE OF (component);
      pressure_in: REAL;
      temp_in: REAL;
      pressure_out: REAL;
      temp_out: REAL;
      massflow: REAL;
      inConnectedTo: OPTIONAL component;
      outConnectedTo: OPTIONAL component;
END_ENTITY;

ENTITY Junction;
      SUBTYPE OF (component);
      ......
END_ENTITY;

.........

ENTITY System;
      propulsion_system: SET[0,?] of component;
      system_mass: REAL;
END_ENTITY;

END_SCHEMA;
```

The exchanged files themselves can either include this EXPRESS information

- in specially formatted ASCII text file formats - EXPRESS Clear Text Encoding, ISO 10303 Part 21,
- or as XML representations compliant to the schemas and data definitions according to ÉXPRESS-X, ISO 10303 Part 28.
  The relatively easy conversion of such type of EXPRESS file information already is obvious when comparing the above notation to the XML file of figure 8.26.
- Also for export of EXPRESS models via an online access to a database (implemented in C++ or Java language) there exists a "Standard Data Access Interface" (SDAI) STEP standard, ISO 10303 Part 22.

Data Exchange STEP/XMI

UML / SysML engineering data in XMI format can be converted from / to the STEP Part 28 XML format. This however is limited to file based exchange. Direct "tool-to-tool" exchange between STEP and UML / SysML tools are not yet possible since STEP does not specify a standardized tool internal information representation as does the Meta Object Facility for UML and SysML tools. Work on an EXPRESS metamodel for UML / SysML mappings to / from EXPRESS however are ongoing at the Object Management Group (OMG).

The STEP Graphical Notation EXPRESS-G

Finally STEP also comprises a graphical representation format for system engineering data - EXPRESS-G. The example system cut out from above figure 10.15 in EXPRESS-G would look somewhat like:

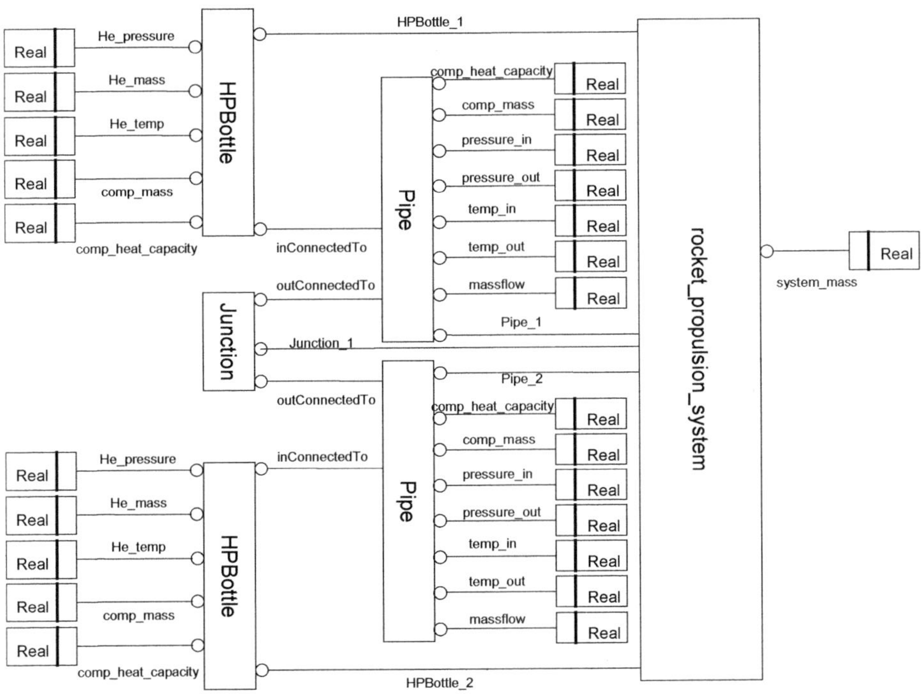

Figure 10.16: Example system representation in Express-G.

The example shows that structure specifications noted in EXPRESS-G quickly tend to become bulky since each characteristic of the system (here e.g. the variables) has to be noted as a separate graphical EXPRESS-G entity. Therefore the spreading of this notation is only very limited.

You think to know when you learn,
are more sure when you can write,
even more when you can teach,
but certain when you can program.

Alan Perlis , Yale University

Part III

Advanced Technologies

11 Service Oriented Simulator Kernel Architectures

FLP © IRS Universität Stuttgart

J. Eickhoff, *Simulating Spacecraft Systems*, Springer Aerospace Technology 1,
DOI 10.1007/978-3-642-01276-1_11, © Springer-Verlag Berlin Heidelberg 2009

In chapter 6.8 on simulator numerics the complexity of a simulator kernel was elaborated. A simulator kernel which is able to import the spacecraft system model topology to be simulated at simulator initialization time and which is able to register the components at the simulator kernel and at the solver is a quite advanced infrastructure and by far these days is not yet state of the art.

These dynamic loading functionalities shall be discussed in detail using first code samples of the issue 4.0 of the academic simulator toolkit *OpenSimKit*, which is currently[28] still under conception. Before diving into technical details a short introduction into the technology of "Service Oriented (Software) Architectures", (SOA), shall be given.

Service Oriented Architectures

Service oriented architectures are a technical approach from information technology to interconnect software systems company-wide. Service oriented architectures in most cases are applied to interconnect business software components. Services with higher levels of abstraction such as stock overview, stocktaking functions, stock value identification, necessary insurance cover, profitability calculation etc. can be created by interconnecting services of abstraction levels. This principle is called "orchestration" in SOA. According lower level functions e.g. can be automatic barcode scanning of parcels at company goods receipt department. Thus computing components are encapsulated into services and coordinated in such a way that their single functions can be integrated into higher services. Or higher abstracted modules can use intermediate results provided by lower services.

In order to interconnect components in an appropriate way, SOA requires an adequate design of the participating components - respectively an appropriate selection of such components, should commercial ones be chosen. Service oriented systems usually are distributed over diverse computing nodes. Each service is registered in a corresponding "directory" and - if applicable - not every service is permanently connected to the overall system. SOA is only a "paradigm" for system architecture. There exists no standard for online component interconnections comparable to a definition language such as UML for software design.

Service Oriented Simulator Kernel Design

The service oriented approach becomes interesting for system simulation if the basic concept and idea are transposed to the internal software architecture of a system simulator. This implies that the simulator as a whole is treated as a SOA in which each module performs specific functions so that the overall functionality of a dynamical system simulation is aggregated. For this the overall simulator system architecture shall be analyzed, as it is depicted in figure 11.1 below. There the basic concept for simulator numerics from figures 6.10 and 6.11 can be recognized.

[28] Spring 2009.

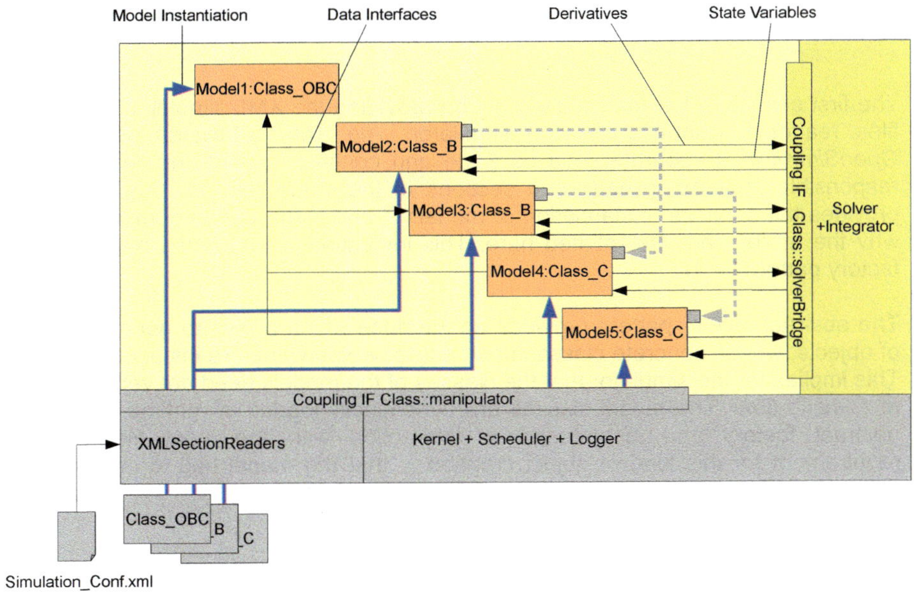

Figure 11.1: Service -oriented simulator architecture - *OpenSimKit* V4.

The program architecture mainly consists of a class hierarchy with

- `XMLSectionReaders` for the modular import of the simulator configuration XML input file,
- the simulator kernel, which is responsible for
  - ◇ creation of model instances as well as port instances,
  - ◇ scheduling of all component computations during the simulation run and for saving of results into log files,
  - ◇ simulation command / control and telemetry interface handling with the control console,
- the model classes, which are loaded and configured according to the entries found in the simulator configuration XML input file, as well as,
- the model and port instances created from the classes by the kernel,
- and finally the numerical solver which comprises all necessary functions to integrate the initial value problem.

Between the kernel and the simulator models a gateway class named `manipulator` is located. The purpose of this class is to control the read / write / set / get access rights of the model variables during initialization and simulation computation. All models are "registered" at the `manipulator`. Regarding numerics a similar registration service between the models and the solver is provided by the class `solverBridge`.

11.1 SOA Implementation of Simulator Initialization

The first service to be analyzed is the one providing the XML simulator configuration files reader functionality for the simulator. The XML subsystem is located in *OpenSimKit* in a separate Java package and contains multiple readers which are responsible for reading dedicated sections of the input files. The XML subsystem provides the services "instantiation of models"[29] and "initialization of models". This is why the models are instantiated here. This instantiation is realized by the abstract factory pattern [101]:

The abstract factory pattern provides an interface which allows to generate a family of objects, whose concrete classes need not to be known before instance generation. This implies for the simulator that the classes of the models need not yet to be known at compile time. The model classes are dynamically loaded at runtime by using the abstract factory and instantiated in interaction with the input file. The only requirement for this kind of object creation is that the model has to implement the service interface `org.opensimkit.Model`.

Instantiation of a Model

The name of the component class is given in the XML file:

```
<model class="org.opensimkit.models.rocketpropulsion.PipeT1" name="model3">
```

The element "`model`" represents a model inside the simulator. The attribute "`class`" specifies the name of the class which shall be used for this model, i.e. it identifies the class of which this model shall be an instance. In this example it is the class "`PipeT1`" which is located in the package `org.opensimkit.models.rocketpropulsion`. The attribute "`name`" specifies the name of the model inside the simulator. In this case it is "`model3`". By using this name the user can access the model via the control console commands.

Below is an extract from the class **`ModelXMLSectionReader`** of the package `org.opensimkit.xml` to illustrate the above mentioned code sequence:

```
try {
    /** Retrieve the Class reference for the name of the class.
     *  The name of the class is located in the variable "type". */
    Class<?> cls = Class.forName(modelType);
    /** Retrieve the constructor with one argument of String. This
     *  must be done, because Model has no default constructor.*/
     Constructor con = cls.getConstructor(new Class[]{String.class});
```

[29] Instantiation = generation of the software instances.

```
        /** Execute this constructor to instantiate a new object from the
         *  originally given name. */
        model = (BaseModel) con.newInstance(name);
        comHandler.addItem(model);
        manipulator.update();
    } catch (Exception ex) {
        System.out.println("Invalid type " + modelType
            + " specified for model " + name);
    }
```

First it is attempted to load the class:

```
Class<?> cls = Class.forName(modelType);
```

Next tried is to retrieve a reference to a constructor of this class with one argument of type `String`:

```
Constructor con = cls.getConstructor(new Class[]{String.class});
```

Then this constructor is invoked. Its return value is a reference to the created / instantiated model. The argument of the constructor is the name the model shall have:

```
model = (BaseModel) con.newInstance(name);
```

This name is read from the XML file:

```
<model class="org.opensimkit.models.rocketpropulsion.PipeT1" name="model3">
```

The reference to the new created model is passed to the component handler:

```
comHandler.addItem(model);
```

These are the steps to instantiate a new model. If errors occur according error messages are submitted. Finally the model variables are initialized. This approach is an implementation of the “abstract factory” design pattern.

Initialisation of Model Variables

Example of a model declaration in the XML input file:

```
<model class="org.opensimkit.models.rocketpropulsion.PipeT1" name="model3">
    <variable name="description"/>
    <variable name="length" unit="m">1.5</variable>
    <variable name="specificMass" unit="kg/m">.6</variable>
    <variable name="innerDiameter" unit="m">.0085</variable>
    <variable name="specificHeatCapacity" unit="J/(kg*K)">500.0</variable>
```

```
    <variable name="surfaceRoughness" unit="m">1.E-6</variable>
    <variable name="temperatures" unit="K" fieldSize="10"
              allFieldEntriesIdentical="true">300.0</variable>
</model>
```

The element “`model`“ can contain an arbitrary number of sub elements (i.e. 0 to n) of the type “`variable`“. Via these it is possible to initialize model variables. Besides scalar values it is also possible to initialize one-dimensional arrays. All primitive Java types as well as strings can be initialized using the input file.

Below an example is given, showing the initialization of a scalar value from the initialization of a model of the class “`org.opensimkit.models.rocketpropulsion.PipeT1`“:

```
<variable name="length" unit="m">1.5</variable>
```

A scalar variable of the name “`length`“ (the name is case-sensitive because Java reflection is being used) is going to be initialized here. The value of the variable is set to 1.5. The declaration of the variable located in the file “`PipeT1.java`“ of the package “`org.opensimkit.models.rocketpropulsion`“ looks like:

```
@Manipulatable private double length;
```

The variable “`length`“ is of type “`double`“ and has the “`private`“ access modifier as well as the annotation “`@Manipulatable`“. This declares the variable to be cleared for both read as well as write access by the user trough set / get functions from the control console. The write access of the manipulator is facilitated by the method `setFromString()` and `getFromString()` respectively. The `setFromString()` method signature looks as follows:

```
public void setFromString(final T instance, final String fieldName,
            final String value)
```

Here “`instance`“ is a reference of a model instance and “`fieldName`“ is a string containing the variable name. The parameter “`value`“ contains the new variable value as a string. If the variable to be altered is not of data type `String` then it is attempted to convert the `String` into the corresponding primitive type. If this is not possible then an exception is thrown. In case the operation was successful, then the variable now has the new value.

Example:

```
manipulator.setFromString(model3, "length", "10");
```

The variable with the name “`length`“ of the model “`model3`“ is assigned the value `10.0`. Here the string “`10`“ is converted into the double value `10.0`.

It is possible to initialize non-scalar model variables like vectors or one-dimensional arrays with the `ModelXMLSectionReader` in a similar way. The models themselves do not provide any `get()` or `set()` methods to retrieve / set variable values. They are completely passive during initialization. The complete initialization service is provided by the `ModelXMLSectionReader`.

This design provides the advantage of being simple to use for the equipment model developer. He is only responsible of declaring settable and readable variables, vectors and arrays. Thus the access methods for these variables need neither to be coded by him nor has the model class to be derived from a base class which offers such a functionality. The simulator design by this means facilitates for the model developer to design his models as coherently as possible. He only is responsible for the proper functional modeling of all relevant physics of the real equipment hardware in the simulator equipment model and he needs a minimum knowledge on kernel implementation details only. An additional advantage is, that in case the *OpenSimKit* kernel design is modified with respect to input read functions and command / control functions (`set()` / `get()`), the model classes themselves are not affected. This makes the design robust to successive simulator kernel improvements and upgrades.

Similar `XMLSectionReaders` are provided for the processing of input file sections for

- the equipment ports,
- the connections between model instances by using functional interfaces,
- the simulation numerics setting parameters (solver accuracy, time step size etc.),
- the specification of variables to be cyclically logged to file,
- and interface packets to be submitted to the control console.

These `XMLSectionReaders` provide in total provide all services needed for a complete initialization of the simulation. By using the abstract factory pattern it is not required to statically link any model class library to the simulator.

11.2 SOA Implementation of the Kernel Numerics

The SOA concept on microscopic level is even more interesting for numerical computations than for the simulator initialization. The equipment models and system models (like space environment model) here serve as as “computation services” for the kernel. All other functions are coded inside the kernel in a strictly model independent manner.

Models Registering at the Solver

For solution of the initial value problem as depicted in figures 6.10 and 6.11, ideally every model needs to inform the solver, which derivatives it can provide. Additionally the solver needs to be informed about which state variables each model needs handed back to perform its internal calculations. This corresponds to the registration of a calculation service in a SOA, please also refer to figure 11.1. The registration of a model at the solver is handled via an according method of the solver. Here again an interface class is involved, which is called `solverBridge` and every model is connected to the solver via this `solverBridge`. During this registration the state variables which each model can provide are queried by the solver, the same applies for the derivatives each model can provide and the state variables which every model needs back from the solver for performing its internal computations. Every model features special markers (annotations like `@ProvidedVariable` , `@RequiredStateVariable` and `@Derivative`) which are read by the `solverBridge` and are provided to the solver on request:

```
public void addModel(Model model) {
    addProvideVariables(solverBridge.getProvidedVariables(model));
    addDerivatives(solverBridge.getDerivatives(model));
    addRequiredStateVariables(solverBridge.getStateVariables(model));
}
```

Specially marked state variables as well as derivatives are declared to the solver. The `solverBridge` subsystem detects the state and derivatives by their annotations. These variables thus then can be queried by the solver.

```
public DemoModel() implements Model {
    @ProvidedStateVariable private double stateVariable1;
    @RequiredStateVariable private double stateVariable2;

    @Derivative private double derivative1;
    @Derivative private double derivative2;
}
```

After all components are registered the solver can verify whether for all registered state variables a "service provider" is available, i.e. an equipment model, which computes the state variable derivative for each time-step respectively for each integration test step.

Computation of Derivatives as Service

During time step integration the models act as service providers for the solver. They compute the state variables and derivatives for each new time or test step. The technical approach used therefore is as follows: Besides access to class global variables the Java reflection API[30] provides access to methods, too. This is applied

[30]In software engineering "reflection" means that a program knows its own structure and is able to modify itself if needed. The "reflection API" is an interface of the Java programming language which provides such functionality. API stands for Application Programming Interface.

also by the `solverBridge` for enabling the kernel to invoke methods of a model. The "callable" annotation can only be used with methods. A method marked by this means can be invoked by the `solverBridge` by using `callMethod()`.

```
@Callable public void computeDiscreteStatechanges() {
  …
}

@Callable public void computeProvidedVars() {
  …
}

@Callable public void computeDerivatives() {
  …
}
```

Here it needs to be emphasized again that the access restrictions are only imposed by the `solverBridge`. The Java reflection API does not provide any such restrictions and does not need any annotation to read / modify variables or invoke methods.

The signature of the method `callMethod()` looks as follows:

```
public Object callMethod(final T instance, final String methodName,
            final Object... parameters)
```

The parameter "`instance`" is already known from the methods `getAsString()` and `SetFromString()` of the manipulator. The parameter "`methodName`" denotes the name of the method which shall be invoked (similar to the "`fieldName`" in the examples above). The parameter "`parameters`" is a so-called variable arguments parameter and represents any number (0 to n) of parameters. The return value of "`callMethod`" is that of the invoked method.

Example:

```
solverBridge.callMethod(model3, "computeDerivatives");
```

Here the method called "`computeDerivatives`" of the model "`model3`" is invoked. As the method "`computeDerivatives`" does not take any parameter, it is not required to pass parameters to the `callMethod()`. This is automatically supported by using the variable arguments parameter in `callMethod()`. The return value of the invoked method is not processed as in this case the return value is "`void`" anyway, see above definition of `computeDerivatives`().

11.3 Orchestration of the Computation and Function Distribution

The coordination of the proper calculation sequence execution in accordance with figure 6.12 is performed by the simulator kernel by calling all the relevant equipment models accordingly via the `solverBridge`. In SOA terminology this coordination of service provider and service user interaction is called "orchestration" of the computations.

Thereby this architecture with registered services and the assembly supported by the abstract factory allows the distribution of the computations of different models over multiple threads inside the simulator, if necessary over multiple CPUs inside one computer or even over multiple CPUs inside multiple computers. The time-step computations inside the models can be distributed over multiple computation nodes similar to Grid-computing, however with equation sets which can be completely different from model class to model class.

Thus with a SOA based simulator kernel architecture, which is still subject to current research, it is possible

- to solve the problem of the dynamical system initialization (including topology information) which was cited in chapter 6.8,
- as well as to relieve the model developer from programming any initialization functions,
- and finally, to provide a highly efficient numerical architecture which offers the numerical performance required by STB and EFM test benches – and this being based on Java as a very user friendly programming language.

For hybrid test benches however, real-time capable Java implementations would be needed to apply this technique like e.g. Java Real-Time System, PERC or the simulation needs to be equipped with real-time capable third party libraries like Javalution (cf. [69] to [71]).

12 Consistent Modeling Technology for all Development Phases

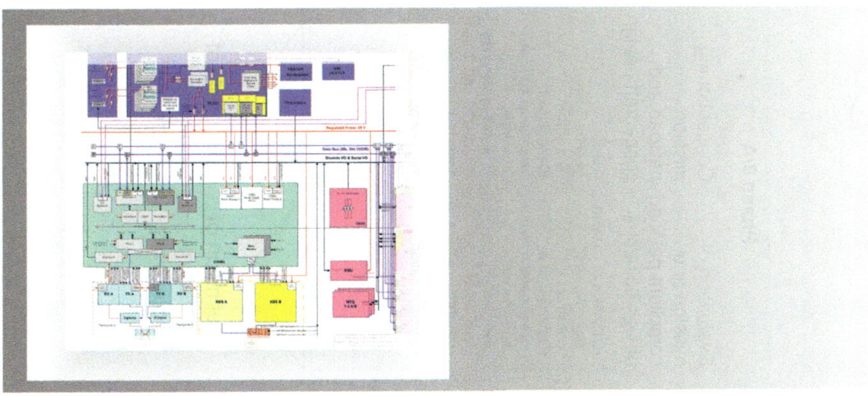

Aeolus Block Diagram © Astrium

J. Eickhoff, *Simulating Spacecraft Systems*, Springer Aerospace Technology 1,
DOI 10.1007/978-3-642-01276-1_12, © Springer-Verlag Berlin Heidelberg 2009

The phases A to E of a spacecraft system development have already been treated in an overview in chapter 3 please also refer to table 3.2. However, especially the early phases of development shall now be covered in a bit more detail.

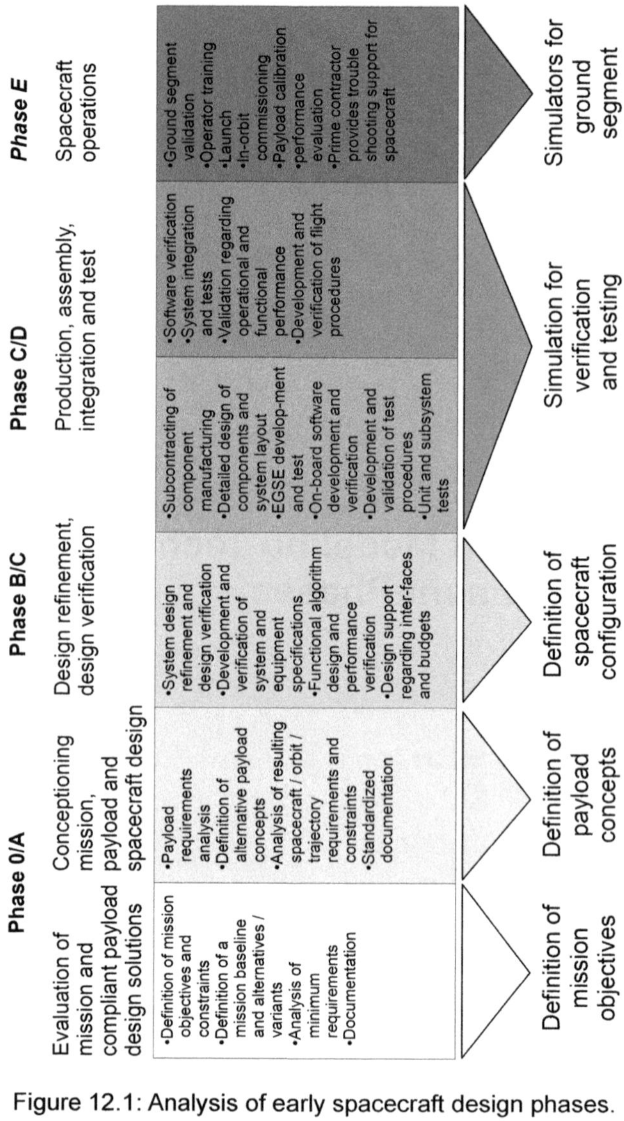

Figure 12.1: Analysis of early spacecraft design phases.

Commercial spacecraft like launchers or telecommunication satellites usually are series productions and a phase 0/A is not performed for each individual satellite. They usually are based on a common platform - for satellites the so-called "satellite bus" - and only vary by number of satellite payload modules, e.g. commercial transponders, their types and used frequency ranges. Phases 0 and A only are passed through when designing an entirely new satellite series platform. In case of Earth observation and science spacecraft usually these are all prototypes or a small series of 3 to 4 instances. So for each such individually tailored mission a new 0/A phase is undertaken. Typically the initiatives for Earth observation and science missions go back to Earth observation proposals which are submitted by scientists to the agency later contracting the spacecraft and which are to be accepted by an agency's board. A first phase 0 mission design step in these phases is in fact carried out between agency and science institutes to define mission objectives, baseline requirements and potential variations.

For a more detailed phase A, an analysis of the mission with payload conception, orbit analysis, operational concept definition and derived spacecraft basic platform design is then carried out by spacecraft manufacturers. As it was already mentioned in chapter 3 such phase A studies usually are subcontracted by the agency to at least 2 spacecraft primes. In the following figure the changing development focus and the changeover from a highly iterative to a more and more linear engineering process over the phases is sketched out.

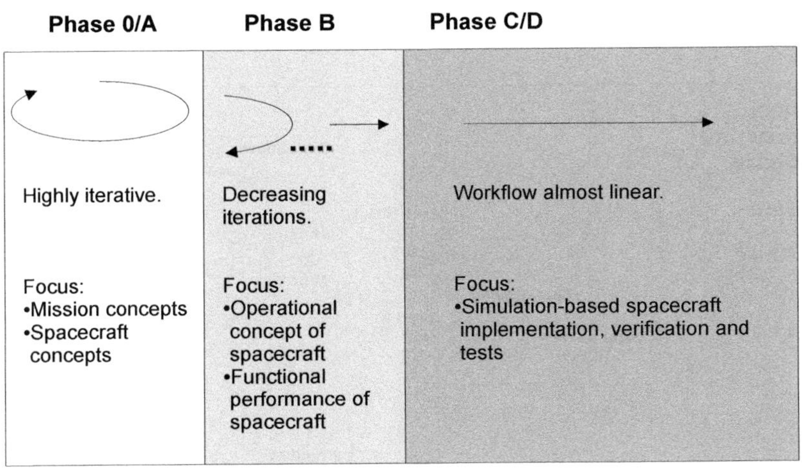

Figure 12.2: Design focus and iteration cycles in spacecraft development phases.

In chapter 3 the modeling and simulation infrastructures where described which are used from conception via design to verification phase and AIT. The ideal situation would be to have available one consistent functional modeling approach and infrastructure for the spacecraft which could be applied from Design Office level modeling via functional simulators (Simulink / SciLab), for AOCS or other control design, down to SVFs and hybrid testbenches. However conceptualization of such an approach requires a closer look at the requirements towards the simulation setups, towards what is currently supported by available setups - and finally what has to be modified for such a consistent cross-phase infrastructure.

12.1 Requirements to a Cross-Phase Design Infrastructure

Considering this background towards a single, cross-phase simulation infrastructure a technical approach shall be evaluated concerning its applicability and extendability for such an infrastructure. As an initial step for each phase, the engineering steps and the required modeling level of detail shall be summed up. The analysis again applies the example of a satellite design.

Budget Engineering (Phase A):

The scope of system modeling at the end of this phase A must provide:

- Defined mission concept including orbits, respectively trajectories
- Specified spacecraft basic structure layout
- Defined on-board equipment
- Identified operational budgets (mass budget, power budget, ground / space data link budgets, required on-board mass memory budget etc.)

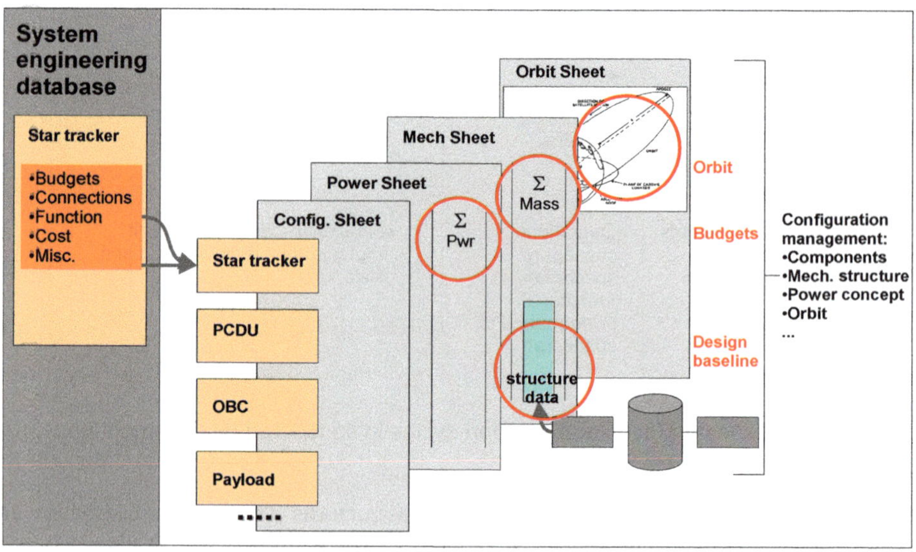

Figure 12.3: Budget Engineering - Phase A.

It can be commented here that this budget engineering work already can be handled appropriately by spreadsheet tools as they are used in Design Offices. At this stage dynamic simulation of spacecraft internal processes is not yet required. Only orbit / trajectory analysis is performed - based on commercial tools like Satellite Toolkit.

Operations Engineering and Control Design (Phase B):

The scope of system modeling at the end of phase B must provide:

- Defined mission concept including orbits respectively trajectories
- Specified spacecraft basic structure layout
- Defined on-board equipment
- Identified operational budgets
- Spacecraft operational modes
- Operational modes of all on-board equipment
- Functional algorithms modeling each equipment's behavior (modeling on engineering unit level only)
- Functional interfaces between spacecraft equipment (not yet identifying real interface types nor protocols)

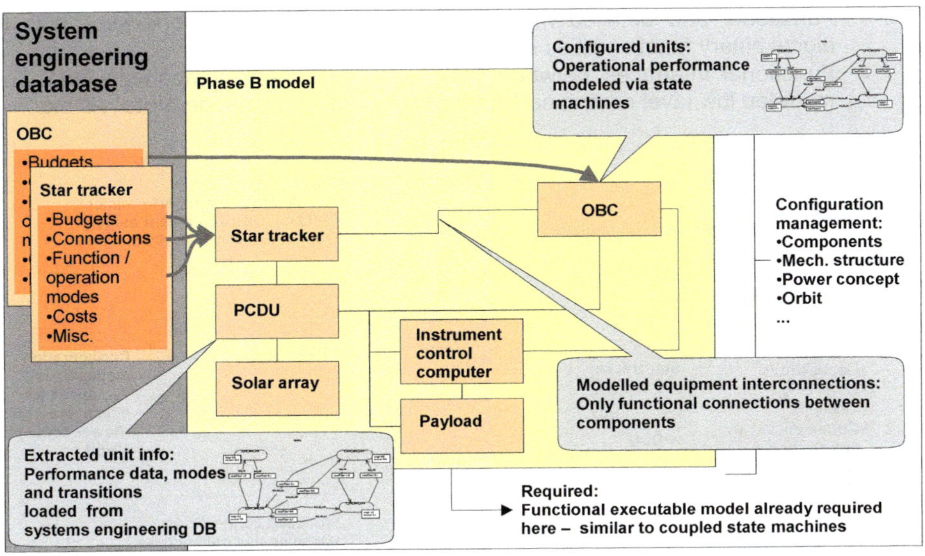

Figure 12.4: Functional modeling for operations engineering and control design.

Here it can be summed up, that the full scope is not fully covered by the functional analysis tools such as Simulink - which in most cases also only are applied to AOCS control design. And even for AOCS control algorithm design in many cases separate setups are implemented - e.g. one for safe mode, one for fine pointing mode. Since the FVB benches contain these derived algorithms - just implemented on the later reference target simulator kernel platform, they in fact show up similar deficiencies.

None of these setups covers the already to be analyzed operational aspects of spacecraft and mission, i.e. operational modes both on spacecraft level as well as on equipment level. A need for reflecting this functionality - represented as user controllable state machines - can clearly be identified here. Furthermore a proper takeover and further evolution of the spacecraft design data set from Design Office level to this first simulation setup is identified as an additional requirement.

Detailed Simulation for Verification and Test (Phases C/D):

The scope of system modeling for simulation in phase C/D must provide:

- Defined mission concept including orbits respectively trajectories
- Specified spacecraft basic structure layout
- Defined on-board equipment
- Identified operational budgets
- Spacecraft operational modes
- Operational modes of all on-board equipment
- Functional and detailed modeling of each equipment's behavior (down to digital binary functionalities in equipment)
- Functional interfaces between spacecraft equipment components including modeled low level command / control details, data exchange protocols etc.

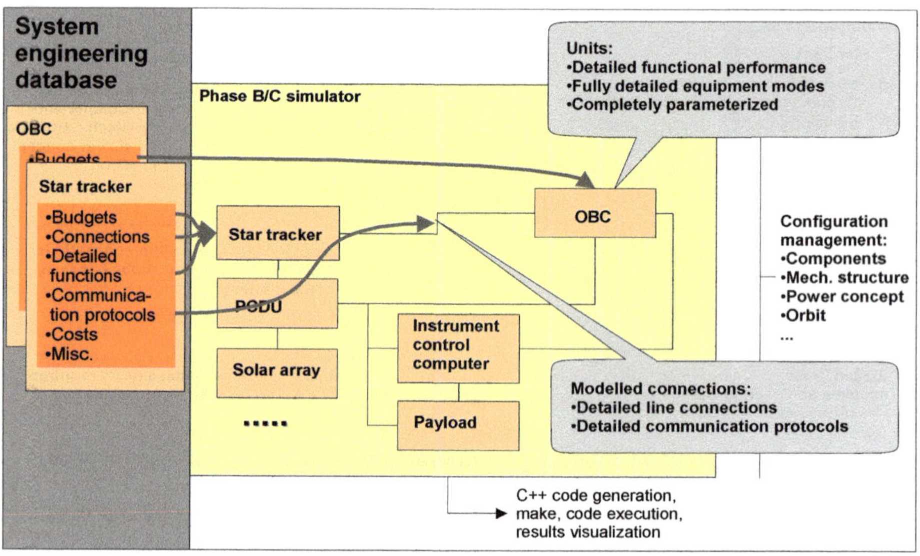

Figure 12.5: Detailed simulation for verification and testbenches.

Analyzing this requirements set it can be concluded, that today's SVF simulators properly fulfill all these requirements, so that no new conceptual invention is required anymore at this point. However also here the need for a proper evolution of the spacecraft design data set and controlled takeover from the previous setup can be identified.

Requirements summary:

When analyzing these three requirement sets and in parallel reviewing the type of toolkits and their configurations currently in use in European spaceflight (cf. table 3.2) then it can be summed up that there exist two fundamental technology breaks in the tool chain today:

One is between the budget and concept oriented toolkits for spacecraft Design Offices and the early functional simulators - e.g for AOCS based on Simulink.

The second break is to bring together again the purely mathematical / functional tools like Simulink - which do not properly reflect spacecraft topology - and operations design approaches to a first complete system simulation. This first complete system simulation is similar to the FVB testbench, but in addition has to include equipment characterizations, equipment / spacecraft operational modes and their commanding - based on state machines.

Thus the ideal sequence would look somehow like:

- Performing budget and basic mission concept engineering based on Design Office tools.
- Then starting control algorithms analysis with dedicated tools like Simulink.
- Setup of an "FVB--" in the form of a simulator which
  - ◇ reflects the spacecraft topology w.r.t components,
  - ◇ reflects equipment interconnection types only on FVB functional engineering unit level,
  - ◇ reflects spacecraft and equipment budget parameters,
  - ◇ reflects equipment and spacecraft modes and functional transitions as well as their commanding based on implemented state machines.

  It will not yet comprise quantitative implementations of control algorithms but will allow first dynamic runs and will be suitable for operations analysis.
- Setup of an "FVB++" including:
  - ◇ The a.m. "FVB--" features
  - ◇ The control algorithms which already today are transferred from the control engineering tools (Simulink etc.) to an FVB,
  - ◇ In addition also control algorithms for other domains than AOCS - such as power control etc.
- Finally extending the FVB to an SVF by adding interface types between equipment models (data buses etc.) and by adding information on equipment interface cross-couplings.
- Then the further process with STB and EFM can continue as described already in chapter 3.

After identifying these requirements and and sketching out a development sequence for such a cross-phase simulation infrastructure also some concepts for its technical implementation shall be presented - always keeping in mind that this definitely still is a research topic also supported by diverse national and European agency studies.

12.2 Cross-Phase Simulation Infrastructure and Engineering Steps

The goal of the applied modeling technique for simulation must be to achieve an executable, functional model of the entire spacecraft including operations aspects already earlier in the design chain. This spacecraft model should be suitable for both design and verification simulations as well as for verification of spacecraft operations procedures. The approach investigated for cross-phase suitability is a slight variation of the simulator implementation technique which was discussed already in chapter 8. The modeling of the spacecraft system and equipment in the described simulator design approach of chapter 8 is based on the software design language UML. The mapping of real equipment features to equipment model UML representation, i.e. to software class structures, state machine representations etc. is performed by the abstraction and modeling expertise of the equipment model programmer. The generated and further instrumented code is based in most cases on the C++ language, since it is platform independent and suitable for real-time implementations. For the cross-phase approach the following enhancements / changes have to be applied to the current approach:

- One is to base the generated code on an object-oriented language which is both able to be interpreted (of course in such cases not offering real-time performance) for prototype code verification and which is compilable for full efficiency and real-time applications. This functionality is provided by Java and the C++ variant "System C", a language originating from digital chip design which can be executed on a runtime kernel. Thus a simulation which was "clicked together" and instrumented with some algorithmic details and characterization data could directly be run. System C in fact a C++ dialect provides a C++ class library (similar to the standard template library) and sophisticated functionalities for modeling of networks and communication processes (buffers, filters, channels etc.) which is essential for representing spacecraft internal equipment interconnections compliant to real topology.

- The second variation to be introduced - compared to simulators for classic FVBs to EFMs - is, to close up the design of the spacecraft model as far as possible to the design of the real system. The spacecraft system design is represented already in SysML in the system engineering database, (SEDB), as it was presented in figure 10.11. The simulation model of the spacecraft can be tightly aligned with this system engineering database representation and software code can be generated from such a derived model representation.

  It shall however be noted explicitly here that the spacecraft / equipment models in the system engineering database and in the simulator only can be similar to a certain extent, but never will be identical. The simulator models e.g. always will comprise certain numerics specific characteristics and the spacecraft representation in the system engineering database e.g. may have AIT specific add-ons or other.

- A third aspect to be changed in the development chain presented in chapter 3 is, that already the budget design results and the elementary system configuration - i.e. the product tree - has to be taken over from the Design Office based phase A study into the system engineering database as starting point for a digital spacecraft design evolution leading to the proposed FVB--, FVB++, SVF, STP, EFM installations.

Starting from these basic insights a stepwise approach for setup of such a cross-phase simulation infrastructure shall be sketched out in the following paragraphs.

Step 1: Phase A mission and spacecraft concept engineering in Design Offices: This task is based on orbit / trajectory simulations and result modeling, based on spreadsheet tools, can be mostly kept untouched compared to the current standard processes. The only add-on has to be an import of the finalized design reference into the system engineering database to be converted to the SEDB data model. To be stored in there are:

- Identified orbit / trajectory characteristics
- Identified necessary spacecraft components and baseline topology (product tree)
- Evaluated design characteristics for entire spacecraft and key equipment respectively (e.g. battery capacity estimations)

Step 2: The first step towards an executable spacecraft baseline is the definition of the spacecraft system topology with respect to "product tree" or decomposition into subsystems and equipment. This can be performed directly in the system engineering model of the SEDB. The availability of the product tree information opens up the representation of a first element-definition based spacecraft representation:

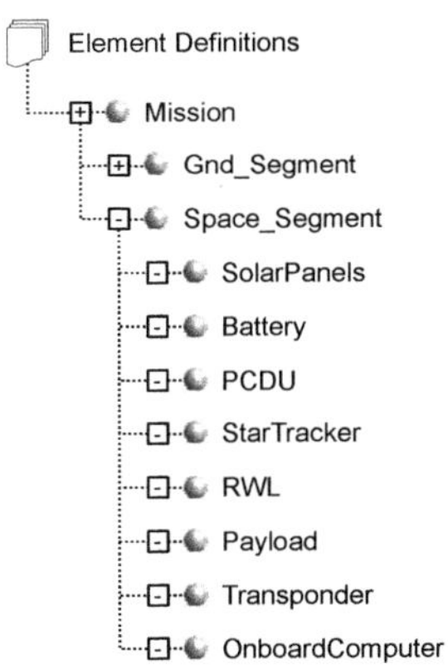

Figure 12.6: Element definitions.

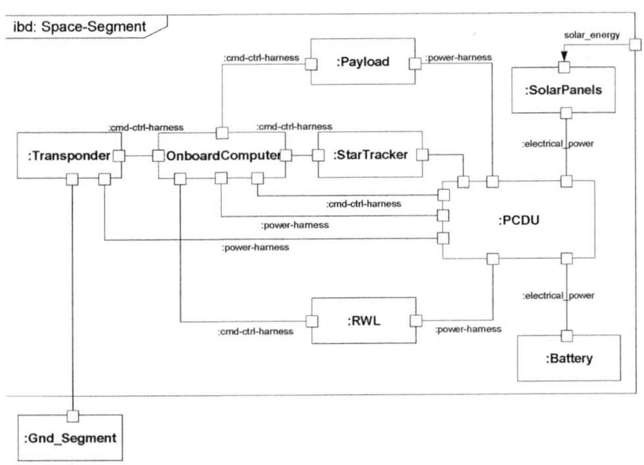

Figure 12.7: System topology based on element definitions.

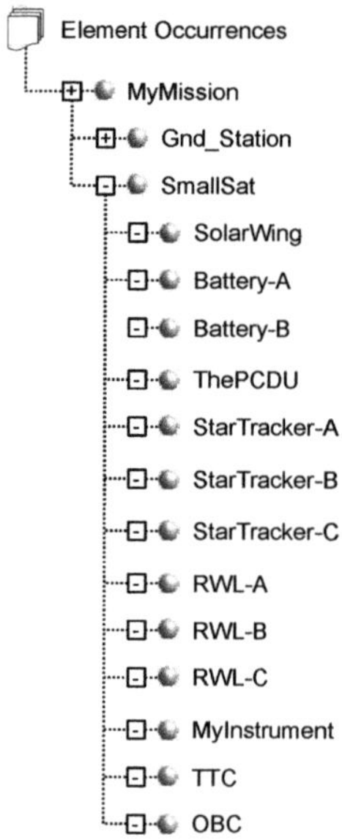

Figure 12.8: Occurrences definition.

So far only the types of equipment on-board have been defined in the element-definitions. Now the definition of the element-occurrences will follow - e.g. the instances of type "Battery" on board the conceptualized spacecraft.

Depending on the development phase, both trees - element definitions and element occurrences - can be rather high level at start of phase B design and can evolve later to be covering all details for a phase C / D simulation like in an SVF. The technical approach does not change with increasing modeling granularity.

Provided with this additional information from the element-occurrence tree and manual definition of equipment interfaces, a first full spacecraft topology can be modeled and stored in the SEDB and a spacecraft simulator model in SysML can be derived accordingly. Such a representation is outlined in figure 12.9.

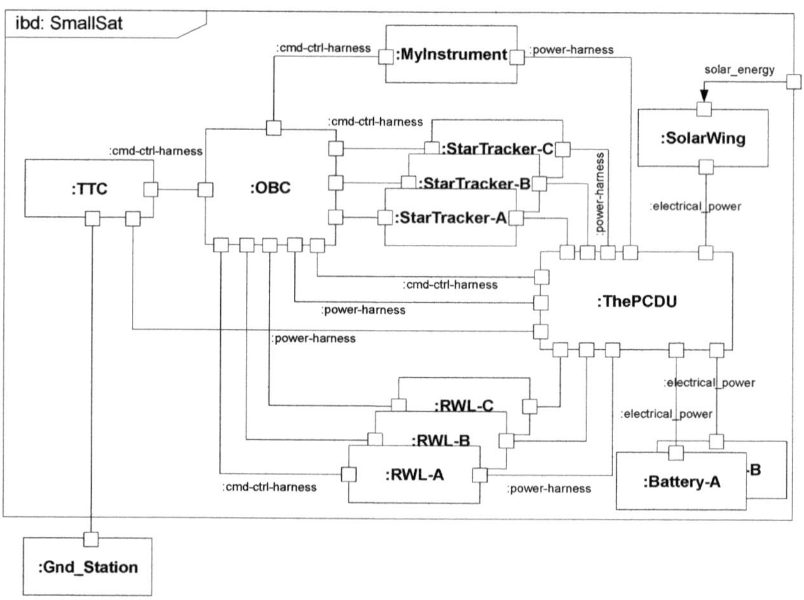

Figure 12.9: System topology based on element occurrences.

Step 3: Thereafter follows the definition of spacecraft system operational modes and the modes for each equipment. Usually the selected notation is to reflect this information in state machine diagrams since this directly allows to note in addition the mode transfer triggers as well as values of mode dependent system or equipment parameters. Also this information flows into the simulator spacecraft design.

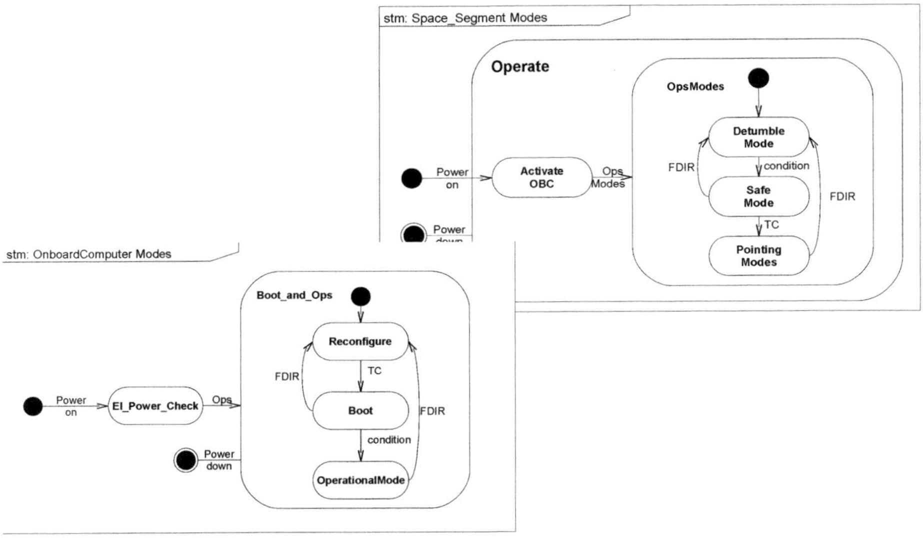

Figure 12.10: Definition of modes and transitions on system and equipment level.

Step 4: A further step then is to align these modes on system and equipment level as it is done by the operations design engineers already in early project phases. Here clearly is allocated which equipment is operating in which equipment mode while the spacecraft is in a certain spacecraft mode. E.g. all payloads will be turned off in spacecraft safe mode. Aligning the spacecraft modes (and implicitly corresponding equipment modes) to the operations sequence of the spacecraft like launcher separation, deployment, de-tumble, safe mode, nominal mode, operational mode (and, in addition, the potential failure handling modes in between), this leads to a sequence often also called "master timeline".

Table 12.1: Definition of spacecraft / equipment mode interactions.

	Boot Mode	Safe Mode	Nominal Mode	Recover Mode
OBC	booting	on	on	reconfigure
PCDU	on	on	on	emergency-on
STR	off	on	on	on
RWL	off	on	on	off
Payload	off	off	operating	off

After this definition of spacecraft and equipment topology, equipment and spacecraft characterization data and mode definitions, all necessary modeling steps for reaching a first executable model have been performed.

Step 5: Thus now from this spacecraft model design, System-C or C++ code can be generated for execution. For System C the state achieved after code generation and instrumentation is somewhat similar to a code in Java language. It has to be compiled for the runtime environment, which however is independent from the target operating system.

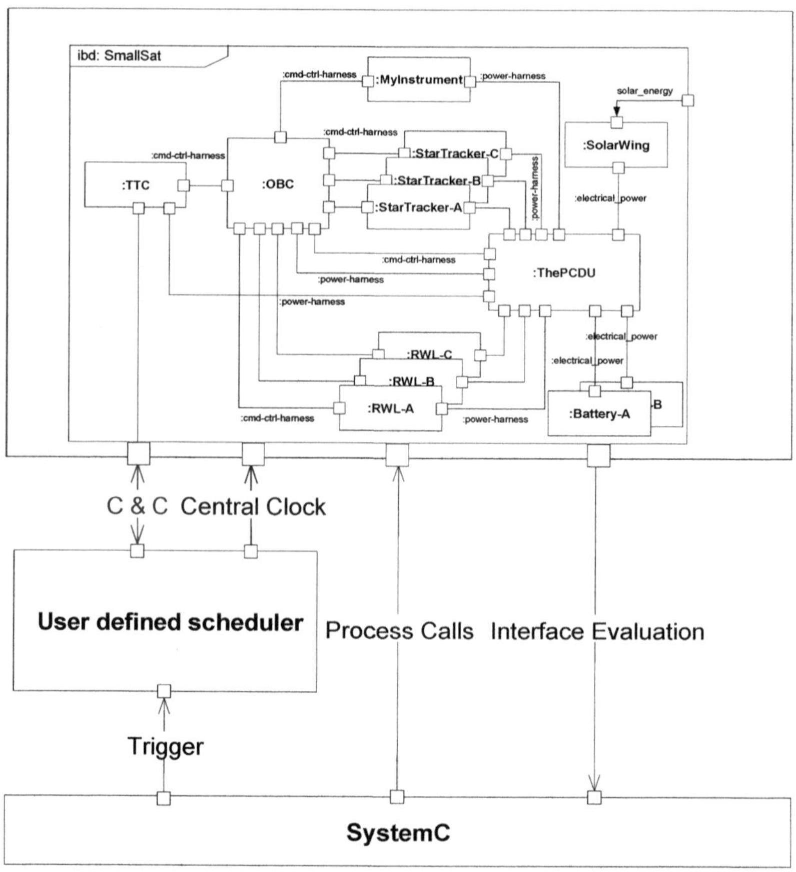

Figure 12.11: Phase B spacecraft system model executable in System C.

The achieved model thus is directly executable, can be commanded and first verification steps for system design, mode switching algorithms or similar can be carried out on this implementation which corresponds to the cited functional verification bench "FVB--". Generally testable are:

- The consistency of operational / mode concept
- The control algorithms w.r.t. mode transfers and affected variables

Step 6: The next step then is to enhance the models of this infrastructure with the results from the control algorithms design in e.g. Simulink - which now is no longer in the main spacecraft model development sequence but provides algorithmic input and performance relevant characterization data as sideline input.

As in Simulink here also the equipment classes and the numerics of designed state machines must be instrumented by additional computational functions. In case numerical analyses for control algorithms have been performed in Simulink, SciLab or a similar tool, C code can be generated from this and can be directly applied for model code instrumentation. Thus this development step provides the control algorithms and based on such an "FVB++" first dynamic spacecraft performance analyses and predictions can be carried out - with some limitations.

Step 7: The final steps now are further refinements of the equipment modeling, algorithmic implementations (e.g. thermal modeling) and especially the implementation of real communication protocols for the modeled equipment connections. Interconnection modeling now is based on models of real data protocols. In this step the functional connections between the identified components are to be refined including all cross-couplings:

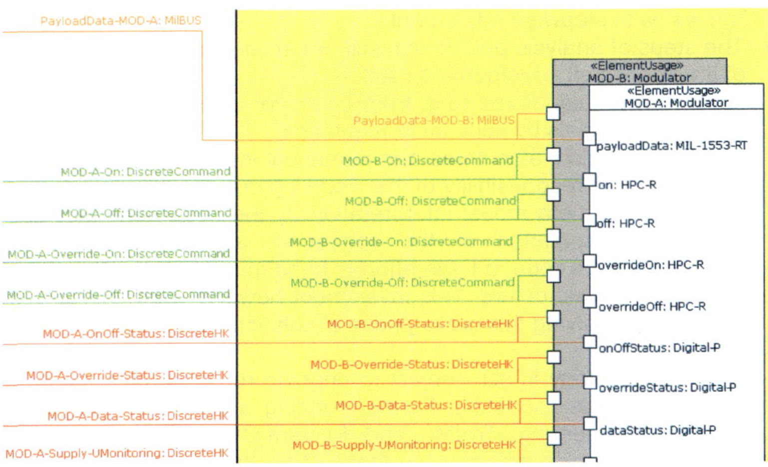

Figure 12.12: Example for equipment interconnections in system.

This in most cases implies the implementation of additional equipment model subelements - e.g. MIL-STD-1553B bus remote terminal functionality - and the entire calibration and failure injection functionality as explained for the model interface layer of simulator components in figure 4.15. Having reached this modeling state now additionally

- detailed simulations with error injections on protocol levels are possible (e.g. for OBSW testing),
- communication budget analyses are possible, and
- usually at this stage the modeling of equipment algorithms has reached the final refinement state and is fully suited for all types of spacecraft performance relevant simulations.

The simulation has achieved a degree of completeness corresponding to a full fledged SVF and is representative for typical phase C / D / E simulators and testbenches as was required in figure 12.5. The presented approach to base all development steps from a first budget analysis architecture down to a full detailed spacecraft simulation on one cross-phase infrastructure approach can be evaluated as follows:

+ The result from Design Office based mission and spacecraft concept analyses are properly taken over into the engineering flow by means of using an SEDB.
+ Equipment / system modeling in SysML, model code generation and instrumentation leads to a functionally executable model. The approach thus provides executable models in very early phases of spacecraft design.
+ The method represents correctly functional aspects as well as system topology.
+ The method allows hierarchical refinement of modeling, i.e. first a high level design which is further detailed later. So the executable spacecraft model "grows" with stepwise refinements.
+ The steps of analysis and design still are in line with the engineering process of the spacecraft development itself.
+ Mathematical algorithms (e.g. from controller design) can be integrated into equipment models by code instrumentation.
- The mathematical algorithms and numerical implementation (applied solvers etc.) are in the responsibility of the modeling engineers.
- For complete meta model interpretation by the code generator to parse the SysML representation of the spacecraft and to generate executable System C / C++ code, a large effort is to be invested once to set up the code generation infrastructure. This investment however then can be reused over all projects applying this cross-phase simulation infrastructure.

The described technology is still subject to ongoing research (see e.g.[100]) and is not only limited to space industry. Further reading and Internet pages on model-based systems engineering, SysML and System C are listed in the according subsection of this book's references annex. For the data model to reflect the spacecraft design in an SEDB, the reader is referred to the ECSS E-TM-10-23 standard [107] and to an ECSS presentation on the model-based system engineering topic in [106].

13 Knowledge-Based Simulation Applications

J. Eickhoff, *Simulating Spacecraft Systems*, Springer Aerospace Technology 1,
DOI 10.1007/978-3-642-01276-1_13, © Springer-Verlag Berlin Heidelberg 2009

Since the beginning of the 70's, a special field of computer science, "Artificial Intelligence", (AI), has made significant progress even though the early hype meanwhile has calmed down. From the early fundamental research during the 80's the first commercial products evolved, the "Expert Systems" or knowledge-based systems. By means of such software tools knowledge is formalizable, storable and processable in the form of objects, cases and rules. The deduction of facts is achieved through instantiation of cases and rules with real object data and evaluation of these rules by a deducing algorithm called "Inference Engine". Before diving into the topic first some terminology definitions shall be given:

Expert system: These are software systems from the artificial intelligence domain, mostly based on object oriented programming techniques. Expert systems comprise an inference engine for processing abstract knowledge. This mechanism uses rules or / and cases to deduce conclusions from existing facts. The rules and respectively cases are stored in a so-called "Knowledge-Base".

Expert system shell: An expert system shell is a development environment for knowledge-based systems. It comprises the essential inference engine and provides for the expert system developer all tools and editors and user interface features to define and visualize rules and cases for knowledge definition. An expert system shell can be compared to a Rich Client Platform for conventional software systems (cf. chapter 8.6).

Inference process: The process of deducing conclusions from initial facts by application of rules / cases.

Inference engine: Also called "Inference Mechanism". Core processing engine of an expert system which is capable of deducing conclusions from facts by instantiation of rules / cases and their evaluation.

Inference chain: The sequence of subsequently applied rules in an inference process including their according intermediate fact sets and results.

Inference strategy: Strategy for identification of the next rule to be applied in the deduction process.

Knowledge-base: "Database" of an expert system where knowledge is stored in form of rules, cases, objects and facts. The overall knowledge of an expert system can be spread over multiple knowledge-bases.

13.1 Modeling of Information for Rule-Based Processing

In rule based expert systems the knowledge is reflected in abstract rules. These rules consist of a conditions part and a conclusions part. In the conditions part all preconditions are listed which are required to fulfill the results in the conclusions part. The generic form of such rules is:

$$P \rightarrow Q \tag{13.1}$$

Which is to be read as:

If preconditions P apply, then the conclusion Q is valid.

Thereby P and Q themselves can be complex logic expressions such as e.g.:

$$P_1 \wedge P_2 \wedge \neg P_3 \rightarrow Q_1 \wedge Q_2 \tag{13.2}$$

Which is to be read as:

When the preconditions P_1 and P_2 apply – but not P_3, then the consequences Q_1 and Q_2 both are valid.

For this rule based processing of system response and output data which may come from system simulations or real system online process data, it is essential that the user is able to enter the rules himself into the expert system. In contrast to assignment operators in conventional programming languages, rules are bi-directionally processable.

In modern expert system shells graphical editors for entering rule preconditions and conclusions are provided, which in most cases follow the outline as given below:

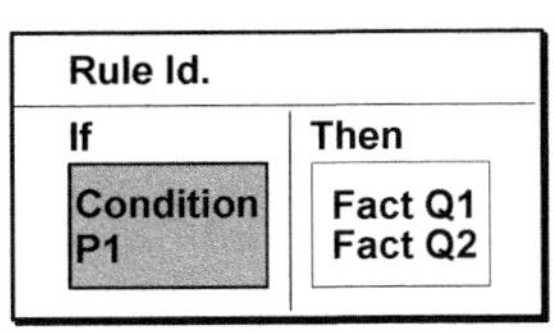

Figure 13.1: Rule format in editor mask.

Rule Processing: Forward Chaining

The typical questions processed via the so-called forward chaining method are of type:

- The facts are identified.
- What are the resulting conclusions?

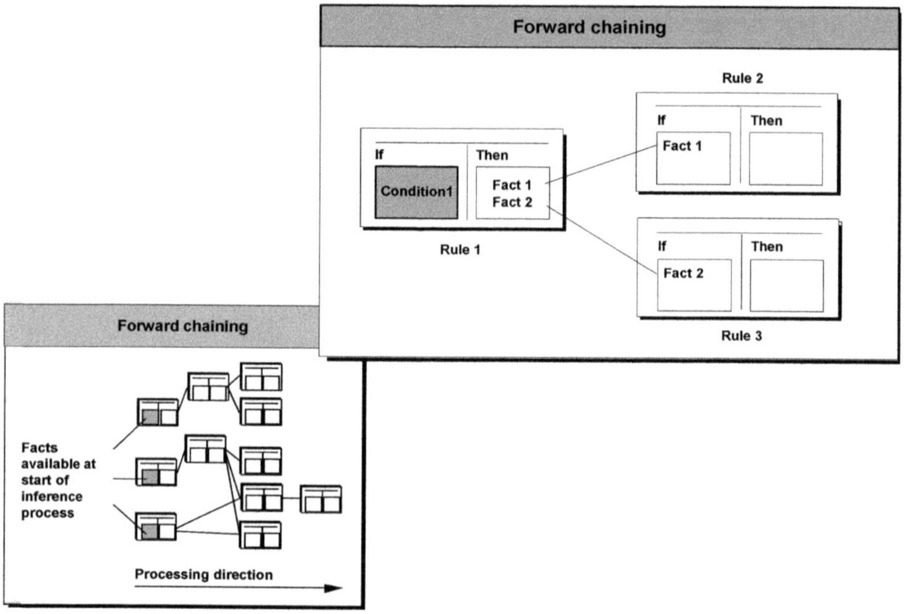

Figure 13.2: Rule network and processing example for forward chaining.

Rule Processing: Backward Chaining

The typical question type processed via the backward chaining method are:

- The facts are identified.
- How did this situation emerge?

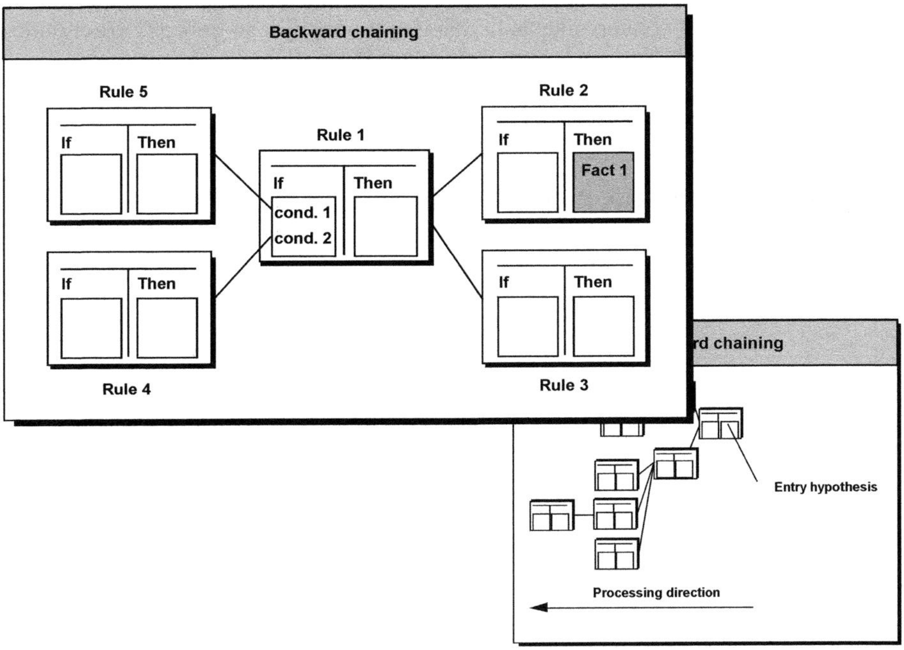

Figure 13.3: Rule network and processing example for backward chaining.

Some additional terminology definitions shall be given gere before further processing:

Inference strategy: An inference strategy is an approach for selection of the next rule(s) in a complex network to be applied ("fired") in an expert system's deduction process. By means of the inference strategy and based on the known facts at a certain point during rule processing, it is selected whether

- first to apply all possible rules on the actual level of known facts ("breadth first"),
- or whether first to process only rules for checking the last rule's precondition in further depth ("depth first"), or,
- whether higher level heuristic approaches shall be applied for selection of the next rule ("gradient method", "hill climbing method" and others).

Truth maintenance system: The "Truth Maintenance System", (TMS), of an expert system is a component permanently checking the consistency of rule instantiations and of deduced facts during the inference process. As soon as contradictory facts or rule conclusions are identified it stops inference processing and alerts the user to fix the situation.

Facts: Are the basic elements processed by rules. For technical applications the fact representation most conveniently is handled in the form of so-called "Object Attribute Value triples" (OAV-triples). An example is given here:

```
(component-C4 temperature 245.8)
```

13.2 Accumulation of Knowledge on a System's Behavior

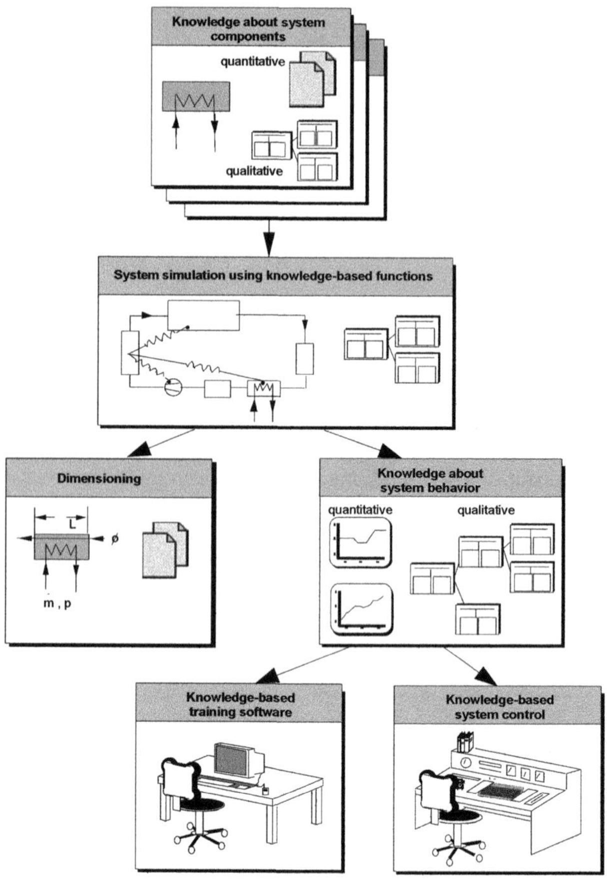

Figure 13.4: Knowledge acquisition on system behavior for user training and system operations.

The objective of knowledge-based simulators is to support acquisition of abstract higher level knowledge on the dynamic behavior of the analyzed system. The aggregation of knowledge starts with information on the system components, their dimensioning, interconnection, their criticality concerning system failure modes, aging effects etc. It is enhanced with knowledge on nominal operations, error symptoms and their propagation in the overall system. Finally this knowledge can be used for:

- System operator training
- System control, and
- Failure diagnosis - in spacecraft engineering called "Failure Detection, Isolation and Recovery" (FDIR)

13.3 Coupling of Knowledge-Processor and simulated / real System

How can knowledge about a system behavior now be made available for the real system's operation? For solving this task a suitable knowledge-processor has to be coupled to a compatible simulator. Diverse system load cases under various operational constraints are simulated and as complement to the symptoms arising, cases and rules are developed for the knowledge-processor. They then allow to identify certain operational cases and failure cases from the according symptomatics. The modeled knowledge on system behavior in its diverse modes evolves with increasing insight of the design engineers in the system's operational behavior and operational and performance constraints.

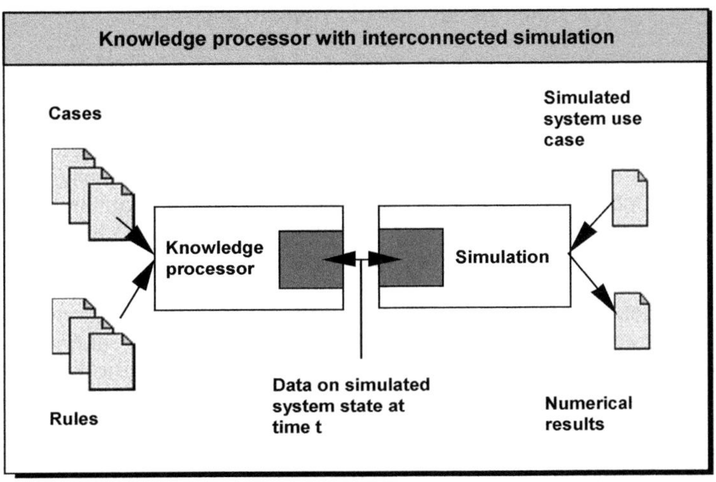

Figure 13.5: Integration of expert system and system simulation.

The functional flow during simulation result analysis is given in the figure below. The sequence of steps over time in the knowledge-processor is depicted in the left column, the functional steps in the simulator are visualized on the right. The information exchange is performed in multiple steps.

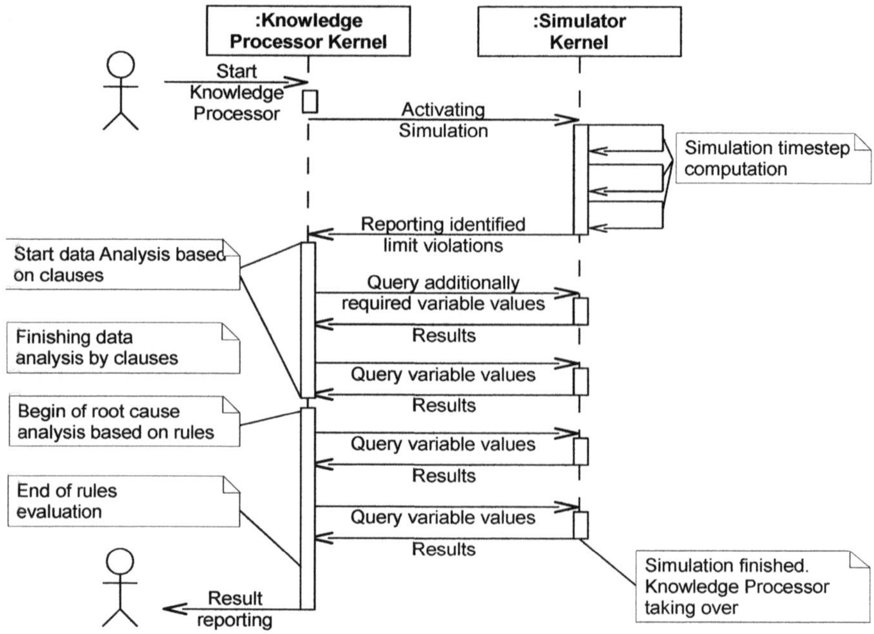

Figure 13.6: Information flow of a system simulation with knowledge-based symptom analysis.

The knowledge-processor is initially inactive and waits for input from the (simulated) system. As soon as parameter limit violations or combined parameter inconsistencies are deduced for the current operational case the data is transferred to the knowledge-processor. The latter immediately starts with data evaluation to identify the situation in more detail. The simulation goes pending. Eventually the knowledge-processor will need further information details to improve quality of its situation analysis and for preparation of a corrective action. In such cases it queries this additional parameter data from the system simulator.

After completion of facts identification the cause identification starts in the knowledge-processor by processing cases and rules. Also during this level of knowledge processing additional facts might need to be queried from the system simulation by the knowledge-processor.

As soon as the knowledge-bases are completed and the system / plant / spacecraft is built, the knowledge-processor can be coupled to the real system via a measurement data acquisition interface which is compatible to the former simulator interface and which allows access to all the system's relevant operational parameters. By this means now the operational case of the system / plant can be monitored, applying the knowledge accumulated during the design phase. A similar expert system based diagnostic - although not online - for the NASA TDRS satellites is described in [119].

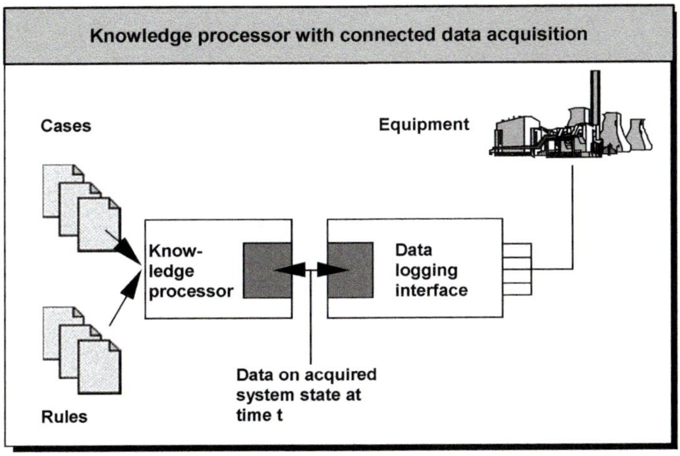

Figure 13.7: Integration of knowledge-processor and system data acquisition.

A further improvement of this method can be achieved if besides the system's data logging interface the system simulator is coupled to the knowledge-processor additionally. Such a synchronously running diagnostic simulation first was presented 1996 in a study of the TU Hamburg-Harburg (cf. [120]). Here the knowledge-processor not only can use system log and measurement data for situation analysis but also can use the connected simulator to further

- identify how the situation would evolve without intervention to the system's operation,
- and the knowledge-processor can deduce quantitatively adequate corrective actions and can analyze their feasibility by simulation before commanding them to the real system.

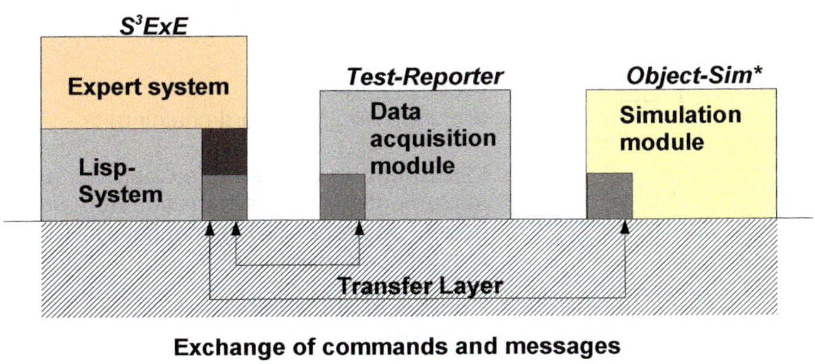

Figure 13.8: Knowledge-processor, data acquisition of system and simulator.

[31] Object-Sim meanwhile has evolved further to the open source simulator "*OpenSimKit*", see [23]

As a demonstration example for such an infrastructure again the well known rocket propulsion system shall be applied.

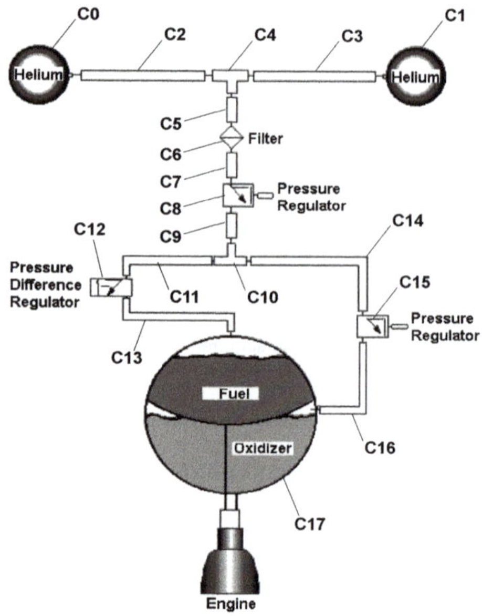

Figure 13.9: Rocket upper stage propulsion system.

The interconnection of numerics and system behavioral knowledge in practice is not established directly via the acquired measurement data from the controlled system to the rule processing. Instead it usually is processed via three levels as depicted below:

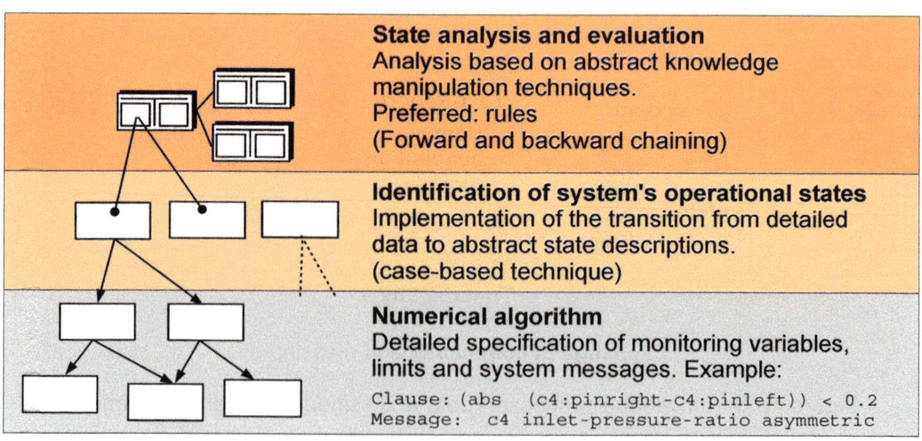

Figure 13.10: Concatenation layers of data and knowledge.

The lowest layer comprises the purely numeric processing of values from real system's data acquisition interface respectively from simulator. For these data which are cyclicly forwarded to the knowledge-processor condition clauses have to be loadable at startup, which permanently check the parameter limit compliance. However it is not sufficient to provide parameter checking against static upper / lower boundary values for each system load case. The full spectrum of processing functionality must be available:

- Absolute upper / lower boundary values for all system equipment parameters
- Relations of multiple parameters towards each other (e.g. pressure relations, pressure differences)
- Relative relations between parameters:
  e.g.: `c16:pout < c13:pout`
  The outlet pressure of component c16 must be below that of c13 (cf. Figure 13.9).
- Relations of complex parameter interdependencies including mathematical functions and relational operators, e.g.:
  `(c71:p / 1.5) < (c27:p - c35:pout * 1.2)`
- Conditional clauses including time dependencies, e.g.:
  `c83:p < (c82:p - 1.387 * t)`

In case such a limit check clause identifies all parameter being within limits, the knowledge-based functions are not activated. The simulation's data acquisition modules continue processing in parallel.

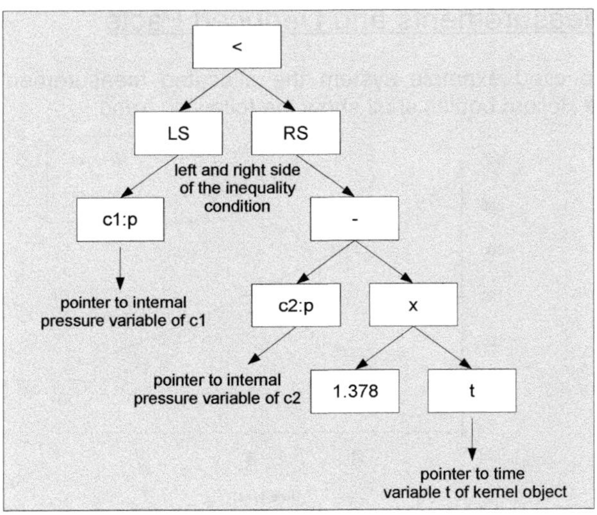

Figure 13.11: Modeling of condition clauses by structure trees. (cf. [120])

This clause based numerical preprocessing should be performed without time consuming data handshake between the knowledge-processor and the data acquisition module simulation. It has to be real-time performance compliant and thus must be implemented based on very efficient techniques. In most cases it will be directly implemented on the data production side, i.e. in the data acquisition module of the connected real system respectively in the simulator in case of a simulated system.

The condition terms for the cited application example of the rocket propulsion system are listed below:

```
// Section: Parameter ranges definition
//-----------------------------------------------------------------
// Please note: Condition expresses normal case - violation leads
to
// msg broadcast.
//
#Error-Condition: ec0
Clause:       (c0:ptotal-c1:ptotal) < 0.2
Message:     c0 pressure-drop-abnormally-low
//
#Error-Condition: ec1
Clause:       (c1:ptotal-c0:ptotal) < 0.2
Message:     c1 pressure-drop-abnormally-low
//
#Error-Condition: ec2
Clause:      abs (c0:ptotal-c1:ptotal)) < 0.2
//pin0 - pressure inlet port 0
//pin1 - pressure inlet port 1
Message:     c4 inlet-pressure-ratio asymmetric
```

Example Measurements and Deduced Facts

Sticking to the cited example system the allocated measurement values for the pressure in the Helium bottles shall show the following trend:

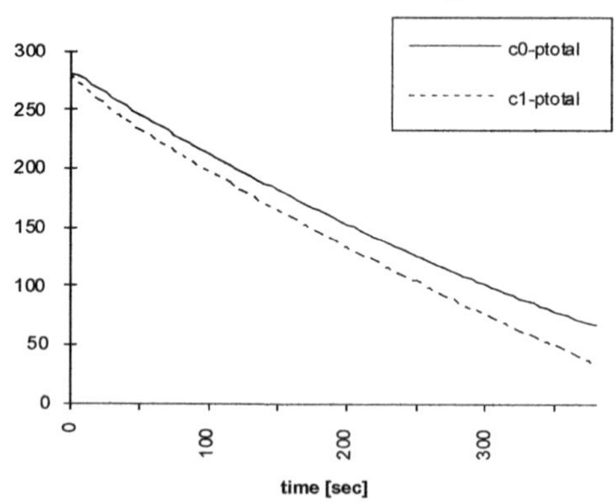

Figure 13.12: Pressure value trends in example system.

Induced through these increasing pressure difference the following "Error Conditions" are triggered:

Table 13.1: Activated error conditions and deduced facts.

EC-Violation	Error-Cond	EC0	c0	PRESSURE-DROP	ABNORMALLY-LOW
EC-Violation	Error-Cond	EC2	c4	INLET-PRESSURE-RATIO	ASYMMETRIC

Case-based Deduction of Initial Facts for Rule Processing

Now according to figure 13.10 the next layer of system state analysis shall be considered, the case-based reasoning level. Here the facts resulting from identified error conditions in the form of the cited OAV-triples are further processed and the initialization information set for the subsequent rule based status analysis and recovery identification is generated. As an example the following cases are defined for a case parser in the LISP programming language:

```
(if (?compx PRESSURE-DROP-ABNORMALLY-LOW)
    (case-assert! (?compx PRESSURE-DROP-IDENTIFIED CRITICAL)))

(if (?compx PRESSURE-DROP-ABNORMALLY-LOW)
    (case-assert!
       ((lisp-function
          get-connection '(?pipe connects (?compx C4)))
        BLOCKING ASSUMED)))

(if (and (C0 PRESSURE-DROP-ABNORMALLY-LOW)
         (C4 INLET-PRESSURE-RATIO ASYMMETRIC))
    (case-assert! (C1 PRESS-GAS-AMOUNT-MIGHT-BE INSUFFICIENT))

(if (and (C1 PRESSURE-DROP-ABNORMALLY-LOW)
         (C4 INLET-PRESSURE-RATIO ASYMMETRIC))
    (case-assert! (C0 PRESS-GAS-AMOUNT-MIGHT-BE INSUFFICIENT))
```

The syntax reads as follows:

Case 1:

If: there exists a component with the symptom PRESSURE-DROP-ABNORMALLY-LOW,

then: assert to the rule based inference engine (uppermost level in figure 1.8) the following fact:
For the identified component applies PRESSURE-DROP-IDENTIFIED CRITICAL

Case 2:

If: there exists a component with the symptom PRESSURE-DROP-ABNORMALLY-LOW,

then: via the function "get-connection", search which pipe connects the affected component with the junction C4, and
assert to the rule based inference engine, the assumption that for the identified pipe component applies BLOCKING ASSUMED.

The syntax of the following cases 3 and 4 can be interpreted in analogy. If the OAV-triples of symptoms identified by the error condition clauses from table 13.1 now are applied to these cases, the case-parser will itself deduce the following results and will assert them to the rule-based inference engine.

```
(C0 PRESSURE-DROP-IDENTIFIED CRITICAL)
(C2 BLOCKING ASSUMED)
(C1 PRESS-GAS-AMOUNT-MIGHT-BE INSUFFICIENT)
```

Rule-based Simulator Control and Result Analysis

At a certain point the knowledge-processor shall have received the pressure profiles from the system as depicted in figure 13.12 and via error condition clause evaluation and the case based processing has asserted the above mentioned fact to its rule based inference engine layer. The facts are identified at this stage and the situation analysis has been performed. Now it is the task of the inference engine to analyze situation criticality, first by means of propagating the further system behavior through simulation and by analysis of simulation results. Precisely for a simple case it shall be detected whether under unchanged continuation of system operation the Helium pressure in one of the bottle drops below 40 bar at foreseen end of rocket stage operation, or not. Therefore first the actual system state is loaded into the simulator, which then simulates the system with the blocked pipe up to stage operations end time of 300 s. Then based on the simulation results rule based evaluation of the situation starts. Of course in real world situations the system continues operations and the simulation has to be fast enough to identify situation criticality and to eventually counteract in time.

```
.....
(rule (simulation-results-match-symptoms ok)
      (and (simulation-results-loading successful)
           (c4 inlet-pressure-ratio asymmetric)
           (lisp-function progn
              '(check-symptoms-match 'inlet-pressure-ratio-asymmetric))))

(rule (simulation-results analyzed)
      (and (simulation-results-match-symptoms ok)
           (?bottle press-gas-amount-might-be insufficient)
           (lisp-function
             multiple-assert
               detect-critical-results
                 'bottle-pressure-below-minimum
                   '(< (table-column ?bottle 'ptotal) 40.0))))
```

```
(rule (information user ((pipe blocking severe)
                          (helium budget in other bottle is not sufficient)
                          (system operation time will be reduced)
                          (shutdown system)))
      (and (simulation-results analyzed)
           (bottle-pressure-below-minimum ?condition ?values)))

(rule (information user ((pipe blocking not critical)
                          (helium budget in other bottle is sufficient)
                          (system operation time will be nominal)))
      (and (simulation-results analyzed)
           (bottle-pressure-above-minimum ?condition ?values))))
```

The LISP rule syntax reads as follows (please note that in LISP always the consequence / result term is placed first in the clause and the conditional terms thereafter):

Rule 1:

If: The system simulation results could successfully be loaded

    simulation-results-loading successful

from the simulator

    and

for component C4 (junction) applies

    inlet-pressure-ratio asymmetric

which was asserted from error condition clause level

    and

via the LISP function "check-symptoms-match“ it could be checked that the simulation results comply to the measured data from the system,

then: the fact shall be asserted that the simulation reproduced properly the system failure case:

    simulation-results-match-symptoms ok

Rule 2:

If: The following applies ("and" conditioning again):
The simulation reproduced properly the system failure case

    simulation-results-match-symptoms ok

and there exists a Helium bottle which is suspected to

    press-gas-amount-might-be insufficient

and when in the simulation result tables for such a bottle at a pressure value below 40 bar can be detected:

    (bottle-pressure-below-minimum < 40)

then: the fact shall be asserted that the simulation results are analyzed (no further involvement of the simulation for the rest of the rule chaining process)

    simulation-results analyzed

Rule 3:

If: The following applies ("and" conditioning again):

```
simulation-results analyzed
```

and there exists a Helium bottle with its

```
bottle-pressure-below-minimum < 40
```

then: inform the user

```
"pipe blocking severe"
"helium budget in other bottle is not sufficient"
"system operation time will be reduced"
```

and shutdown the system in this case[32]

```
shutdown system
```

Rule 4:

If: The following applies ("and" conditioning again):

```
simulation-results analyzed
```

and there exist only Helium bottles with

```
bottle-pressure-above-minimum < 40
```

then: inform the user

```
"pipe blocking not critical"
"helium budget in other bottle is sufficient"
"system operation time will be nominal"
```

with no further actions for system control.

In the example case the Helium bottle pressure also for the emptier one stays above 40 bars at simulated stage operation time so that at the end of the inference chain rule 4 is fired and no stage shutdown is commanded.

The example deliberately showed a very elementary implementation of such knowledge processing in a low level AI language, namely LISP. More modern expert system shells of course allow rule definition in a much more elegant to read layout, however the caveats remain the same. Essential is the very careful stepping from simpler identification rules (such as rule 1 and 2 in above example) to higher decision rules (see rules 3 and 4) building on top of each other, without doubling, conflicts and overdefinitions. As a summary the following criteria and requirements for rules can be summed up:

- The conditions or antecedences of a rule are similar to IF-statements in higher programming languages, however mind that rules are propagateable by forward chaining (example above) and backward chaining (not demonstrated in above example, but being the basline method for [119]).

[32] Shutdown is used here as simplified end of demonstration. Of course for a real rocket stage this would not be a permissible solution.

- As a consequence rules must allow the formulation of complex conditions, and thus have to provide a full spectrum of logical operators as do the case definitions.

- The preprocessing of data and messages from the data acquisition tool respectively the simulator by a case based inference, allows to deduce selected symptoms from low level data already. These higher level system states and symptoms then can be used for abstract rule based analysis.

- The rules and cases need to be formulated as generically as possible, see e.g. cases 1 and 2 in the above example which work independent of whether the right or left pipe between bottle and junction is affected by blockage.

- In practice the situation often arises, that during inference process additional low level system data values are required which so far have not been processed by the error condition clauses nor by the case based pre-processing.

- This leads to the requirement for rules or the inference engine to provide all necessary functionality to query such additional data from the monitored system and to process them inside the rules.

- And this leads further to the requirement for rules and the inference engine to provide mathematical comparison operations and all sorts of mathematical functions to be usable in the condition terms and conclusion terms of a rule. Rules here have to provide the same full scope of mathematical functionality as the cases.

- As further requirement, variables must be definable withing rules. By use of variables rules of much higher flexibility can be implemented and computed data can be handed over more easily between rules. This allows computation on rule level to be more independent from the variable set of acquired real-system or simulated-system data.

In the presented example additional functions for rule prioritization, for rule exclusion and for rule contradiction avoidance are not considered. For these topics and for the techniques of cyclic reprocessing of rules with updated data and facts the reader is referred to the corresponding specialist literature in the artificial intelligence domain.

To avoid inference rule chains to becoming too complex, unmaintainable and untestable, it has proven good practice to generate groups of rules for dedicated tasks and to arrange the overall system architecture such that the inference engine always only processes one group of rules. Input facts, deduced result facts and data are shared in a common knowledge-pool over all rules. The following figure depicts such a structuring of the overall knowledge-base and groups of rules being processed stepwise.

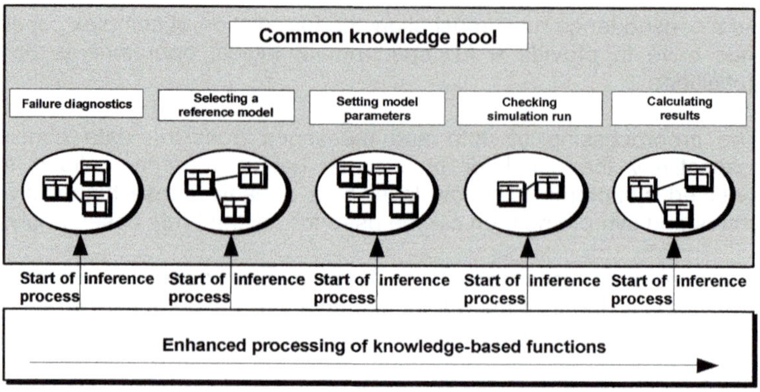

Figure 13.13: Groups of rules for dedicated processing steps.

Expert System, Measurement Data Processing and Simulation

In the field of knowledge-based applications the space domain is by far not the most innovative. Process engineering and plant control in chemical industry, electricity generation and heavy industries are partly much further advanced here. A first system which applied the above mentioned approach of plant data acquisition, knowledge-based data evaluation and parallel running "best-process" simulation is the Thyssen-Krupp coke oven plant in Duisburg-Schwelgern, Germany, which went into operation in 2001.

Figure 13.14:
Coke oven plant in Duisburg-Schwelgern.
© Thyssen-Krupp

It provides a sophisticated process control with synchronously running best-process simulation (with coke "battery control" and "gas control") to avoid undercooked coke with gaseous remainders which would lead to cavities when used in steel production. The simplifying factor here is, that a coke burning cycle in an oven is an hours long slow process, so that no sophisticated requirements are imposed here towards expert system and parallel running simulation with respect to their numerical performance.

The realization of operational state input data for the expert system is performed cyclically. The best process control selects settings for an assumed optimum control state for the process. Then it activates the simulation to compute the progress of the coke burning process over time using the assumed process control settings. As soon as new acquired measurement data are available from the real process, these are reanalyzed by the expert system, process configuration parameters are readjusted and again reverified by parallel simulation and simulation result analysis. Please also refer to [123] to [124].

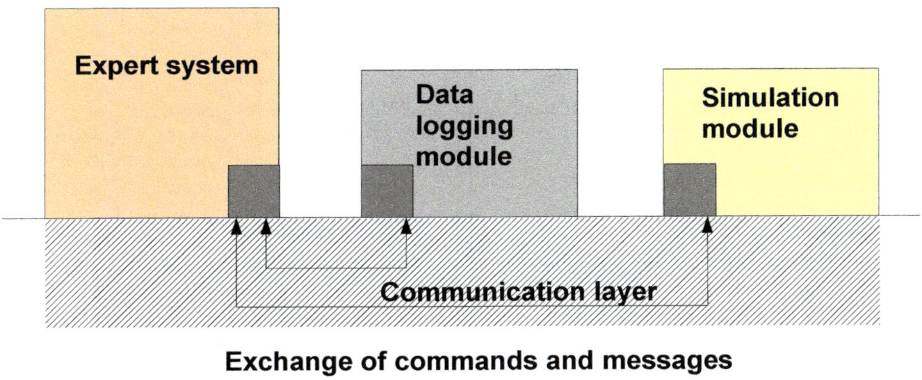

Figure 13.15: Expert system with coupled data acquisition module and simulation.

Thus the expert system cyclically verifies whether the results fit with the desired target values. Figure 13.16 depicts the flows. In summary it can be concluded that:

- Such a knowledge-based plant / system diagnosis and failure correction system is a significant support for operation of complex systems such as chemical plants, oil rigs, but also potentially for spacecraft.
- On the other hand it becomes obvious, that such knowledge-based systems are not in the position to control a plant / system preemptively avoiding operational errors or failures.
- Therefore the knowledge-based diagnosing and simulation on the one side and the cybernetic control on the other side are two fundamentally complementing techniques.

- The system regulation works actively controlling until detection of deviations or failure symptoms. The knowledge-based diagnosis and recovery induces corrective actions with quantitatively correct interaction in error cases.

It is an opportunity for the space domain to extend operations simulators for spacecraft as they are used in ground control stations and were discussed in chapter 3.2.5 by these knowledge-based techniques.

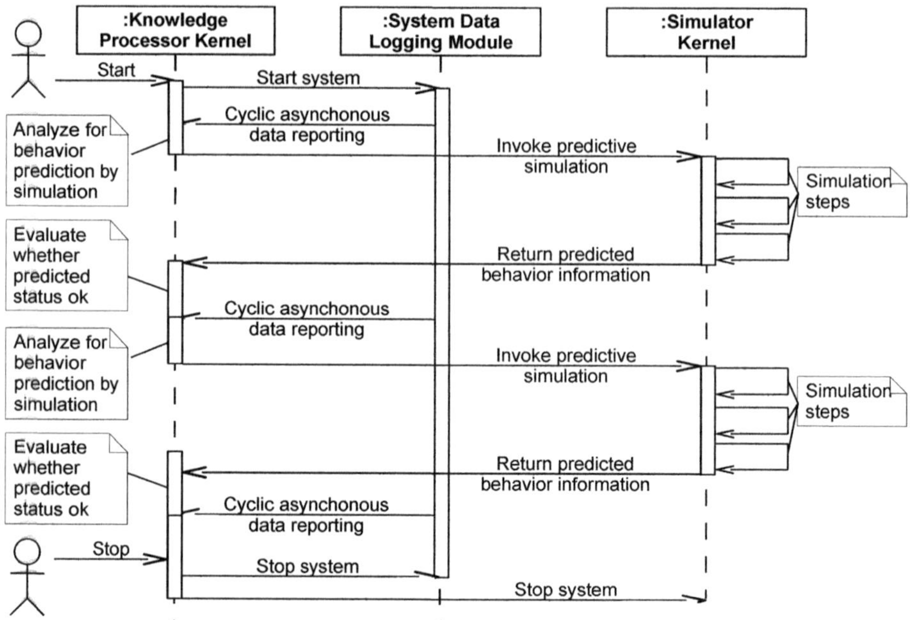

Figure 13.16: Functional flow for cyclic analysis of best-process simulation results.

13.4 Application of Expert Systems for User Training

It already was mentioned that such kinds of formalized knowledge about the system's behavior, failure cases, symptoms and limits can be reused for system operator training. For modern training software so-called "courseware" systems are used to generate training material. They present the lecture content to the candidate - later being the system operator - in the form of lessons and scripts, sorted according to topics and supported by multimedia techniques.

The abstract formulation of knowledge on system behavior as it is implemented in expert system rules (assuming a more handy notation than the presented LISP examples of course) is optimally suited for reuse in such courseware systems. Figure 13.17 shows which information about the system and its behavior flows into a courseware and how consecutive lessons for the later system operator are derived. These lessons can comprise simulations of system nominal and failure cases.

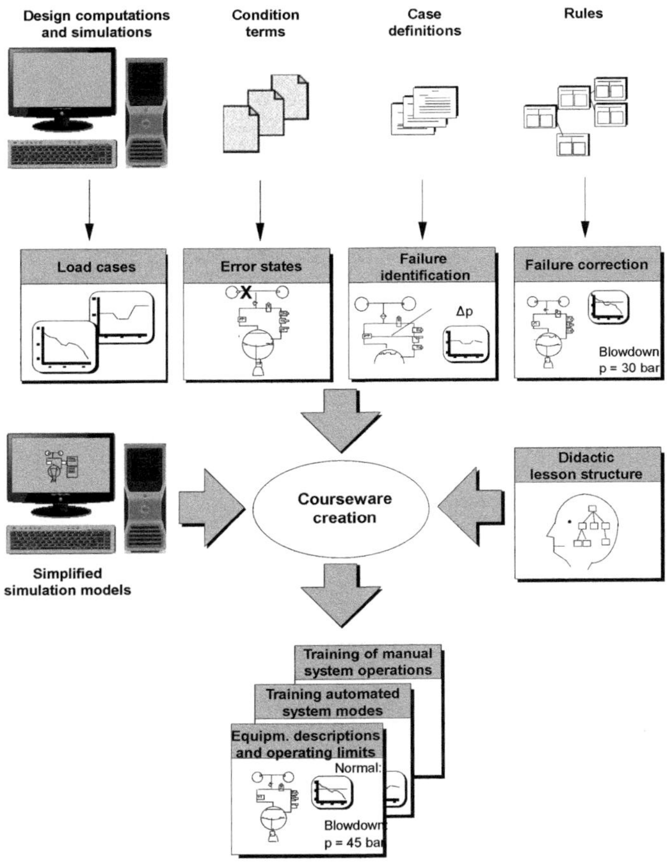

Figure 13.17: Knowledge on system behavior taken over into courseware.

13.5 Implementation Technology: Rules as Fact Filters

Finally the implementation technique of such rules in expert systems shall be elaborated in a bit more detail. The implementation technique for rules as demonstrated in the example above is based on data structures called "streams". Streams simplified are a sort of list structures which represent a continuous data flow.

In an expert system such data streams comprise facts to be evaluated and variable bindings. Rules of an expert system can be implemented as filters for such streams. A rule then filters away all facts respectively parameter bindings from a stream, which are identified not to be applicable. On the other hand rules can identify bindings and can impose new facts, added to the data stream thereafter. Rule chains as they appear in forward or backward chaining can be implemented by arranging such data stream filters in sequential order. Figure 13.18 shows a simple example for rule implementation based on filters for streams.

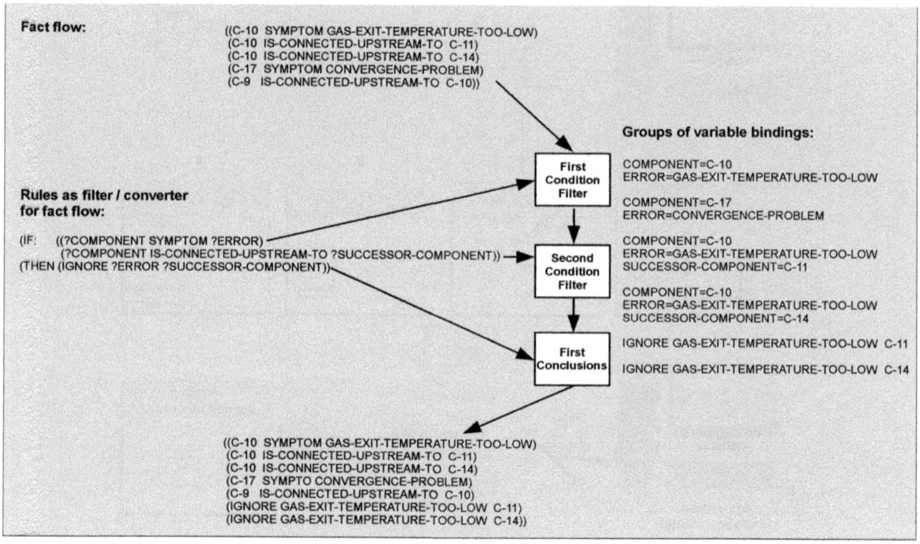

Figure 13.18: Rule implementation as data stream filter.

Specific for the processing of complex rule networks by inference engines is the fact, that larger fact bases have to be checked against each of the conditional terms of each rule. The resulting facts from one rule processing step plus the old, already known facts once again have to be checked against each of the conditional terms of the next rule and so forth. The fact sets are growing and as a result the memory resource allocation requirements are growing. In the example of figure 13.18 first all facts are input for the conditional terms of the rule. Not before having passed the first filter which identifies components indicating an error (here C-10 and C-17) any variable bindings are made which limit the scope of facts to be processed further. An additional effort increase is induced by rules applying variables in their clauses, eventually even multiple variables in one rule. For these variables then a combinatorial number of variable bindings can arise, at least for processing by some of the conditional term filters.

The most complex fact sets however arise during backward chaining of rules while checking validity of hypotheses. Therefore first one hypothesis is assumed, i.e. one combination of facts with dedicated variable bindings, and by processing these via rules - usually much more than only one - the hypothesis can be proven or disproved.

In the latter case the loop starts again until a successful hypothesis, i.e. variable binding has been found. This process is extremely resource consuming when being based on the a.m. clear text fact notation and filter implementation.

Thus the approach for processing rules by an inference engine with variable bindings in clear list notation as depicted in figure 13.18 is not a practicable approach for memory resource consumption reasons. Also the permanent changes in allocated memory can be difficult to handle. For higher performance expert systems a different implementation technique of the fact streams therefore is applied, based on the so-called "delayed evaluation" of streams. The technical principle behind it works as follows:

Instead of making available all elements of a fact stream in a list as stack in memory, which anyway is processed sequentially, the delayed evaluation technique always only stores one data tuple. The first part of the stream is available in clear notation, i.e. in directly evaluatable form. The second element contains a pointer to a data structure. This structure comprises a function pointer and variable bindings for the function arguments. By this function and variable bindings, the next element of the fact stream can be computed. Such a pointer to an executable function with fixed variable bindings in LISP is called a "closure". For an example of a fact chain with delayed evaluation please refer to figure 13.19. The example presented there depicts the fact set of the 2nd condition term filter from figure 13.18, once as clear list implementation (2 facts) and below as tuples with one clear text element and closure for computing the next element in the fact list. This technique does not save numeric processing effort, but limits memory resources and avoids problems with combinatorial numbers of variable bindings in backward chaining steps.

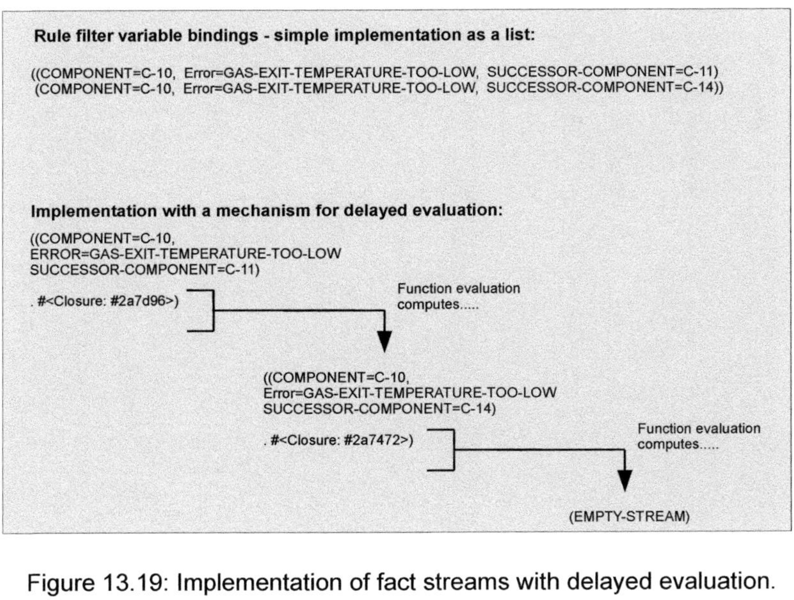

Figure 13.19: Implementation of fact streams with delayed evaluation.

This technique also is applied in the implementation cited from [120]. It is used for the implementation of fact sets which are to be processed, such as the initial fact set depicted in figure 13.18. Furthermore it is applied for parameter bindings for variables in rules which have to be kept fixed during successive processing of all condition terms of one rule (see figure 13.19) and finally for handling of entire bindings for rule setting during backward chaining. By this means an efficient and fast mechanism for rule processing is implemented which imposes only moderate requirements concerning needed memory resources.

Further reading and Internet pages concerning knowledge-based systems are provided in the according subsection of this book's references annex. Readers with specific interest in informatics of knowledge-based systems are referred to [118].

14 Simulation of Autonomous Systems

New Horizons © NASA

J. Eickhoff, *Simulating Spacecraft Systems*, Springer Aerospace Technology 1,
DOI 10.1007/978-3-642-01276-1_14, © Springer-Verlag Berlin Heidelberg 2009

Testing on-board software, (OBSW), is one of the main use cases for system simulators throughout spacecraft development. Simulation based testing has been proven as successful approach both for tests of simple OBSW data management functions as a well as for higher control and monitoring layers.

A modern on-board software – using the example of satellites again – typically consists of the following major building blocks:

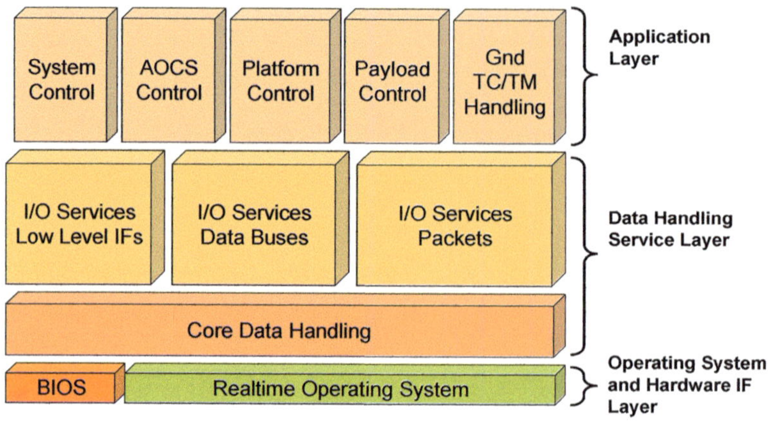

Figure 14.1: Building blocks of a modern satellite on-board software.

14.1 Testing Conventional on-board Software Functions

On-board software needs to be verified regarding its elementary functions which include:

- The on-board computer "Basic I/O-System", (BIOS), which provides access to the memory, processor and controller chips. It usually is qualified by the OBC hardware provider.
- The real-time operating system, which forms the basis of the OBSW application and which needs to be completely verified by the OBSW supplier.
- A further layer is formed by the "Input Output Services", (I/O-Services), which also are to be verified at OBSW supplier level. These I/O-services in the OBSW interface the I/O-controller hardware chips - handling data from / to the harness lines connecting the OBC to external equipment. Depending on the interface type, this can cover the full spectrum from simplest analog line acquisitions to complex on-board equipment data packets transferred via high level data bus protocols.
- Finally, on the highest level, the functions of spacecraft control are implemented and need to be verified. These functions cover the system monitoring, the attitude and orbit control, satellite platform control (power,

thermal etc.) and last but not least the payload control. This layer comprises also the ground communication functions during radio contact. This OBSW layer includes all the management of the spacecraft operational modes as well as the submodes of all subsystems and equipment.

These conventional on-board software functions include already very limited features falling into the domain of of system autonomy. E.g. a typical Earth observation satellite, which has no permanent ground station contact throughout segments of its polar orbit, executes part of its nominal system control functions as so called "time tagged commands". These time-tagged commands are grouped in sequences and are typically used to control payload operations eventually combined with platform maneuvers. As an example the following on-board function sequence for an Earth observation satellite could be controlled via time tagged commands:

- Shortly before arriving at the acquired target to be observed, the command is executed to switch PCDU power on for the payload.
- Then the payload is switched on.
- The payload is calibrated.
- Eventually the satellite AOCS is commanded to a specific pointing precision mode or roll-over mode.
- The observation itself is performed by the instrument.
- An according payload shut-down follows.

All the time tags for these steps have been precomputed on ground in this case before procedure uplink. This functionality of time-tagged command handling provides the feature to execute fixed sequences. Any non conformity during work-down of the sequence will lead to either the payload or even the entire spacecraft to switch back to a safe mode.

In addition some OBSW designs allow the upload of so-called "On-Board Control Procedures", (OBCPs), into the on-board computer from the ground. OBCPs are scripts in a kind of a macro language, which is interpreted and executed (a time-coding is also possible) on-board. With these OBCPs also a limited reaction to certain events (target identification etc.) is possible automatically. This functionality also can be applied for automatic docking and rendezvous maneuvers and is not limited to satellites.

These two kinds of simplest system autonomy in any case are subject of OBSW tests carried out on simulation based test benches. The verification of the correct processing of the OBCPs for each nominal and failure case already is a very demanding task for the verification engineer.

14.2 Testing Failure Management Functions

The next type of to-be-verified autonomous functions are "Failure Detection, Isolation and Recovery", (FDIR), functionalities. FDIR functions are implemented in all OBSW layers from the lowest level of monitoring I/O-line data traffic to the highest level of attitude error determination or payload failure detection. FDIR functions need to intervene autonomously, reacting to the various error cases. As such they cover all spacecraft operational modes. They are mostly organized as follows:

- The lowest level covers hardware errors - to identify e.g. an incorrect answering data bus controller or sensor). The errors on this level can be easily handled, e.g. by executing a data acquisition retry or a short reset of the controller device (not to be mixed with the equipment the controller is built in). The spacecraft in such simple cases keeps its main operating mode.
- The next hierarchy level comprises already errors on which the spacecraft overall operating mode cannot be retained – e.g. on a satellite the payload can no longer be operated. The system changes into a more secure operating mode – may be even down to safe-mode.
- The third level comprises such severe errors that the on-board computer needs to be restarted – eventually parts of it might need to be switched to redundant side (redundant processor board, I/O-board or other).
- The fourth level comprises hard wired hardware alarms which are issued directly to the TTR-board of the OBC or to the transponder - e.g. in case of short circuits of the power supply). These alarms can be detected on ground in case of such a contingency and as the case may be (partly) correctable, even if no OBSW is running anymore.

The testing of these highly security relevant and autonomous functions is an essential part of the OBSW developers testing task and simulation based testbenches greatly ease the work since failures much simpler can be injected, debugging of the OBSW is possible and the real spacecraft OBC hardware is kept available for AIT test purposes and is not blocked.

14.3 Testing Higher Levels of System Autonomy

For satellite spacecraft it can be stated that practically each example of today's standard satellites already features certain on-board autonomy functionality. The same applies e.g. for transfer vehicles like ATV with their autonomous docking functions etc. For satellites autonomy is largely focused to perform operations during orbit periods without ground contact. Furthermore it has to manage the mentioned failure cases without ground station contact even if the payload needs to be switched off. At least a transition in a stable safe mode has to be guaranteed. But there exist even higher levels of autonomy which can be achieved - and should be tested. However before discussing these, some terminology definitions shall be made to define functionalities and features - also because the term “autonomy” sometimes is used in imprecise contexts.

Table 14.1: Key terminology definitions concerning autonomy.

Autonomy	Autonomy is a system feature based on various functionalities and technologies. Autonomy can be implemented on the basis of automatic functions and / or autonomous functions.
Automatic functions	They run directly and straightforward and are initiated according to a master schedule or by a control program, (OBCP executor). They are checked against a schedule (operations timeline).
Autonomous functions	They implement the decision making procedure and a reaction to anomalies, (events), which occur during the execution of an automatic function.
Event	Events are triggered either through anomalies (violation of limits, error flag activations) or through the occurrence of a predefined status modification (e.g. position reached, attitude reached).
Autonomous spacecraft system	Such systems involve both spacecraft (e.g. satellite) plus ground segment and are distinguished by a wide independence from permanent human interventions. The distribution of intelligent functions for autonomy between space segment and ground segment is not prescribed.
Autonomous spacecraft	A spacecraft which is characterized by being largely independent from ground support and ground contact. Its intelligent functions achieving the autonomy are implemented aboard and serve to achieve essential parts of the mission objectives without ground intervention.

In addition to these definitions diverse levels of autonomy are to be distinguished. A break down into tree categories is given in the following table:

Table 14.2: Levels of autonomy.

Level	Characteristics
0	• Automatic monitoring of physical parameters like attitude, position, battery state of charge, temperatures • Autonomous navigation technologies (GPS) • Autonomous failure handling based on: ◇ Redundancy switches ◇ Spacecraft mode switching (survival-mode / safe-mode)
1	• Precise error identification by low level intelligent functions • On-board fault diagnosis by voting mechanisms and logic based functions • Based thereupon an operational autonomy for gradually adapted degraded mode or failure mode handling
2	• Flexible, knowledge-based fault diagnosis (manages also problems which have not been predicted) • Knowledge-based on-board planning of actions, reactive planning in failure / degradation cases • Autonomous optimization of spacecraft operations plans / procedures

Autonomy is a key system level technology for spacecraft and it can have two basic different characteristics:

"Enabling Technology" - enables a certain mission: • Enables survival of the spacecraft without radio contact • Enables spacecraft maneuvers without radio contact • Enables mission product generation without radio contact	e.g. for: • Interplanetary missions • Military missions
"Process Improvement Technology" - simplifies or cheapens the mission: This can be achieved by autonomy allowing: • Single-shift spacecraft operations in the control center • Mission product generation without radio contact • Data recording and processing focused on user requests	e.g. for: • Earth observation spacecraft • Telecom satellites • Navigation missions • Deep space probes

14.4 Implementations of Autonomy and their Focus

High level of autonomy: "Enabling Technology" Typical focus: On-board autonomy, e.g. for space probes, landers, rovers, transfer vehicles.

Basic characteristics of this type of spacecraft systems are:

- The level of autonomy typically being adjustable between
  - ◊ simple execution of macro command sequences, and
  - ◊ on-board available mission planner.
- The technical implementation inside the on-board software is focused towards:
  - ◊ Modular on-board software concepts
  - ◊ OBSW based on real operating systems

- ◇ Multi CPU board architectures, eventually separate payload OBCs
- ◇ Higher control software layers being independent from the target hardware

These autonomous system architectures have to remain maintainable and have to be verified and validated. For this purpose spacecraft testbenches capable of simulating detailed scenarios are necessary. They need to provide error injection mechanisms on different levels of the simulated spacecraft. If necessary they must provide to model complex error symptoms through parallel manipulation of multiple failure symptom relevant parameters to test the on-board software's error identification mechanisms.

Moderate level of autonomy:
"Process Improvement Technology"

Typical focus: Autonomy of the entire ground / space system, e.g. for satellites.

Basic characteristics of this type of spacecraft systems are:

- System control for the mission product of a satellite being supported by so-called "user requests":
  - ◇ This means - e.g. taking the example of an Earth observation satellite again - a mission product customer no longer specifies when the on-board instrument shall be switched on with which settings,
  - ◇ but instead he specifies the geometrical observation target, the spectral characteristics or other mission product setting he desires and the delivery date of the mission product.

- Supported by a mixture of sources of archive data in the ground segment from previous observations and by identification of still missing information an intelligent mission planning system can generate the command sequence for on-board execution for the satellite
  - ◇ to observe the mission product parts not available in archive,
  - ◇ to downlink the data, and
  - ◇ to merge on ground the latest observation data with archive data

  for to finally deliver the requested mission product to the customer.
- Such an infrastructure opens the door towards a semi-automatic single shift operation of spacecraft platform and payload control.

Only the precise testing of the overall scenario can prove the "process improvement". Required here are again system environments to simulate detailed scenarios. However besides pure spacecraft simulation here the functionalities of the ground segment elements - including the user request based mission planning - are to be included in such verification scenarios.

14.4.1 Improvement Technology – on-board SW / HW Components

Example 1: Level of Automation in on-board-SW

In October 2001 ESA launched the first satellite of the PROBA series - "Project for On-board Autonomy". With these satellites new technologies heading for higher levels of on-board autonomy or higher automation levels in satellite operations were tested.

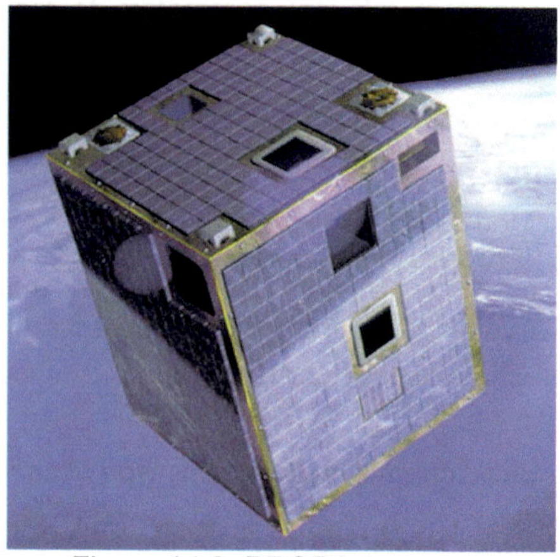

Figure 14.2: PROBA 1 © ESA

PROBA 1 served for in-flight testing of following technologies:

- First in orbit use of Europe's 32bit radiation hard space application microprocessor - the ERC32 chip set
- First use of a digital signal processor (DSP) as instrument control computer ICU
- First in orbit application of a new designed "autonomous" star sensor
- Use of on-board GPS for the first time
- And following innovations in the on-board software:
    - ◇ ESA for the first time flying an OBSW coded in C instead of Ada.
    - ◇ ESA for the first time flying an OBSW based on an operating system (VxWorks) instead of a pure manual coded implementation.
    - ◇ The GNU C compiler for the ERC32 target finally was validated though flying an OBSW running on the ERC32.

The achieved new on-board functionalities were:

- For the first time having an ESA satellite with position determination in orbit by means of GPS
- Attitude determination through an active star sensor automatically identifying star constellations
- Autonomous prediction of navigation events (target flyover, station flyover)
- Thereupon based a minimum of on-board "mission planning"

Example 2: Complexity of Operational Modes in on-board-SW

A comparable level of on-board autonomy was implemented in the ESA Moon probe SMART-1. Here the focus was less targeting for active on-board equipment - like active star sensor - or the software implementation strategies, but rather on consistency of operating modes and mode transitions during Moon approach and maneuvers.

Figure 14.3: SMART-1 © ESA

SMART-1 for the first time in ESA missions used a ion engine which implied strong cross-influences between the power supply system and thermal influences induced by the ion engine. Furthermore the use of the ion propulsion largely influenced the AOCS design.

Therefore the attitude adjustment to the Sun / to the Earth / to the Moon was optimized according to thermal and power supply criteria. Ground station visibility and data handling were subordinate and needed to be covered by corresponding autonomous functions. System simulation played an essential role for the verification of the avionics system as well as for the verification of the on-board software requirements.

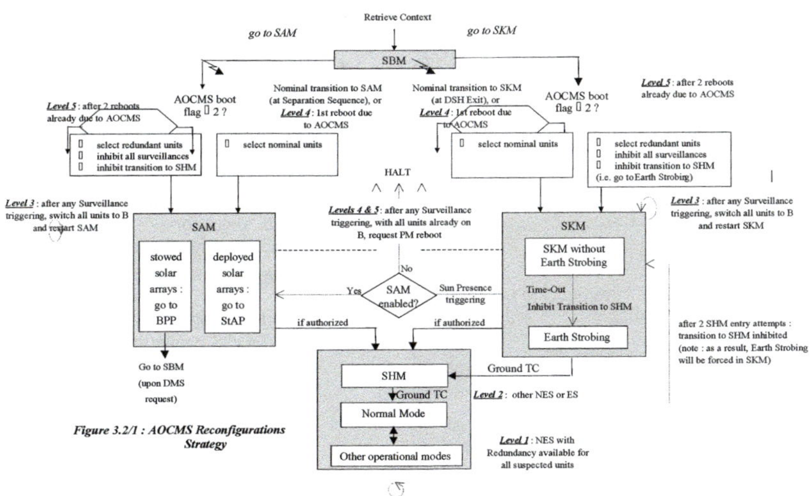

Figure 14.4: Operating modes of the ESA lunar probe SMART-1 (extract). © ESA

14.4.2 Improvement Technology – Optimizing the Mission Product

Example 3: Development Infrastructure: "Autonomy Testbed"

This example depicts a combined ground / space architecture of the ESA study "Autonomy Testing" where the design of a potential on-board mission planning for payload operation was analyzed. The idea behind this is that a user "only" needs to transmit his observation request ("user request") to the satellite. The user defines

- by which payload,
- in which operating mode,
- with which settings,
- he wants to have which target area observed.

The satellite constantly collects user requests from the various sequentially visible ground stations and is equipped with an intelligent mission planning system. This system generates a detailed timeline comprising all commands for attitude control (orientation), payload, power system etc.

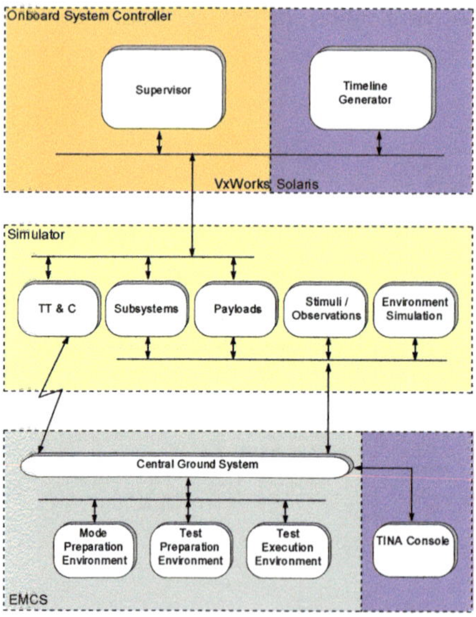

- ❑ **Autonomous On-board Architecture:**
  - System Supervisor (from DLR MARCO study) Level of autonomy scaleable from simple macro-command execution via onboard control procedures processing up to onboard timeline execution
  - TINA Timeline Generator, providing onboard generation of directly executable mission timelines from user requests and platform service requests.
- ❑ **Test Infrastructure:**
  - Simulated Satellite and Space Environment:
    - SSVF simulator
    - Spacecraft model, and environment models derived from SSVF
  - Ground segment/checkout system:
    - SSVF/CGS configuration
    - TINA console for user-request definitions

Figure 14.5: On-board autonomy test infrastructure: "Autonomy Testbed" (cf. [125]).

© Astrium

The prototype from the ESA "Autonomy Testing" study consisted of:

- A Core EGSE acting as a fictitious ground station
- A satellite simulator
- An on-board computer board as VMEbus board
- An on-board software with a macrocommand interface running on this board
- A mission planning algorithm which created an activity timeline from the cited user requests including all macrocommands to the on-board software.

The on-board software executed the spacecraft macrocommands in the generated mission timeline and thus controlled the simulated satellite. In this autonomy testbed complex scenarios were tested which comprised

- nominal operating cases in which user requests were uplinked, processed and the results were downlinked at the next ground station contact,
- furthermore scenarios which lead to planning conflicts on-board and where the user requests could only be partially satisfied within the operating period,
- and finally scenarios during which equipment manually injected failures occurred and where at initially a suitable error recovery needed to be identified and to be performed, followed by a replanning of the activities since after error recovery the satellite already had missed some of the observation targets.

Figure 14.6: Autonomy testbed setup. © Astrium

Such scenarios imposed extremely high requirements towards

- the mission planning algorithms, and
- the on-board software (which needs to intercept any potentially erroneous commands, which might be created by the mission planning tool),
- and to the spacecraft simulation infrastructure which has to reflect sufficiently realistically the overall scenario including payload operations.

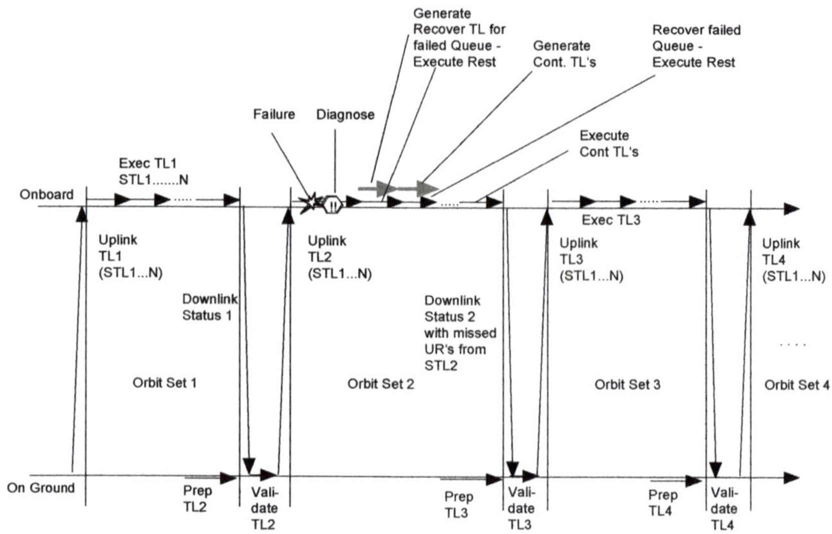

Figure 14.7: Autonomous recovery scenario on board. © Astrium

14.4.3 Enabling Technology – Autonomous OBSW for Deep Space Probes

Example 4: OBSW Concept of the NASA New Horizons Probe

Figure 14.8: New Horizons Probe. © NASA

In spring 2006 NASA launched the deep space probe "New Horizons" to explore the trans Neptunian objects Pluto and Charon. It represents probably the highest level of on-board autonomy ever flown to date.

The on-board software of New Horizons is based on a case and rule interpreter similar to the algorithm described in chapter 13 (however not coded in LISP). In place of on-board control procedures as used in conventional satellites here intelligent structures are implemented to control the nominal approach maneuvers as well as the error recovery. Cases are implemented on the lower processing level to identify abstract symptoms from parameter measurements and above these cases a rule network is implemented for situation analysis and system control.

The following figure provides a sketch of a small extract from the overall rule network – here for the handling of an error during Pluto approach. The failure either can be handled or results in the space probe going to safe mode - depending on the detailed conditions. The rule network approximately matches the technique depicted in figure 13.2, however only the forward chaining method is used for processing here.

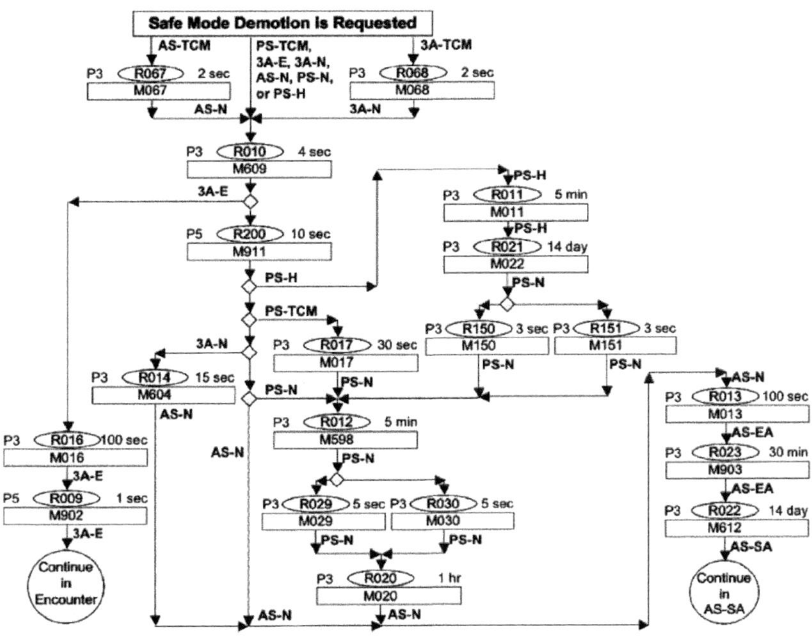

Figure 14.9: Extract of a rule-based mode-transition network of an OBSW (from [126]) © NASA

For explanation of the figure:

- The Rxxx-identifiers represent rules.
- The Myyy-identifiers represent macros which are executed by the activated rules.

- All spacecraft commands initiated by rules are encapsulated in such macros.
- The transition times for the rules / macro execution are depicted as well (some cover several days due to spacecraft coast or approach phases).
- For the rules / macros the on-board processor executing them is shown (in this extract from the rule network P3 and P5 are cited)
- and in the rule identification information is contained (for details see [126]):
  - ◇ The rule priority
  - ◇ The rule persistence,
  - ◇ The methodology how the rule result is to be handled by the inference system, when the rule result is obviously outdated
  - ◇ The state during the loading of the rule into memory (active / inactive).

Testing such rule networks for correctness and completeness is by far more difficult than the testing of conventional on-board software. Accordingly more complex will be the spacecraft simulators which need to provide injections of all sorts of error combinations and which need to be able to model such long lasting remote planet approach maneuvers. The simulations furthermore are required to test the on-board software with respect to its proper prioritization of error recoveries in multi-failure scenarios.

Further reading and Internet pages on on-board autonomy are listed in the according subsection of this book's references annex.

15 References

J. Eickhoff, *Simulating Spacecraft Systems*, Springer Aerospace Technology 1,
DOI 10.1007/978-3-642-01276-1_BM2, © Springer-Verlag Berlin Heidelberg 2009

References on Design Offices

[1] Scheuble, M.; Mager, R.:
The Satellite Design Office (SDO) Capability,
3rd European System Engineering Conference,
Toulouse, 21.-24. May 2002

[2] ESA/ESTEC:
Concurrent Design Facility,
http://www.estec.esa.nl/pr/facilities/cdf.php3

[3] ESA/ESTEC:
Students Design a Space Station,
http://www.esa.int/export/esaCP/ESALCWUTYWC_index_0.html

[4] Jet Propulsion Laboratory Project Design Centre:
http://pdc.jpl.nasa.gov

[5] Goddart Space Flight CentreMission Design Centre:
http://imdc.gsfc.nasa.gov

[6] California Institute of Technology, Laboratory for Spacecraft & Mission Design:
http://www.lsmd.caltech.edu

[7] Satellite Toolkit:
http://www.stk.com

References on Simulation Tools for Control Engineering

[8] The MathWorks Inc.:
http://www.mathworks.com

[9] The Modelica Association:
http://www.modelica.org

[10] SCILAB Organization:
http://scilabsoft.inria.fr/ und
http://www.scicos.org

[11] PSpice:
http://www.electronics-lab.com

[12] Sinda/Fluint
http://www.crtech.com

References on Verification Testbeds

MDVE (EADS Astrium GmbH - Satellites, Friedrichshafen):

[13] N.N:
MDVE – Model-based Development & Verification Environment,
Technology flier of Astrium GmbH, Friedrichshafen, January 2004

[14] Eickhoff, Jens; Hendricks, Reinhard; Flemmig, Jörg:
Model-based Development and Verification Environment,
54th International Astronautical Congress,
Bremen, September 29th - October 1st, 2003

[15] Hendricks, Reinhard; Eickhoff, Jens:
The significant role of simulation in satellite development and verification,
Die Schlüsselrolle von Simulationstechnik in Satellitenentwicklung und Test,
Aerospace Science and Technology
Volume 9, Issue 3 , April 2005, Pages 273-283

[16] Eickhoff, Jens; Falke, Albert; Röser, Hans-Peter:
Model-Based Design and Verification – State of the Art from Galileo Constellation down to small University Satellites,
25. International Astronautical Congress, Valencia, Spain,
October 2.-5., 2006
see also
Acta Astronautica, Vol. 6, Issues 1-6, June - August 2007, Pages 383-390
http://dx.doi.org/10.1016/j.actaastro.2007.01.027

SimWare (EADS Astrium S.A.S. - Satellites, Toulouse):

[17] Vatan, Bruno:
A Generic Product based Infrastructure for Simulators in Space Area,
Proceedings of the Conference on Data Systems in Aerospace,
DASIA 1997, Sevilla, Spain, 26.-29. May1997

Eurosim (EADS Astrium - Dutch Space BV.):

[18] http://www.eurosim.nl/

Simulator 3rd Generation (MDVE Successor, EADS Astrium - Satellites):

[19] Eisenmann, Harald; Cazenave, Claude:
SimTG: Successful Harmonization of Simulation Infrastructures,
10th International Workshop on Simulation for European Space Programmes,

SESP 2008, October 7th - 9th 2008, ESA/ESTEC, Noordwijk, Netherlands

References on Simulation Tools used in Ground Stations

SIMSAT (ESA / ESOC):

[20] http://www.esoc.esa.de/external/mso/simsat.html

[21] http://www.dlr.de/dlr/Raumfahrt/RF-Management/Ind/vega.pdf

BASILES (CNES)

[22] N.N:
BASILES - Distributed Simulators & 3D Visualization,
TSP Workshop, 2007,
http://download.savannah.gnu.org/releases/tsp/events/First_TSP_Workshop_27march2007/BASILES_HLA_TSP.pdf

References on Open Source Simulation Tools

[23] http://www.opensimkit.org
and
http://www.opensimkit.de

References on Equipment Modeling

[24] Hyder, A.K.; Wiley, R.L. et. al.:
Spacecraft Power Technologies,
Imperial College Press, 2000,
ISBN: 1860941176

[25] Hallmann, W.; Ley, W.:
Handbuch der Raumfahrttechnik,
Carl Hanser Verlag, München, 1988,
ISBN: 3-446-15130-3

[26] Wertz, J.R (Ed.):
Spacecraft Attitude Determination and Control,
Kluwer Academic Publishing, 9th Edition, 1997,
ISBN: 90-277-0959-9

On-board Computer Modeling:

[27] N.N:
TSIM ERC32/LEON simulator,
Gaisler Research,
http://www.gaisler.com/cms/index.php?option=com_content&task=view&id=38&Itemid=56

AOCS Components Modeling:

[28] www.irs.uni-stuttgart.de/lehre/v_kleinsatellitenentwurf_selfstudy_online/pdf/Kap5_KSE_AOCS_www.pdf

[29] Nicolini, D.:
LISA Pathfinder FEEP Subsystem,
6th LISA Symposium, ESA/ESTEC, Noordwijk, Netherlands, June 22., 2006

Electric Components Modeling:

[30] www.irs.uni-stuttgart.de/lehre/v_kleinsatellitenentwurf_selfstudy_online/pdf/Kap4_KSE_Leistungselektronik_www.pdf

[31] Kordesch, K.V.:
Brennstoffbatterien,
Springer-Verlag, Wien, 1984
ISBN: 3211818197

[32] Linden, D.:
Handbook of Batteries and Fuel Cells,
Mc Graw Hill, 1984

[33] Regenhardt, P.A.:
F-Cell: The ASPEN Fuel Cell Model,
US Department of Energy, March 1985,
NASA TM-103210

Fuel Sloshing:

[34] http://www.esa.int/SPECIALS/Launchers_Home/SEMNFZ0XDYD_0.html

Life Support Systems:

[35] World Spaceflight News:
Inside the International Space Station: Environmental Control and Life Support System (ECLSS),
Astronaut Training Manual (Ring-bound),
ISBN: 1-893472-19-1

[36] Melton, Robert G.:
Space Station Technology (Progress in Technology)

Society of Automotive Engineers, November 1996
ISBN: 978-1560917465

References on Algorithms for Attitude and Orbit Control

[37] http://en.wikipedia.org/wiki/List_of_orbits

[38] Montenbruck, O.; Gill, E.:
Satellite Orbits - Models, Methods, Applications,
Springer-Verlag, Berlin, 2001,
ISBN-13: 978-3-540-67280-7

[39] Chobotov, Vlaidimir A.:
Spacecraft attitude dynamics and control,
Krieger Publishing, Malabar, 1991,
ISBN-13: 978-0894640315

[40] Wertz, J.R (Ed.):
Spacecraft Attitude Determination and Control,
Kluwer Academic Publishing, 9th Edition, 1997,
ISBN: 90-277-0959-9

[41] Sidi, M.J.:
Spacecraft Dynamics and Control:
A practical Engineering Approach,
Cambridge University Press, 2nd Edition, 2000

[42] Fehse, W.:
Automated Rendezvous and Docking of Spacecraft,
Cambridge University Press, 2003,
ISBN: 0-521-82492-3

[43] Larson, W.J.; Wertz, J.R:
Space Mission Analysis and Design,
Kluwer Academic Publishing, 2nd Edition, 1993
ISBN: 0792319982

[44] Fortescue, P.W.; Stark, J.; Swinert, G.:
Spacecraft Systems Engineering,
J. Wiley & Sons, 3rd Edition, 2003,
ISBN: 0470851023

References on Modeling Satellite Power Systems

[45] Hyder, A.K.; Wiley, R.L. et. al.:
Spacecraft Power Technologies,
Imperial College Press, 2000,
ISBN: 1860941176

[46] Hallmann, W.; Ley, W.:
Handbuch der Raumfahrttechnik,
Carl Hanser Verlag, München, 1988,
ISBN: 3-446-15130-3

References on Simulation Numerics

[47] Engeln-Müllges, G.; Reutter, F.:
Numerische Mathematik für Ingenieure,
Bibliographisches Institut, BI Wissenschaftsverlag, 1973

[48] N.N.:
ESRF - The European Synchrotron Radiation Facility ScientificLibraries
http://www.esrf.fr/UsersAndScience/Experiments/TBS/SciSoft/
http://www.esrf.eu/UsersAndScience/Experiments/TBS/SciSoft/Links/Fortran

[49] N.N.:
Numerical Recipes in C: The Art of scientific Computing,
Cambridge University Press, 1988 – 1992,
ISBN: 0-521-43108-5

[50] Urban, Karsten; Lehn, Michael:
Übungen zur Vorlesung Numerik 1b
Fakultät für Mathematik der Universität Ulm
http://www.mathematik.uni-ulm.de/numerik/teaching/ws07/NUM1b/prog/anleitung04/
and
http://www.mathematik.uni-ulm.de/numerik/teaching/ws07/NUM1b/prog/blatt4/newton2d.cc

[51] Westerberg, A.W.; Hutchison, H.P.; Motard, R.L.; Winter, P.:
Process Flowsheeting
Cambridge University Press, 1979,
ISBN: 0 521 22043 2

[52] http://de.wikipedia.org/wiki/CG-Verfahren

[53] Burden, L.B.; Faires, J.D.:
Numerical Analysis,
Prindle, Weber and Schmidt, Boston 1989, 3rd Edition

[54] Deuflhard, P.; Hohmann, A.:
Numerische Mathematik,
W. de Gruyter Verlag, Berlin, 1991,
ISBN: 3110171813

[55] Meis, Th.; Marcowitz,U.:
Numerische Behandlung partieller Differentialgleichungen,
Springer-Verlag, Berlin, 1978

[56] N.N.:
Numerical Recipes in C: The Art of scientific Computing,
Cambridge University Press, 1988 – 1992,
ISBN: 0-521-43108-5

[57] N.N.:
International Mathematical and Statistical Library, User's Manual,
IMSL Math/Library, Houston 1989

[58] Engeln-Müllges, G.; Reutter, F.:
Formelsammlung zur num. Mathematik mit Standard FORTRAN77 Programmen,
Bibliographisches Institut, BI Wissenschaftsverlag, 1989,
ISBN: 3411031859

Please also refer to the substantial literature on this topic in the field of chemical process engineering.

References on Model Driven Architecture

[59] N.N.:
MDA Information Page
http://www.omg.org/mda/

[60] Soley, Richard M.:
Model Driven Architecture: Executive overview,
http://www.omg.org/mda/executive_overview.htm

[61] Kleppe, Anneke; Warmer, Jos; Bast, Wim:
MDA Explained. The Model Driven Architecture: Practice and Promise.
Addison-Wesley Professional, 2003,
ISBN-13: 978-0-321-19442-8

[62] Gruhn, Volker; Pieper, Daniel; Röttgers, Carsten:
MDA - Effektives Software-Engineering mit UML2 und Eclipse,
Springer-Verlag, 2006,
ISBN-13: 978-3-540-28744-5

References on Object Oriented Programming Languages

C++:

[63] Stroustroup, Bjarne:
The C++ Programming Language
Addison Wesley, Reading, Massachussetts,
2nd Edition, 1993,
ISBN: 0-201-53992-6

[64] Eckel, Bruce:
Using C++, Covers C++ Version 2.0,
Osbourne McGraw Hill, Berkeley 1989,
ISBN: 0-07-881522-3

[65] Ellis, Margaret A.; Stroustroup, Bjarne:
The Annotated C++ Reference Manual
Addison Wesley, Reading, Massachussetts, 1990,
ISBN: 8131709892

[66] Coplien, James O.:
Advanced C++, Programming Styles and Idioms
Addison Wesley, Reading, Massachussetts, 1992,
ISBN: 0-201-54855-0

[67] Coplien, James O.; Schmidt, Douglas C. (Eds.):
Pattern Languages of Program Design
Addison Wesley, Reading, Massachussetts, 1995,
ISBN: 0-201-60734-4

Java and real-time Java:

[68] Krüger, Guido:
Handbuch der Java-Programmierung
Addison Wesley, 4th Edition, 2006,
ISBN-13: 978-3-8273-2361-3

[69] Sun Java™ Real-Time System 2.1 Technical Documentation
http://java.sun.com/javase/technologies/realtime/reference/rts_productdoc_2.1.html

[70] N.N.:
AONIC PERC – Virtual Machine
A Brief Synopsis of Real-Time Java
http://www.aonix.com/real_time_java.html

[71] N.N.:
javalution - the Java Solution for Real-Time and Embedded Systems
http://javolution.org/

References on UML

[72] Erler, Thomas:
Das Einsteigerseminar UML 2,
Verlag Moderne Industrie Buch AG & Co KG, Bonn, 2004,
ISBN: 3-8266-7363-8

[73] Booch, Grady; Rumbaugh, James; Jacobson, Ivar:
The Unified Modelling Language User Guide,

Addison Wesley Longman, Reading, Massachussetts, 1999
ISBN: 0-201-57168-4

[74] James Rumbaugh, Ivar Jacobson, Grady Booch:
The Unified Modeling Language Reference Manual,
Addison Wesley Longman, 1999,
ISBN: 020130998X

[75] Sinan Si Alhir:
Learning UML
O'Reilly, 2003,
ISBN: 0-596-00344-7

[76] N.N.:
UML Notation Overview
Compendium of the diagram types on two pages
http://www.oose.de/uml

[77] N.N.:
UML for managers
Short abstract on concepts and diagrams
http://www.parlezuml.com

[78] N.N.:
OpenAmeos – The OpenSource UML Tool
http://www.openameos.org/

[79] Jeckle, Mario; Rupp, Chris; Hahn, Jürgen; Zengler, Barbara; Queins, Stefan:
UML2 glasklar,
Carl Hanser Verlag, München Wien, 2004

References on XML

[80] Mintert, Stefan (Ed.):
XML & Co. Die W3C-Spezifikationen für Dokumenten- und Datenarchitektur,
Addison-Wesley, 2002,
ISBN: 3827318440

[81] Ray, T.; Siever, E. (Ed.):
Learning XML,
O'Reilly, 2001,
ISBN: 0596000464

[82] Eckstein, Robert; Casabianca, Michel:
XML kurz & gut,
2nd Edition, O'Reilly, 2002,
ISBN: 3-89721-235-8

[83] http://xml.apache.org/ ("Xerces" Parser)

[84] http://www.saxproject.org/ ("SAX" Parser)

[85] W3C Document Object Model
http://www.w3.org/DOM/

References on Eclipse

[86] Wieland, Thomas:
C++ mit Eclipse,
www.cpp-entwicklung.de/downld/Cpp-mit-Eclipse.pdf

[87] Künneth, Thomas:
Einstieg in Eclipse 3.3,
Einführung, Programmierung, Plug-In-Nutzung
Galileo Computing, 2007,
ISBN-13: 978-3-89842-792-0

[88] Daum, Berthold:
Java Entwicklung mit Eclipse 3.2
dpunkt.verlag, Heidelberg, 4.Ed, 2006,
ISBN: 3-89864-426-X

[89] Burnette, Ed:
Rich Client Tutorial Part1, http://www.eclipse.org/articles/Article-RCP-1/tutorial1.html
Rich Client Tutorial Part2, http://www.eclipse.org/articles/Article-RCP-2/tutorial2.html
Rich Client Tutorial Part3, http://www.eclipse.org/articles/Article-RCP-3/tutorial3.html

References on System Engineering Standards and Tools

[90] Object Management Group:
http://www.omg.org/uml

[91] http://syseng.omg.org

[92] http://www.omg.org/mda

[93] http://de.wikipedia.org/wiki/Model_Driven_Architecture

[94] http://www.sysml.org

[95] International Council on System Engineering:
http://www.incose.org

[96] N.N.:
OMG Systems Modeling Language (OMG SysML™), V1.1,

May, 2008,
http://www.sysml.org/docs/specs/OMGSysML-v1.1-AS-ptc-08-05-16.pdf

[97] ISO:
ISO 10303-1, Industrial automation systems and integration,Product Data Representation and Exchange,
Part 1 – Overview and fundamental principles, ISO, 1994

[98] http://www.steptools.com/library/standard/index.html

[99] Grose, T.J.; Doney, G.C; Brodsky, S.A.:
Mastering XMI,
Java Programming with XMI, XML and UML,
Wiley Computer Publishing, 2002,
ISBN: 0-471-38429-1

[100] Fuchs, J.:
Der virtuelle Entwurfsprozess
(Virtual Spacecraft Design –VSD)
http://www.dlr.de/sc/Portaldata/15/Resources/dokumente/WS_120607/Presentation_VSD_070612_Fuchs.pdf

References on Software Design Patterns

[101] Gamma, Erich; Helm, Richard; Johnson, Ralph; Vlissides, John M.:
Design Patterns: Elements of Reusable Object-Oriented Software,
Addison-Wesley Professional, 1994,
ISBN: 0-201-63361-2
ISBN-13: 978-0201633610

[102] Shalloway, Alan;. Trott, James R.:
Design Patterns Explained: A New Perspective on Object-Oriented Design (2nd Edition),
Addison-Wesley Professional, 2001,
ISBN-13: 978-0201715941

References on Model-based System Engineering

[103] SysML Partners:
Systems Modeling Language (SysML) Specification, version 0.9 DRAFT, May 2005

[104] Kress, Martin:
Applying SysML/UML 2.0 to Functional System Engineering for Space Applications,
Diploma Thesis, Berufsakademie Ravensburg, September 2005

[105] Open SystemC Initiative:
SystemC 2.0.1 Language Reference Manual,
Revision 1.0, Jan Jose, 2003

[106] De Koning, H.P.; Eisenmann, H.:
ECSS standard supporting MBSE across the whole space product life cycle
INCOSE Model-based System Engineering Workshop,
Albuquerque, NM, USA, 24. / 25. January 2008

[107] N.N.:
ECSS E-TM-10-23, Issue 0, Rev. 20,
ECSS, February, 2009

[108] N.N.:
Exchange of engineering analysis models and results for space,
Part 1: Application protocol: Network-model and results format (STEP-NRF),
6.1 intermediate release,
European Space Agency (ESA), August, 2008

[109] Eickhoff, Jens; Herpel, Hans-Juergen; Steinle, Tobias; Birn, Robert; Steiner, Wolf-Dieter; Eisenmann, Harald; Ludwig, Tobias:
System Engineering Infrastructure evolution - Galileo-IOV and the steps beyond,
DASIA 2009, Istanbul, Turkey, 26. - 29. May 2009

References on Software Standards

ECSS Standards:

[110] http://www.ecss.nl/

DO-178B:

[111] RCTA/EUROCAE:
Software Considerations in Airborne Systems and Equipment Certification, DO-178B/ED-12B, December 1992

[112] http://www.rtca.org/downloads/ListofAvailable_Docs_WEB_NOV_2005.htm

Galileo Software Standard:

[113] Montalto, Gaetano:
The Galileo Software Standard as tailored from ECSS E40B/Q80,
European Satellite Navigation Industries SPA,
BSSC Workshop on the Usage of ECSS Software Standards For Space Projects,
http://www.estec.esa.nl/wmwww/EME/Bssc/BSSCWorkshopProgrammev5.htm

NASA Standards (Top level view):

[114] http://sw-eng.larc.nasa.gov/process/documents/wddocs/LaRC_Local_Version_of_SWG_Matrix.doc

MIL Standards:

[115] MIL-STD-2167A,
Military Standard, Defense System Software Development,
Department of Defense, Washington, D.C., February 29, 1988.

References on Knowledge-Based Systems

[116] Jackson, P.:
Introduction to Expert Systems
Addison Wesley, Menlo Parc, California, USA 1994
ISBN: 0-201-17578-9

[117] Winston, P. H.; Horn, B. K. P.:
Lisp
Addison-Wesley, Bonn 1987
ISBN: 0-201-08319-1

[118] Abelson H.; Sussmann, G. J.; Sussmann, J.:
Structure and Interpretation of Computer Programs
The MIT Press, Cambridge, Massachusetts, USA 1985

[119] http://ntrs.nasa.gov/archive/nasa/casi.ntrs.nasa.gov/19890017221_1989017221.pdf

[120] Eickhoff, Jens:
Modulare Programmarchitektur für ein wissensbasiertes Simulationssystem mit erweiterter Anwendbarkeit in der Entwicklung und Betriebsüberwachung verfahrenstechnischer Anlagen,
VDI Verlag, Reihe 20 Rechnerunterstützte Verfahren, Nr. 196, 1996

[121] Sturm, F.A.:
Wissensbasierte Prozessführung,
BWK Band 55, Nr. 9, S. 42-45
Springer-Verlag, 2003

[122] http://www.scheckreiter.de/TKEC/Berichte_3.htm

[123] http://www.uhde.eu/cgi-bin/byteserver.pl/pdf/broschueren/Kokerei/Schwelgern_engl.pdf

References on on-board Autonomy

[124] http://www.esa.int/SPECIALS/Proba_web_site/SEMHHH77ESD_0.html

[125] Eickhoff, Jens:
System Autonomy Testbed
Produktbroschüre der Dornier Satellitensysteme GmbH,
D-88039 Friedrichshafen, 1997

[126] Moore, Robert C.:
Autonomous Safeing and Fault Protection for the New Horizons Mission to Pluto
The Johns Hopkins University Applied Physics Laboratory, Laurel, Maryland, USA,
57th, International Astronautical Congress,Valencia, Spain,
October 2.-6., 2006

[127] http://www.nasa.gov/mission_pages/newhorizons/main/index.html

[128] http://smart.esa.int/science-e/www/area/index.cfm?fareaid=10

References Diverse

[129] Eickhoff, Jens; Schwarzott, Walter:
Thermo-fluiddynamische Simulation unterstützt durch wissensbasierte Systeme,
DGLR Jahrestagung, Friedrichshafen, 1990

[130] Eickhoff, Jens; Bisanz, Reinhold:
Thermophysical Simulation in Aerospace Supported by Object-oriented Software Technologies,
Proceedings of the European Simulation Multiconference 1994,
Universitat Politechnica de Catalunia,
Barcelona, 1994

[131] http://en.wikipedia.org/wiki/Tcl

[132] http://www.satellite-links.co.uk/directory/rheasystem.html

[133] Krafzig Dirk; Banke, Karl; Slama, Dirk:
Enterprise SOA: Service Oriented Architecture Best Practices
Prentice Hall PTR, 2005,
ISBN-13: 9780131465756

Index

9783642012754